RELIABILITY AND MAINTENANCE

NETWORKS AND SYSTEMS

RELIABILITY AND MAINTENANCE
NETWORKS AND SYSTEMS

Frank Beichelt
Peter Tittmann

CRC Press
Taylor & Francis Group
Boca Raton London New York

CRC Press is an imprint of the
Taylor & Francis Group, an **informa** business
A CHAPMAN & HALL BOOK

CRC Press
Taylor & Francis Group
6000 Broken Sound Parkway NW, Suite 300
Boca Raton, FL 33487-2742

First issued in paperback 2019

ISBN-13: 978-1-4398-2635-5 (hbk)
ISBN-13: 978-1-138-38198-8 (pbk)

Library of Congress Cataloging-in-Publication Data

Beichelt, Frank, 1942-
  Reliability and maintenance : networks and systems / Frank Beichelt, Peter Tittmann.
    p. cm.
  Includes bibliographical references and index.
  ISBN 978-1-4398-2635-5 (hardcover : alk. paper)
  1. Reliability (Engineering)--Mathematical models. I. Tittmann, Peter. II. Title.

TA169.B45 2012
620'.00452--dc23   2012010143

Visit the Taylor & Francis Web site at
http://www.taylorandfrancis.com

and the CRC Press Web site at
http://www.crcpress.com

I dedicate this book to the memory of my former student and colleague

Dr. sc. techn. Lutz Sproß.

Lutz has considerably contributed to reliability engineering. More important to me, he was a man of utmost integrity.

Frank Beichelt

Contents

Preface

Since the 1960s, hundreds of books and thousands of papers have been published in the field of mathematical modeling for reliability and maintenance. Hence, no book of this size can claim to give a comprehensive survey on all the areas dealt with. The topics of this book have on the one hand been chosen by the interests of the authors and on the other hand by their intention to present a book which is interesting to engineers, computer scientists, operations researchers, and mathematicians, both with regard to education and application. Every part of the book starts with mathematical foundations and basic models, leads to more sophisticated models and frequently to the state-of-the-art. Usually, hints at some relevant recent literature are given. But we are well aware of the fact that the list of references is far from complete and refer sometimes to more comprehensive surveys. We did try to be correct in the citations of pioneering papers.

Part I on Binary Coherent Systems, possibly complemented by fault tree analysis, can serve as the basis for a one-semester introductory course in structural reliability theory for students in all the fields listed above. This part, moreover, provides theoretical foundations for the reliability analysis of communication networks, which is the subject of Part II. Texts on network reliability analysis are currently scarce, so the lecturers may welcome this book as a suitable base for preparing courses for students of computer science and mathematics in this field. Based on the tools presented in Part II, theoreticians and network operators will be in a position to do network reliability analyses on their own. Part III may be a suitable base for an introductory course in maintenance theory for engineers and operations researchers. Because most of the maintenance policies presented are simply structured and, therefore, easy to implement, maintenance engineers in industry may find them particularly suitable for application in their respective companies. Generally, the text as a whole can serve as a basis for self-study for readers who have mastered their undergraduate mathematical training as engineers or scientists. However, some additional knowledge in graph theory and combinatorics for Part II would be helpful.

Comments and suggestions regarding the book are most welcome and should be sent to Frank.Beichelt@wits.ac.za (Parts I and III) and peter@hs-mittweida.de (Part II).

Frank Beichelt and *Peter Tittmann*

Part I

Coherent Binary Systems

Chapter 1

Fundamental System Structures

1.1 Simple Systems

1.1.1 Basic Concepts

A system is called *simple* if it can be considered a unit with regard to all relevant reliability and maintenance aspects, in particular with regard to the development, occurrence, and localization of failures, as well as with regard to scheduling, materializing and accounting maintenance measures.

This explanation does not exclude that simple systems in practice may be fairly complicated structures or products. In a large computer network, for instance, the individual computers normally can be regarded as simple systems. In the context of the following sections, simple systems are the components, elements, parts or subsystems of *structured (complex) systems*. Knowledge of reliability parameters of their underlying simple systems is a prerequisite for the reliability analysis and maintenance planning of structured systems. In this section, let S be a simple system. We will assume that S can only be in one of the following two states:

state 1 (*available, operating, functioning*),

state 0 (*unavailable, failed, down*).

Hence, the indicator variable for the state of the system at time t is a *binary* or *Boolean variable* with range $\{0,1\}$:

$$z(t) = \begin{cases} 1 & \text{if } S \text{ is available at time } t \\ 0 & \text{otherwise} \end{cases}. \tag{1.1}$$

This assumption implies that the transition of a system from state 1 into state 0 occurs in negligibly small time, i.e. the system is subjected to *sudden failures*. The chosen binary range $\{0,1\}$ is, of course, arbitrary, but later it will prove a very convenient choice.

3

If S starts operating at time $t = 0$, then its *lifetime* L is the time from $t = 0$ to the time point of its failure. *Distribution function (failure probability)*, *survival function (survival probability)*, and *probability density* of L are denoted as $F(t)$, $\overline{F}(t)$, and $f(t)$, i.e.

$$F(t) = \Pr(L \le t), \quad \overline{F}(t) = 1 - F(t) = \Pr(L > t), \quad f(t) = \frac{dF(t)}{dt}.$$

If the system is not replaced or repaired after a failure, then

$$\Pr(z(t) = 1) = \Pr(L > t) = \overline{F}(t).$$

Expected value (mean value) $\mathbb{E}(L)$ and *variance* $Var(L)$ of L are

$$\mathbb{E}(L) = \int_0^\infty t\, f(t)dt, \quad Var(L) = \int_0^\infty (t - \mathbb{E}(L))^2 f(t)dt.$$

Since L is a nonnegative random variable, its expected value is also given by

$$\mathbb{E}(L) = \int_0^\infty \overline{F}(t)dt. \tag{1.2}$$

The *residual lifetime* $L_x = L - x$ given $L > x \ge 0$ is the random time span from $t = x$ to the system failure at time L. Hence, its distribution function is

$$F_x(t) = \Pr(L \le x + t \,|\, L > x) = \frac{\Pr(x < L \le x + t)}{\Pr(L > x)}.$$

Thus, $F_x(t)$ and $\overline{F}_x(t) = 1 - F_x(t)$ are

$$F_x(t) = \frac{F(x + t) - F(x)}{\overline{F}(x)}, \quad \overline{F}_x(t) = \frac{\overline{F}(x + t)}{\overline{F}(x)}. \tag{1.3}$$

The *failure rate (hazard rate)* $\lambda(x)$ of S is defined as the limit

$$\lambda(x) = \lim_{\Delta x \to 0} \frac{1}{\Delta x} F_x(\Delta x) = \frac{1}{\overline{F}(x)} \cdot \lim_{\Delta x \to 0} \frac{F(x + \Delta x) - F(x)}{\Delta x}. \tag{1.4}$$

Thus,

$$\lambda(x) = \frac{f(x)}{\overline{F}(x)}. \tag{1.5}$$

The value $\lambda(x)$ is *not* a probability (not even an *instantaneous probability*), but a (failure) probability rate, namely the probability of a system, which has survived interval $[0, x]$, to fail in $(x, x + \Delta x]$, referred to the length Δx of this interval and letting $\Delta x \to 0$. Thus, $\lambda(x)$ is a measure for the propensity of a system of *age* x to fail.

The definition (1.4) of the failure rate can be written in the equivalent form

$$\Pr(L - x \le \Delta x \,|\, L > x) = \lambda(x)\Delta x + o(\Delta x), \tag{1.6}$$

where $o(x)$ is the *Landau order symbol* with regard to the limit $x \to 0$, i.e. any function $o(x)$ of a variable x satisfying

$$\lim_{x \to 0} \frac{o(x)}{x} = 0, \tag{1.7}$$

i.e. $o(x)$ "tends faster to 0" than the function $f(x) = x$. Therefore, for small Δx, the product $\lambda(x)\Delta x$ is a good approximation to the probability that a system, still operating at age x, will fail in $(x, x + \Delta x]$. Thus, if a number n of identical systems start operating at $t = 0$ and n_x of them have survived the interval $[0, x]$ and $n(\Delta x)$ of these n_x fail in $(x, x + \Delta x]$, then $n(\Delta x)/n_x$ is a suitable point estimate for $\lambda(x)$. Usually, due to *early failures*, the function $\lambda(x)$ will decrease for a while after starting up a system, and, due to wearing out, will increase after a sufficiently large time point x. In between, frequently a nearly constant failure rate can be observed (*bath tub curve* of the failure rate). In other words, a system is *rejuvenating* within a certain time span after taking up its operation, and, from a certain time point later in its life, it will start *aging*. In practice, however, there is usually an overlap between these three phases of system life.

By integrating both sides of (1.5) from $x = 0$ to $x = t$, we obtain $\int_0^t \lambda(x)dx = -\ln \overline{F}(t)$, or, equivalently,

$$F(t) = 1 - e^{-\Lambda(t)}, \quad \overline{F}(t) = e^{-\Lambda(t)}, \quad f(t) = \lambda(t) \, e^{-\Lambda(t)}, \quad t \geq 0, \tag{1.8}$$

where

$$\Lambda(t) = \int_0^t \lambda(x)dx \tag{1.9}$$

is the *integrated failure rate*. Thus, if L has a probability density, then its probability distribution is fully characterized by its failure rate and vice versa. By combining formulas (1.3) and (1.8),

$$F_x(t) = 1 - e^{-[\Lambda(x+t) - \Lambda(x)]}, \quad \overline{F}_x(t) = e^{-[\Lambda(x+t) - \Lambda(x)]}. \tag{1.10}$$

Obviously, all the concepts introduced also refer to lifetimes of organisms. In survival analysis, for instance, the failure rate belonging to the lifetime distribution of human beings is called the *force of mortality*.

1.1.2   Nonparametric Classes of Probability Distributions

The concept of the failure rate has given rise to the definition of classes of nonparametric probability distributions, which model the aging behavior of systems under different aspects.

Definition 1.1 $F(t)$ *is IFR (DFR) if its residual lifetime distribution function* $F_x(t)$ *is increasing in x for any fixed $t > 0$ (decreasing in x for any fixed $t > 0$).*

From (1.10): If $F(t)$ has a density, then $F(t)$ is *IFR (DFR)* iff $\lambda(x)$ is increasing (decreasing) in x. *IFR (DFR)* stands for *increasing (decreasing) failure rate*. Here and in what follows, *increasing (decreasing)* means *nondecreasing (nonincreasing)*. It also is quite common, and maybe even more intuitive, to say that *a system has an increasing (decreasing) failure rate* if its lifetime distribution function is *IFR (DFR)* (even if $f(t)$ does not exist).

Technical systems may have periods in their life in which they do not age, for instance in periods of no or reduced stress, or when maintenance actions like repairs actually rejuvenate the system, i.e. the failure rate of the system immediately after a repair is lower than the failure rate immediately before a failure. In case of human beings, the force of mortality of somebody having overcome a serious disease will be lower than at the onset of the disease. To model situations like that from the aging point of view, distribution functions of type *IFRA (increasing failure rate average)* or *DFRA (decreasing failure rate average)* will be more suitable than *IFR (DFR)*.

Definition 1.2 $F(t)$ *is* IFRA (DFRA) *if the function* $-(1/t)\ln\overline{F}(t)$ *is increasing (decreasing) in t.*

The notation *IFRA (DFRA)* is motivated by the fact that according to (1.8), given the existence of the density $f(t) = F'(t)$,

$$\overline{\lambda}(t) = \frac{1}{t}\int_0^t \lambda(x)dx = -\frac{1}{t}\ln\overline{F}(t).$$

Thus, $\overline{\lambda}(t) = -(1/t)\ln\overline{F}(t)$ is the average failure rate in $[0,t]$.

Obviously, *IFR (DFR)* implies *IFRA (DFRA)*. Note that probability distributions may belong to classes *IFR, DFR, IFRA, and DFRA* only in certain time intervals and then may switch to another class.

For recent comprehensive surveys on the stochastic modeling of the aging phenomenon of systems, in particular on the construction of lifetime distributions with bathtub-shaped failure rates, see [192] and [124].

1.1.3   Parametric Probability Distributions

In this section, some important probability distributions of nonnegative random variables such as life- and repair times are listed.

Exponential Distribution L has an *exponential distribution* with parameter λ if it has distribution function

$$F(t) = 1 - e^{-\lambda t}, \quad t \geq 0; \; \lambda > 0.$$

Density, failure rate and expected value of L are

$$f(t) = \lambda e^{-\lambda t}, \; t \geq 0; \quad \lambda(t) \equiv \lambda, \; E(L) = 1/\lambda.$$

Since in this case the failure rate is constant, systems with exponentially distributed lifetimes do not age. In particular, if such a system has survived interval $[0, x]$, then, from the point of view of its failure behavior, it is at time $t = x$ *as good as new*. More exactly, this means that

$$F(t) \equiv F_x(t) \text{ for all nonnegative } x. \tag{1.11}$$

Thus, $F(t)$ is both *IFR* and *DFR*. By (1.3), this equation is equivalent to

$$\overline{F}(x + t) = \overline{F}(x)\overline{F}(t), \quad x \ge 0, \ t \ge 0. \tag{1.12}$$

It is well known that (1.12) is valid iff $\overline{F}(t)$ has structure $\overline{F}(t) = e^{at}$, $t \ge 0$, for any constant a. Thus, the exponential distribution is the only probability distribution satisfying (1.12).

Weibull distribution L has a *Weibull distribution* with shape parameter β and scale parameter θ if its distribution function is given by

$$F(t) = 1 - e^{-(t/\theta)^\beta}, \ t \ge 0; \ \ \theta > 0, \ \beta > 0. \tag{1.13}$$

Density and failure rate are

$$f(t) = \frac{\beta}{\theta} \left(\frac{t}{\theta}\right)^{\beta-1} e^{-(t/\theta)^\beta}, \ t \ge 0,$$

$$\lambda(x) = \frac{\beta}{\theta} \left(\frac{x}{\theta}\right)^{\beta-1}, \ x > 0.$$

The expected value of L is

$$E(L) = \theta\Gamma(1/\beta + 1), \tag{1.14}$$

where

$$\Gamma(x) = \int_0^\infty u^{x-1} e^{-u} du, \ x > 0, \tag{1.15}$$

is the *Gamma function*. For $\beta > 1$, the failure rate strictly increases to ∞ if $x \to \infty$, whereas for $\beta < 1$ the failure rate strictly decreases to 0 if $x \to \infty$. For $\beta = 1$, the failure rate is constant: $\lambda(x) \equiv \lambda$; i.e. in this case the Weibull distribution becomes the exponential distribution with parameter $\lambda = 1/\theta$. Thus, depending on the value of its scale parameter, the Weibull distribution can be *IFR* or *DFR*, respectively.

Frechét distribution L has a *Frechét distribution* with shape parameter β and scale parameter θ if it has distribution function

$$F(t) = e^{-(\theta/t)^\beta}, \quad t > 0; \ \theta > 0, \ \beta > 0. \tag{1.16}$$

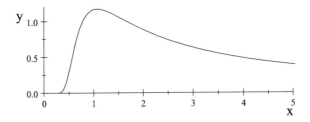

Figure 1.1: Failure rate of the Frechét distribution for $\beta = 2$ and $\theta = 1$

Density and failure rate are

$$f(t) = \beta\theta^\beta t^{-(\beta+1)} e^{-(\theta/t)^\beta}, \ t > 0,$$

$$\lambda(x) = \frac{\beta\theta^\beta x^{-(\beta+1)}}{e^{+(\theta/x)^\beta} - 1}, \quad x > 0.$$

If $\beta > 1$, the failure rate increases to an absolute maximum λ_m, and then decreases to 0. Thus, if $\lambda(x_m) = \lambda_m$, the Frechét distribution is *IFR* in $[0, x_m]$ and *DFR* in $[x_m, \infty)$. For instance, if $\theta = 1$ and $\beta = 2$, then $\lambda_m = \lambda(1.0695) \approx 1.17$ (Figure 1.1). The expected value of L is

$$E(L) = \theta \ \Gamma(1 - 1/\beta). \tag{1.17}$$

Gamma distribution L has a *gamma distribution* with parameters α and λ if its distribution function is given by

$$F(t) = \frac{\lambda^\alpha}{\Gamma(\alpha)} \int_0^t x^{\alpha-1} e^{-\lambda x} dx, \ \ t \geq 0, \ \alpha > 0, \ \lambda > 0. \tag{1.18}$$

Density and expected value are

$$f(t) = \frac{\lambda^\alpha}{\Gamma(\alpha)} t^{\alpha-1} e^{-\lambda t}, \ \ t \geq 0, \quad E(L) = \frac{\alpha}{\lambda}.$$

If $\alpha = n \geq 1$ is an integer, then the gamma distribution is called an *Erlang distribution* with parameters n and λ. Since $\Gamma(n) = (n - 1)!$, distribution function, density and expected value of the Erlang distribution are:

$$F(t) = 1 - e^{-\lambda t} \sum_{k=0}^{n-1} \frac{(\lambda t)^k}{k!}, \quad f(t) = \lambda \frac{(\lambda t)^{n-1}}{(n - 1)!} e^{-\lambda t}, \quad E(L) = \frac{n}{\lambda}. \tag{1.19}$$

(Note that the sum of n independent, exponentially with parameter λ distributed random variables has an Erlang distribution with parameters λ and n.) If $\alpha = 1$, the gamma distribution becomes the exponential distribution. The gamma distribution is *IFR* for $\alpha \geq 1$ and *DFR* for $0 < \alpha \leq 1$.

1.2 Structured Systems

1.2.1 Introduction

Let S be a system consisting of the *elements (subsystems, components)* e_1, $e_2,..., e_n$. The lifetime of e_i is L_i. Distribution function, survival function, density, and failure rate of L_i are denoted as $F_i(t)$, $\overline{F}_i(t)$, $f_i(t)$, and $\lambda_i(t)$; $i = 1, 2,..., n$. The corresponding notation for the system S is $F_s(t)$, $\overline{F}_s(t)$, $f_s(t)$, and $\lambda_s(t)$. In this section, the $L_1, L_2, ..., L_n$ are generally assumed to be independent random variables. The relationship between the availability (reliability) of the system and the availabilities (reliabilities) of its elements is graphically represented by so-called reliability block diagrams. A *reliability block diagram* is a graph (see Chapter 4, Definitions 4.1 and 4.2) with an *entrance node (entrance vertex)* s and an *exit node (exit vertex)* t. In such a diagram, all elements of the system are represented by one or more edges of the graph. But, depending on the way the graphical represention is being done, there may be edges, which do not symbolize elements, but which are necessary to fully determine the structure of the system and which are thought to be always available (see Figure 1.3, right). The system is available (operating) if and only if there is at least one path from s to t with all elements, symbolized by the edges of this path, being available (operating). In a reliability block diagram, edges are generally directed, but some may be undirected. The directions of the edges need not be explicitely marked if they are obvious. Undirected edges allow for paths from s to t which pass these edges in opposite directions. In Chapters 1 and 2, if necessary, the direction of edges is indicated by an arrow.

Fault trees are another graphical way of illustrating the reliability behavior of binary systems, see e.g. [18, 187, 158].

1.2.2 Series Systems

A *series system* is available iff all of its n elements are available. (Obviously, this explanation only makes sense if all elements are active.)

Hence, the reliability block diagram of a series system (Figure 1.2) only consists of one path from s to t. A series system fails as soon as one of its elements fails. Thus, its lifetime is the lifetime of that element which fails first:

$$L_s = \min(L_1, L_2, ..., L_n).$$

Figure 1.2: Series system

Thus, $\Pr(L_s > t) = \Pr(L_1 > t, L_2 > t, ..., L_n > t)$ so that

$$\overline{F}_s(t) = \overline{F}_1(t) \cdot \overline{F}_2(t) \cdots \overline{F}_n(t). \tag{1.20}$$

Hence, the survival probability of a series system with independent elements is equal to the product of the survival probabilities of its elements. By (1.8), Formula (1.20) can equivalently be written as

$$e^{\int_0^t \lambda_s(x)dx} = e^{\int_0^t [\lambda_1(x)+\lambda_2(x)+ \cdots +\lambda_n(x)]dx}.$$

Thus, the failure rate of a series system with independent elements is equal to the sum of the failure rates of its elements:

$$\lambda_s(t) = \lambda_1(t) + \lambda_2(t) + \cdots + \lambda_n(x).$$

In particular, if the elements have exponential lifetime distributions with parameters (failure rates) $\lambda_1, \lambda_2, ..., \lambda_n$, then the lifetime distribution of the series system is exponential with parameter (failure rate) $\lambda_s = \lambda_1 + \lambda_2 + \cdots + \lambda_n$. Hence, the expected system lifetime is

$$\mathbb{E}(L_s) = \frac{1}{\lambda_1 + \lambda_2 + \cdots + \lambda_n}.$$

1.2.3 Redundant Systems

An increase in the survival probability of a series system can only be achieved by increasing the survival probabilities of its elements. However, with increasing n, increasing the survival probabilities of the elements to meet a required level for the survival probability of series system may soon become unfeasible for technological or economic reasons. Consider a simple numerical example. Let the series system consist of n identical elements with joint survival probabilities 0.999 in the interval $[0, t_0]$. Table 1.1 shows the decrease of the system survival probability $\overline{F}_s(t_0)$ with increasing n. If $n \geq 100$, for most applications $\overline{F}_s(t_0)$ is unacceptably low.

n	1	10	100	1000
$\overline{F}_n(t_0)$	0.9990	0.9900	0.9048	0.3677

Table 1.1: Survival probabilities of a series system

The availability of a series system requires the availability of all its elements. Systems which do not have this property are called (*structurally*) *redundant*. In redundant systems, a failure of an element may not lead to a system breakdown. Thus, when a redundant system is working, some of its elements are active, whereas other of its elements may be not or only partially included in keeping the system operating. Hence, at a given point in time, elements of a redundant system can be classified as follows:

1. *Active* or *working elements*.

2. *Redundant elements.* They are called *passive* if they are not subject to any stress and, therefore, cannot fail. Passive elements are also called elements in *cold redundancy* or elements in *cold standby*. Redundant elements are called *partially active* if they are exposed to a lower stress level than active elements. They may fail, but with lower probability than when being (fully) active. Partially active elements are also called elements in *warm redundancy* or elements in *warm standby*. Elements in *hot redundancy* (*standby*) fully participate in the work of the system and, hence, they are exposed to the same stress level as active elements. Thus, there is no real difference between active elements and elements in hot redundancy in an operating system. (Sometimes the term "standby" is used in the sense of "cold standby".) Obviously, in an operating series system, all elements are active.

Parallel Systems A *parallel system* is available if at least one of its n elements is available (Figure 1.3).

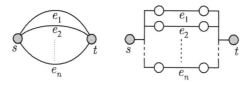

Figure 1.3: Parallel system — two graphical representations

We assume that if a new parallel system starts operating, all of its elements are active. The system fails as soon as the last of its elements fails. Therefore, the lifetime of a parallel system is

$$L_s = \max(L_1, L_2, ..., L_n),$$

so that $\Pr(L_s \leq t) = \Pr(L_1 \leq t, L_2 \leq t, \cdots, L_n \leq t)$. Hence, in case of independently operating elements, the distribution function of L_s is equal to the product of the distribution function of the lifetimes of its elements:

$$F_s(t) = F_1(t)F_2(t) \cdots F_n(t). \tag{1.21}$$

The corresponding survival probability of the system is

$$\overline{F}_s(t) = 1 - F_1(t)F_2(t) \cdots F_n(t).$$

Example 1.3 *A parallel system consists of* n *elements with independent, identically exponentially distributed lifetimes:*

$$F_i(t) = \Pr(L_i \leq t) = 1 - e^{-\lambda t}, \quad i = 1, 2, ..., n, \quad t \geq 0.$$

Then, by (1.21),
$$F_s(t) = (1 - e^{-\lambda t})^n, \quad t \geq 0. \tag{1.22}$$

Hence, from (1.2), the expected system lifetime is
$$E(L_s) = \int_0^\infty [1 - (1 - e^{-\lambda t})^n] dt.$$

By substituting $x = 1 - e^{-\lambda t}$ one gets
$$E(L_s) = \frac{1}{\lambda}\left(1 + \frac{1}{2} + \cdots + \frac{1}{n}\right).$$

What is the smallest integer $n = n^*$ with the property that the system survives the interval $[0, 1/\lambda]$ with probability 0.95? In view of (1.22), n^* is the smallest n satisfying
$$\overline{F}_s(1/\lambda) = 1 - (1 - e^{-\lambda t})^n \geq 0.95 \quad \text{or, equivalently,} \quad \frac{\ln 0.05}{\ln(1 - e^{-1})} \leq n.$$

It follows $n^* = 7$.

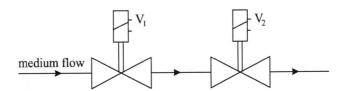

Figure 1.4: Series connection of two valves

In practice, technological and reliability theoretic system structures need not coincide. Consider, for instance, the technological series connection of two magnetic valves for controlling the flow of a medium (liquid, gas) through a pipe (Figure 1.4). Each of these valves can completely interrupt the flow of the medium, when closed. Hence, if the system failure is defined as "*not being able to interrupt the flow of the medium,*" this technological series system behaves like a reliability theoretic parallel system. However, if the system failure is "*interrupted medium flow*" (*blocked pipe*) because at least one of the valves fails to open, this technological series system needs to be modeled by a reliability theoretic series system.

To include redundancy in a system, engineers frequently design *series-parallel-systems*, i.e. parallel systems connected in series or *parallel-series-systems*, i.e. series systems connected in parallel (Figure 1.5). But couplings of parallel and series systems may also arise when the system itself can be modeled as a pure parallel or series system, respectively. As a simple example, consider a parallel system the elements of which are subjected to a so-called

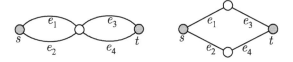

Figure 1.5: Series-Parallel and Parallel-Series System

common cause failure. The root causes of such failures are usually external (natural disasters, software failures, power cuts, human errors et al.). If such a common cause failure disables all the elements of the parallel system, then it can be included in the reliability model by a dummy element e_c which is in series with the parallel system. If e_c fails (i.e. a common cause failure occurs), then the whole system (parallel system with e_c in series) fails. Given that no common cause failure occurs, the reliability of the whole system is that of the parallel system.

k-out-of-n Systems A *k-out-of-n-system* is available if and only if at least k out of its n elements are available.

Sometimes, a k-out-of-n system defined in this way is called a k-out-of-n *system:G* (available $\simeq$ good). Similarly, a *k-out-of-n system:F* is unavailable if and only if at least k elements are unavailable (failed). Thus, a k-out-of-n system:G is equivalent to an $(n-k+1)$-out-of-n system:F. Hence, there is no real need to separately consider k-out-of-n systems:G and k-out-of-n systems:F.

In what follows, all elements are assumed to be active. Then, a series system is an n-out-of-n-system and a parallel system is a 1-out-of-n system. If $(L_1^*, L_2^*, ..., L_n^*)$ denotes the ordered sample of the lifetimes $(L_1, L_2, ..., L_n)$ of the elements, i.e. $L_1^* < L_2^* < \cdots < L_n^*$, then the lifetime of a k-out-of-n system is

$$L_s = L_{n-k+1}^*.$$

To determine the distribution function of L_s, let us next assume that the L_i are iid with distribution function $F(t)$. Then the probability that at time t exactly j selected elements are available and $n-j$ are not, is $[F(t)]^{n-j}[\overline{F}(t)]^j$. Since there are $\binom{n}{j}$ different possibilities to select j available elements from n, the probability that any j elements are available and $n-j$ are not, is

$$\binom{n}{j}[F(t)]^{n-j}[\overline{F}(t)]^j.$$

Hence, failure and survival probability of the system with regard to interval $[0, t]$ are

$$F_s(t) = \sum_{j=0}^{k-1}\binom{n}{j}[F(t)]^{n-j}[\overline{F}(t)]^j,$$

and

$$\overline{F}_s(t) = \sum_{j=k}^{n} \binom{n}{j} [F(t)]^{n-j} [\overline{F}(t)]^{j}, \tag{1.23}$$

respectively. The expected system lifetime is

$$\mathbb{E}(L_s) = \sum_{j=k}^{n} \binom{n}{j} \int_0^{\infty} [F(t)]^{n-j} [\overline{F}(t)]^{j} dt.$$

In particular, for the exponential distribution $F(t) = 1 - e^{-\lambda t}$, $t \geq 0$, one gets

$$\mathbb{E}(L_s) = \frac{1}{\lambda} \left(\frac{1}{k} + \frac{1}{k+1} + \cdots + \frac{1}{n} \right). \tag{1.24}$$

Now let us consider the more general case that the L_i are independent, but not necessarily identically distributed. If $\mathbf{y} = (y_1, y_2, ..., y_n)$ with $y_i = 0$ or 1 and $\overline{y}_i = 1 - y_i$, then the survival probability of the system with regard to the interval $[0, t]$ is

$$\overline{F}_s(t) = \tag{1.25}$$

$$\sum_{\substack{(\mathbf{y}, \ y_1+y_2+\cdots y_n=j) \\ j=k,k+1,\cdots,n}} F_1^{y_1}(t) F_2^{y_2}(t) \ \cdots \ F_n^{y_n}(t) \overline{F}_1^{\overline{y}_1}(t) \overline{F}_2^{\overline{y}_2}(t) \ \cdots \ \overline{F}_n^{\overline{y}_n}(t).$$

Note that the condition $y_1 + y_2 + \cdots + y_n = j$ implies that j elements are available and $n - j$ are not. There exist $\binom{n}{j}$ different vectors $\mathbf{y}$ which satisfy this condition.

k-out-of-n systems are widely used in military and industry to make systems *fault-tolerant*. A hydraulic control system, for instance, may be designed in such a way that it will operate fully satisfactorily if only 4 out of its 5 pumps are operating. Multidisplay systems in the cockpit of an aircraft are frequently designed as 2-out-of-3 systems. A communication system may designed in such a way that it functions normally if at least k out of n installed transmitters are available. *Majority systems* are special k-out-of-n systems with property $k > n/2$, i.e. a majority system is available iff more than a half of its elements are available. As a simple example, if n channels in parallel transmit signals (supposed to be identical) to a decision-making device and this device received identical signals from more than $n/2$ channels, then this signal will be considered the true one. (The input into the channels may be obtained from measurement devices that measure one and the same physical quantity.)

For surveys on k-out-of-n systems see, e.g., [190] and [264].

Example 1.4 *The pressure in a high pressure tank is monitored by 3 identical, independent sensors. If at least 2 sensors show the same pressure, then it is considered to be the true one. The probability that a sensor indicates the correct pressure at time t be* $\overline{F}(t)$. *Then, by (1.23), the probability that at least 2 sensors show the same pressure is*

$$\overline{F}_s(t) = [\overline{F}(t)]^2 [3 - 2\overline{F}(t)].$$

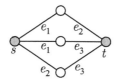

Figure 1.6: 2-out-of-3-system

Let the sensors have the constant failure rate $\lambda = 0.0001 \ [h^{-1}]$. Then the probability that a sensor indicates the correct pressure (within given tolerance limits) after 500 h is

$$e^{-0.05} = 0.951.$$

But a 2-out-of-3-system would indicate the correct pressure after 500 h with probability

$$e^{-0.1}(3 - 2e^{-0.05}) = 0.993.$$

The use of 2-out-of-3 (majority) systems is quite popular in reliability engineering. The additional effort is still acceptable and the reliability gain is usually sufficient. Figure 1.6 shows a reliability block diagram of a 2-out-of-3 system.

A special class of k-out-of-n systems are the *consecutive k-out-of-n systems*. Their elements are linearly ordered, more seldom circular (from the technological point of view). They fail as soon as k consecutive elements are not operating. As an example, consider a conveyor belt. If up to $k-1$ consecutive cylinders out of a total number of n fail, then the neighboring cylinders can carry their load, and no system failure arises. Two other examples were discussed by *Chiang and Niu* [88]:
1) Along a pipeline, n pumps are installed to maintain the required flow of oil. If up to $k-1$ consecutive pumps are down, then the neighboring pumps have the capacity to take over the full load of the failed pumps. But this is no longer possible if k or more successive pumps are down.
2) As part of a telecommunication system, n microwave stations have to transmit information from point s to point t. Each station has the capacity to transmit information over a distance of k stations. Thus, no transmission of information between s and t is possible if at least k successive microwave stations are not operating.
Despite their technological series structure, linear consecutive k-out-of-n-systems are systems with redundancy. Hence, they frequently have a high *a priori* availability. To achieve this level of availability with a parallel system would usually require a much higher technological input. Imagine in the above pipeline-pump-example the additional technological and economic effort necessary for replacing one pipeline with several pipelines in parallel.
For recent surveys on consecutive k-out-of-n systems see e.g. [226, 115].

1.3 Exercises

Exercise 1.1 *A nonnegative random variable X has a* uniform distribution *over the interval $[0,T]$ if it has the distribution function*

$$F(t) = t/T, \quad 0 \le t \le T.$$

Show that $F(t)$ is IFR.

Exercise 1.2 *A nonnegative random variable X has a* Pareto distribution *over the interval $[\alpha, \infty)$ with positive parameters α and β if it has the distribution function*

$$F(t) = 1 - (\alpha/t)^{\beta}, \quad \alpha \le t; \ \beta > 0.$$

Show that $F(t)$ is DFR.

Exercise 1.3 *A nonnegative random variable X has a* logarithmic normal *distribution with parameters μ and σ if it has the distribution function*

$$F(t) = \Phi\left(\frac{\ln t - \mu}{\sigma}\right), \quad t > 0,$$

where $\Phi(\cdot)$ is the distribution function of a standardised normally distributed random variable $N(0,1)$.
a) Show that X has structure $X = e^Y$, where $Y = N(\mu, \sigma^2)$.
b) Show that $\mathbb{E}(X) = e^{\mu + \sigma^2/2}$ and $Var(X) = e^{2\mu + \sigma^2}\left(e^{\sigma^2} - 1\right)$.
c) Investigate the aging behavior of a system with lifetime X.

Exercise 1.4 *A nonnegative random variable X has a* triangle distribution, *the density of which is depicted in Figure 1.7. Determine the failure rate of this distribution.*

Exercise 1.5 *Let X be a positive random variable with range $\{1, 2, ...\}$ and probability distribution $\Pr(X = k) = (1-p)^{k-1}p, \ 0 < p < 1$, i.e. X has a* geometric distribution *with parameter p. (Sometimes the range $\{0, 1, 2, ...\}$ is adequate. In this case, $\Pr(X = k) = (1-p)^k p, \ k = 0, 1, ...$)*

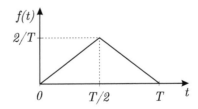

Figure 1.7: Triangle distribution

a) Show that $\Pr(X > n + k \,|\, X \geq k) = \Pr(X > n), \quad k = 1, 2, ...$ *Compare this formula to (1.11) and interpret it.*

b) The failure rate of any discrete random variable X with range $\{1, 2, ...\}$ is defined as

$$\lambda(k) = \frac{\Pr(X = k)}{\sum_{i=k}^{\infty} \Pr(x = i)}, \quad k = 1, 2, ...$$

Show that $\lambda(k) \equiv p$ for the geometric distribution.

Exercise 1.6 *The probability density $f(t)$ of a nonnegative random variable X is given by the mixture of two exponential distributions with parameters $\lambda_1 = 1$ and $\lambda_2 = 2$: $f(t) = 0.6 \cdot e^{-t} + 0.4 \cdot 2 \cdot e^{-2t}$, $t \geq 0$. Check whether this distribution is IFR.*

Exercise 1.7 *The failure rate of a system has the following bath tub shape:*

$$\lambda(x) = \begin{cases} -2x^2 + 4, & 0 \leq x \leq 1, \\ 2, & 1 \leq x \leq 3, \\ 2(x - 3)^2 + 2, & 3 \leq x. \end{cases}$$

a) Sketch this function.
b) Determine the distribution function of the system lifetime.

Exercise 1.8 *Figure 1.8 shows the reliability block diagram of a system consisting of 6 independent elements with entrance node s and exit node t. All elements have availability p. Determine the availability of the system.*

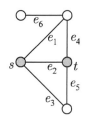

Figure 1.8: System with 6 elements

Exercise 1.9 *A parallel system consists of 4 independent elements, each with availability p, $0 < p < 1$. As soon as the first element fails, the availabilities of the 3 elements still operating decrease in view of increasing stress by a quarter of p. If the second element fails, the availabilities of the two remaining operating elements decrease by a third of their current level, and if the third element fails, the availability of the last operating element decreases to half of its current level. Let $p = e^{-0.01t}$. Determine the probability that the system survives the interval $[0, 200]$.*

Exercise 1.10 *A system consists of one active element and one identical spare element in cold redundancy. The active element has an exponentially distributed lifetime with parameter $\lambda = 0.01 \ [h^{-1}]$. As soon as the active element fails, the spare element becomes active.*

a) Determine the probability that the system survives 200 hours if change-over switching from the redundant to the active mode is absolutely reliable.

b) Determine the same probability on condition that the change-over switch device fails with probability 0.05.

c) Under otherwise the same conditions, compare the survival probability determined under a) with the one of a 2-element parallel system.

Exercise 1.11 *A parallel system consists of two identical active elements, each with survival probability $\overline{F}(t) = e^{-\lambda t}$, $t \geq 0$. To each active element a spare element in cold redundancy is assigned. When a spare element becomes active, it also has survival probability $\overline{F}(t) = e^{-\lambda t}$, $t \geq 0$. Let $\lambda = 0.01$. Determine the probability that the system survives the interval $[0, 300]$.*

Exercise 1.12 *Apart from Figure 1.6, plot three other different reliability block diagrams for a 2-out-of-3 system.*

Exercise 1.13 *The elements of a 2-out-of-4 system have availabilities 0.9, 0.8, 0.7, and 0.6.*

a) Develop a reliability block diagram for this system.

b) Determine the availability of this system.

Exercise 1.14 *The elements of a consecutive 2-out-of-4 system have the availabilities 0.9, 0.8, 0.7, and 0.6.*

a) Develop a reliability block diagram for this system.

b) On condition that the elements operate independently, determine the availability of this system.

c) How many series systems consisting of 4 independently operating elements with the same availabilities do you have to connect in parallel to achieve at least the same availability?

Chapter 2

Complex Systems

2.1 Foundations

In this chapter we introduce a general class of binary systems, which contain the ones dealt with in Chapter 1 as special cases. The reliability structures of most real-life binary systems can be modeled by a member of this class. We use the same basic notation as the one proposed in Section 1.2.1, but consider the system at a fixed time point or in the stationary regime. The foundations of the theory developed in this chapter have been laid by *Birnbaum et al.* [52].

The indicator variables of the system S and its elements e_i, $i = 1, 2, ..., n$, are denoted as

$$z_s = \begin{cases} 1 & \text{if } S \text{ is available} \\ 0 & \text{otherwise} \end{cases}, \quad z_i = \begin{cases} 1 & \text{if } e_i \text{ is available} \\ 0 & \text{otherwise} \end{cases}.$$

For z_s and the z_i being random variables, there exist probabilities p_s and p_i with

$$z_s = \begin{cases} 1 & \text{with probability } p_s \\ 0 & \text{with probability } 1 - p_s \end{cases}, \quad z_i = \begin{cases} 1 & \text{with probability } p_i \\ 0 & \text{with probability } 1 - p_i \end{cases},$$

where p_s is called the *system availability (reliability)* and the p_i are the *element availabilities (reliabilities)*. Since z_s and z_i are $(0, 1)-$variables,

$$p_s = \Pr(z_s = 1) = \mathbb{E}(z_s), \quad p_i = \Pr(z_i = 1) = \mathbb{E}(z_i).$$

The following assumption is essential in what follows: *The states of the elements uniquely determine the state of the system.* Hence, for different systems S there exist different functions φ with

$$z_s = \varphi(z_1, z_2, ..., z_n). \tag{2.1}$$

φ is called the *structure function* or the *system function* of S, and n is the *order* of the system or the *order* of φ. Function φ characterizes the structural dependence between the state of the system and the states of its elements. The system availability $p_s = \Pr(\varphi = 1)$ is the expected value of φ:

$$p_s = \mathbb{E}(\varphi(z_1, z_2, ..., z_n)). \tag{2.2}$$

Thus, knowledge of the structure function is crucial for determining the system availability. In particular, if the $z_1, z_2, ..., z_n$ are independent random variables, then we will see that p_s is a function of the element availabilities $p_1, p_2, ..., p_n$. In this case, letting $\mathbf{p} = (p_1, p_2, ..., p_n)$, formula (2.2) is written as

$$p_s = h(\mathbf{p}) \text{ or simply as } p_s = h(p) \text{ if } p = p_1 = p_2 = \cdots = p_n. \tag{2.3}$$

Given a reliability block diagram of a system (or its fault tree), there are computerized algorithms for determining its structure function. Unfortunately, the computation time for obtaining the structure function is generally exponentially increased with increasing order n. Thus, developing computationally efficient algorithms for determining φ is a main problem in reliability theory. In achieving this goal, the formalism of Boolean algebra will play an important role. Hence, we will next summarize its terminology and basic rules.

Let x and y be any two $(0, 1)$−variables, i.e. two *Boolean* or *binary* variables, which assume values 0 or 1. The key relations between x and y are *conjunction, disjunction,* and *negation.*

$$\text{conjunction: } x \wedge y = \begin{cases} 1 & \text{if } x = y = 1 \\ 0 & \text{otherwise} \end{cases}$$

$$\text{disjunction: } x \vee y = \begin{cases} 0 & \text{if } x = y = 0 \\ 1 & \text{otherwise} \end{cases}$$

$$\text{negation: } \quad \overline{x} = \begin{cases} 1 & \text{if } x = 0 \\ 0 & \text{if } x = 1 \end{cases}$$

An advantage of assigning the range $\{0, 1\}$ to Boolean variables is that these three relations can be expressed by arithmetic operations:

$$\text{conjunction: } x \wedge y = xy$$
$$\text{disjunction: } x \vee y = x + y - xy$$
$$\text{negation: } \overline{x} = 1 - x$$

A third and, as we will see later, very useful way of equivalently defining conjunction and disjunction is:

$$\text{conjunction: } x \wedge y = \min(x, y)$$
$$\text{disjunction: } x \vee y = \max(x, y)$$

Definition 2.1 *Two Boolean variables x and y are* disjoint *or* orthogonal *if $xy = 0$. In this case, $x \vee y = x + y$.*

In particular, x and its negation $\bar{x}$ are disjoint: $x\bar{x} = 0$.

For any $(0,1)$-variables x, y, and z, it is easy to verify the following properties:

$$
\begin{aligned}
\text{Idempotency:}\quad & x \wedge x = x, \quad x \vee x = x \\
\text{Commutativity:}\quad & x \wedge y = y \wedge x, \quad x \vee y = y \vee x \\
\text{Associativity:}\quad & (x \wedge y) \wedge z = x \wedge (y \wedge z) \\
\text{Associativity:}\quad & (x \vee y) \vee z = x \vee (y \vee z) \\
\text{Distributivity:}\quad & x \wedge (y \vee z) = (x \wedge y) \vee (x \wedge z) \\
\text{Distributivity:}\quad & x \vee (y \wedge z) = (x \vee y) \wedge (x \vee z) \\
\text{Absorption:}\quad & x \vee (x \wedge y) = x, \quad x \wedge (x \vee y) = x \\
\text{Double negation:}\quad & \bar{\bar{x}} = x \\
\text{de Morgan rules:}\quad & \overline{x \wedge y} = \bar{x} \vee \bar{y}, \quad \overline{x \vee y} = \bar{x} \wedge \bar{y}
\end{aligned}
\tag{2.4}
$$

As we will see later, a particularly important property is the following one:

$$
\text{Orthogonalization:}\quad x \vee y = x + \bar{x}y = x + y - xy.
\tag{2.5}
$$

Now let $x_1, x_2, ..., x_n$ be $(0,1)$-variables. Their *conjunction* and *disjunction* are defined as

$$
\bigwedge_{i=1}^{n} x_i = x_1 \wedge x_2 \wedge \cdots \wedge x_n = \begin{cases} 1 & \text{if } x_1 = x_2 = \cdots = x_n = 1 \\ 0 & \text{otherwise} \end{cases},
$$

$$
\bigvee_{i=1}^{n} x_i = x_1 \vee x_2 \vee \cdots \vee x_n = \begin{cases} 0 & \text{if } x_1 = x_2 = \cdots = x_n = 0 \\ 1 & \text{otherwise} \end{cases}.
$$

Equivalent representations are

$$
\bigwedge_{i=1}^{n} x_i = \prod_{i=1}^{n} x_i = \min(x_1, x_2, ..., x_n),
\tag{2.6}
$$

$$
\bigvee_{i=1}^{n} x_i = \coprod_{i=1}^{n} x_i = \max(x_1, x_2, ..., x_n),
\tag{2.7}
$$

where

$$
\coprod_{i=1}^{n} x_i = 1 - \prod_{i=1}^{n} \bar{x}_i .
\tag{2.8}
$$

The *de Morgan rules* are

$$
\overline{\bigwedge_{i=1}^{n} x_i} = \bigvee_{i=1}^{n} \bar{x}_i, \quad \overline{\bigvee_{i=1}^{n} x_i} = \bigwedge_{i=1}^{n} \bar{x}_i .
\tag{2.9}
$$

A Boolean variable y, which is a function of other Boolean variables $x_1, x_2, \ldots,$ x_n, is called a *Boolean function of order* n:

$$y = y(\mathbf{x}), \quad \mathbf{x} = (x_1, x_2, \ldots, x_n) \in V_n,$$

where V_n is the set of all state vectors $\mathbf{x} = (x_1, x_2, \ldots, x_n)$ with property that the x_i can take on only values 0 or 1. Thus, V_n has cardinality 2^n, i.e. it contains 2^n different vectors. Special Boolean functions of order n are the conjunction $y = \bigwedge_{i=1}^{n} x_i$ and the disjunction $y = \bigvee_{i=1}^{n} x_i$ as well as the structure function φ given by (2.1). The representation of a Boolean function in dependence on their arguments x_i is usually not unique. Hence, two Boolean functions y_1 and y_2 of order n are called *(logically) equivalent* if $y_1(\mathbf{x}) = y_2(\mathbf{x})$ for all $\mathbf{x} \in V_n$. In particular, every Boolean function $y = y(\mathbf{x})$ can be represented by a logically equivalent disjunction of Boolean functions C_k, which are conjunctions of some x_i and $\overline{x}_j$:

$$y = \bigvee_{k=1}^{c} C_k = C_1 \vee C_2 \vee \cdots \vee C_c, \tag{2.10}$$

and by a conjunction of Boolean functions D_k, which are disjunctions of some x_i and $\overline{x}_j$:

$$y = \bigwedge_{k=1}^{d} D_k = D_1 \wedge D_2 \wedge \cdots \wedge D_d. \tag{2.11}$$

The representations (2.10) and (2.11) are called *disjunctive normal form* and *conjunctive normal form* of y, respectively. If the C_k are pairwise disjoint, i.e $C_i C_j = 0$ for $i \neq j$, then the disjunctive normal form (2.10) simplifies to

$$y = \sum_{k=1}^{c} C_k = C_1 + C_2 + \cdots + C_c. \tag{2.12}$$

In what follows, any disjunctive normal form of y with pairwise orthogonal C_k is called a *disjoint sum form* (*orthogonal form*) of y.

Note that two conjunctions (products) C_k and C_l of some x_i and $\overline{x}_j$ are disjoint iff there exists an x_i, which is a factor of C_k, and at the same time its negation $\overline{x}_i$ is a factor of C_l, or vice versa.

2.2　Coherent Systems

2.2.1　Definition, Properties and Examples

In this section we discuss properties of the structure function (2.1) and methods for its generation. Usually we will write (2.1) in the form

$$z_s = \varphi(\mathbf{z}), \quad \mathbf{z} \in V_n. \tag{2.13}$$

Then we have for the basic systems introduced in Chapter 1,

$$\text{Series system:} \quad \varphi(\mathbf{z}) = \prod_{i=1}^{n} z_i = \min(z_1, z_2, ..., z_n)$$

$$\text{Parallel system:} \quad \varphi(\mathbf{z}) = \coprod_{i=1}^{n} z_i = \max(z_1, z_2, ..., z_n)$$

$$\text{k-out-of-n system:} \quad \varphi(\mathbf{z}) = \begin{cases} 1 & \text{if } z_1 + z_2 + \cdots + z_n \geq k \\ 0 & \text{if } z_1 + z_2 + \cdots + z_n < k \end{cases} \quad (2.14)$$

As mentioned before, Boolean functions are not unique. The aim must be to construct structure functions for a system with low complexity (with regard to the number of its terms, the number of variables involved in the terms and their mutual connections within the terms). The first step in this direction is to eliminate unnecessary variables. To identify such variables, let

$$(r_i, \mathbf{z}) = (z_1, z_2, ..., z_{i-1}, r, z_{i+1}, ..., z_n), \quad (2.15)$$

where $\mathbf{z} = (z_1, z_2, ..., z_n) \in V_n$ and $r = 0$ or $r = 1$. This notation means that state of element e_i is fixed.

Definition 2.2 *Element e_i is called* irrelevant *if*

$$\varphi((0_i, \mathbf{z})) = \varphi((1_i, \mathbf{z})) \quad \text{for all } \mathbf{z} \in V_n.$$

Otherwise e_i is called relevant, $i = 1, 2, ..., n.$

Thus, whether an irrelevant element is available or not has no influence on the system availability. Just as requiring that all system elements are relevant, it also makes sense to assume that the structure function does not decrease if a nonavailable element becomes available, or equivalently, that a nonavailable system becomes available when one of its element fails. This leads to the following definition:

Definition 2.3 *A binary system S with structure function $\varphi(\mathbf{z})$, $\mathbf{z} \in V_n$, is called* coherent *or* monotone *if*

a) every element is relevant, and
b) $\varphi((0_i, \mathbf{z})) \leq \varphi((1_i, \mathbf{z}))$ for all $\mathbf{z} \in V_n.$

Let $\mathbf{0} = (0, 0, ..., 0)$ and $\mathbf{1} = (1, 1, ..., 1)$. Then a) and b) imply two intuitively obvious properties of coherent systems. The first one is

$$\varphi(\mathbf{0}) = 0, \quad \varphi(\mathbf{1}) = 1. \quad (2.16)$$

To verify (2.16), let us assume that $\varphi(\mathbf{0}) = 1$. Then, since φ is nondecreasing, $\varphi(\mathbf{0}) = \varphi(\mathbf{1}) = 1$. But this implies that all elements of S are irrelevant, contrary to a). The proof of $\varphi(\mathbf{1}) = 1$ is analogous. The second property is

$$\prod_{i=1}^{n} z_i \leq \varphi(\mathbf{z}) \leq \coprod_{i=1}^{n} z_i . \quad (2.17)$$

Since φ is nonnegative, the left-hand side of (2.17) is trivial if $\prod_{i=1}^{n} z_i = 0$. If $\prod_{i=1}^{n} z_i = 1$, then $\mathbf{z} = \mathbf{1}$ so that, by (2.16), $\varphi(\mathbf{1}) = 1$. The right-hand side of (2.17) is proved analogously.

The inequalities (2.17) express that any coherent system is *structurally stronger (weaker)* than a series (parallel) system with the same number of elements.

Definition 2.4 *Let φ be the structure function of S. The* dual structure function *to φ is defined as*

$$\varphi_d(\mathbf{z}) = 1 - \varphi(\mathbf{1} - \mathbf{z}), \tag{2.18}$$

where $\mathbf{z} = (z_1, z_2, ..., z_n)$ and $\mathbf{1} - \mathbf{z} = (\overline{z}_1, \overline{z}_2, ..., \overline{z}_n)$. $\varphi_d(\mathbf{z})$ is the structure function of S_d, the dual system *to S.*

The dual system to a series (parallel) system is a parallel (series) system. More generally, the dual system to a k-out-of-n system is a $(n-k+1)$-out-of-n system. Dual systems frequently occur with systems which have two failure modes (see valve system, Figure 1.4).

Example 2.5 *Let S be the series-parallel system shown in Figure 1.5 (left). With $\mathbf{z} = (z_1, z_2, z_3, z_4)$, its structure function is*

$$\varphi(\mathbf{z}) = (z_1 \vee z_2) \wedge (z_3 \vee z_4)$$

or, equivalently,

$$\varphi(\mathbf{z}) = (1 - \overline{z}_1 \overline{z}_2)(1 - \overline{z}_4 \overline{z}_4).$$

The structure function dual to φ is

$$\begin{aligned}
\varphi_d(\mathbf{z}) &= 1 - (1 - z_1 z_2)(1 - z_3 z_4) \\
&= z_1 z_2 + z_3 z_4 - z_1 z_2 z_3 z_4 \\
&= z_1 z_2 \vee z_3 z_4 \ .
\end{aligned}$$

Thus, $\varphi_d(\mathbf{z})$ is the structure function of a system which is a parallel connection of two series systems (Figure 1.5, right).

It should be mentioned that there are real systems consisting only of relevant elements, which, however, from the reliability point of view, are not coherent, i.e. their structure functions do not increase in each argument. Example 2.6 presents a system of this kind. Others have been e.g. analyzed by *Inagaki* and *Henley* [159] and *Zhang* and *Mei* [313].

Example 2.6 *A pump P feeds acid to a tank T (Figure 2.1). Due to power fluctuations the pump is subject to, the acid level L in the tank varies. However, this is only cause for concern if the acid level is equal to or exceeds a*

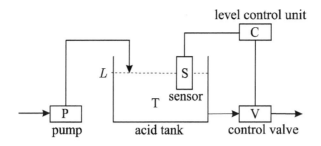

Figure 2.1: Acid level control system

critical level L_c. The level is monitored by a sensor S, which continuously transmits the actual acid level to a control unit C. If $L \geq L_c$, C makes a control valve V to increase the acid draining off. The control valve is assumed to be absolutely reliable. If $L < L_c$, then S always correctly responds to this situation. However, S may indicate an admissable acid level although $L \geq L_c$. Hence, the following indicator variables are relevant:

$$z_1 = \begin{cases} 1 & \text{if } L < L_c \ (P \text{ pumps normally}) \\ 0 & \text{if } L \geq L_c \ (P \text{ pumps too much}) \end{cases}$$

$$z_2 = \begin{cases} 1 & \text{if } S \text{ indicates correctly} \\ 0 & \text{otherwise} \end{cases}$$

$$z_3 = \begin{cases} 1 & \text{if } C \text{ responds correctly to the signal from } S \\ 0 & \text{otherwise} \end{cases}$$

The system is considered to be functioning if $L < L_c$ or, if $L \geq L_c$, then this state has been identified and is being removed (acid draining off is taking place). Hence, the system is functioning if it is in one of the following three states:

$$(1,1,1), \ (0,1,1), \ (0,0,0). \tag{2.19}$$

It is amazing that in this example the system is functioning even in state $(0,0,0)$. This is due to the fact that C may wrongly respond to a wrong signal from S. (On condition $z_1 = 0$, the wrong signal from S is $L < L_c$.) The structure function of the system is

$$\varphi(\mathbf{z}) = z_1 z_2 z_3 \vee \overline{z}_1 z_2 z_3 \vee \overline{z}_1 \overline{z}_2 \overline{z}_3 \ , \tag{2.20}$$

since φ assumes value 1 iff $\mathbf{z} = (z_1, z_2, z_3)$ is one of the vectors (2.19). Note that there is no need to include the control valve V in (2.20), since, by assumption, it does not fail. The three conjunctions in (2.20) are obviously disjoint so that φ can be written as

$$\varphi(\mathbf{z}) = z_1 z_2 z_3 + \overline{z}_1 z_2 z_3 + \overline{z}_1 \overline{z}_2 \overline{z}_3 \ = z_2 z_3 + \overline{z}_1 \overline{z}_2 \overline{z}_3 \ .$$

φ is not monotone, since $\varphi(0,0,1) = 0$ and $\varphi(0,0,0) = 1$. Thus, if the system is in the faulty state $(0,0,1)$, then the failure of the control unit C will render the system functioning.

Level of parallel redundancy If in the design of a (sub-) system redundancy has to be included to meet the required availability level, then the designer has two basic options (if not excluded by technological or economic reasons): Either two systems each with structure function φ are connected in parallel, or the individual elements of the system are replaced by two elements in parallel. (In both cases, the total number of elements in the enhanced system is $2n$.) What option is preferable from the availability point of view? To answer this question, let $\mathbf{x} = (x_1, x_2, ..., x_n) \in V_n$, $\mathbf{z} = (z_1, z_2, ..., z_n) \in V_n$, and

$$\mathbf{x} \vee \mathbf{z} = (x_1 \vee z_1, x_2 \vee z_2, ..., x_n \vee z_n)$$

Then

$$\varphi(\mathbf{x}) \vee \varphi(\mathbf{z}) \leq \varphi(\mathbf{x} \vee \mathbf{z}) \tag{2.21}$$

To verify this inequality, note that $x_i \leq \max(x_i, z_i) = x_i \vee z_i$ for all $i = 1, 2, ..., n$. Hence, since φ is increasing in every variable, $\varphi(\mathbf{x}) \leq \varphi(\mathbf{x} \vee \mathbf{z})$. Analogously, $\varphi(\mathbf{z}) \leq \varphi(\mathbf{x} \vee \mathbf{z})$. Thus,

$$\varphi(\mathbf{x}) \vee \varphi(\mathbf{z}) = \max(\varphi(\mathbf{x}), \varphi(\mathbf{z})) \leq \varphi(\mathbf{x} \vee \mathbf{z}).$$

In other words: Redundancy at element level is preferable to redundancy at system level.

Figure 2.2 illustrates the situation for $n = 2$: By replacing the elements e_1 and e_2 in Figure 2.2 a) with two elements in parallel, one gets Figure 2.2 b), and by connecting two series systems of type Figure 2.2 a) in parallel one gets Figure 2.2 c). The inequality (2.21) makes sure that for all element availabilities p_1, p_2, p_3, p_4, a system with reliability block diagram Figure 2.2 b) has a higher availability than a system with reliability block diagram Figure 2.2 c).

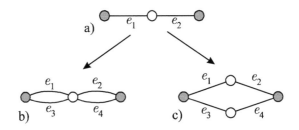

Figure 2.2: Redundancy options

Level of series switch   Now, to do the analogous thing for a series connection (although this has nothing to do with redundancy), let

$$\mathbf{x} \wedge \mathbf{z} = (x_1 \wedge z_1, x_2 \wedge z_2, ..., x_n \wedge z_n) = (x_1 z_1, x_2 z_2, ..., x_n z_n).$$

(Note that $\mathbf{x} \wedge \mathbf{z}$ is the scalar product of vectors $\mathbf{x}$ and $\mathbf{z}$.) Then,

$$\varphi(\mathbf{x} \wedge \mathbf{z}) \leq \varphi(\mathbf{x}) \wedge \varphi(\mathbf{z}). \tag{2.22}$$

The proof is similar to the one of (2.21). This inequality implies that connecting two systems, each with structure function φ, in series is from the availability point of view superior to replacing each element by a series connection of the two corresponding system elements.

Pivotal decomposition   By making use of the notation (2.15), every structure function $\varphi(\mathbf{z})$ of the order n, $n > 1$, can be represented as a sum of two disjoint Boolean functions of the order $n - 1$ in the following way:

$$\varphi(\mathbf{z}) = z_i \varphi((1_i, \mathbf{z})) + \overline{z}_i \varphi((0_i, \mathbf{z})); \quad i = 1, 2, ..., n. \tag{2.23}$$

Element e_i is called a *pivotal element*. Principally, every element can be selected as the pivotal element. However, one will select the pivotal element in such a way that the Boolean functions $\varphi((1_i, \mathbf{z}))$ and $\varphi((0_i, \mathbf{z}))$ become "as simple as possible." Then applying the *(pivotal) decomposition formula* (2.23) may drastically reduce the computational effort for obtaining $\varphi(\mathbf{z})$. Of course, (2.23) can also be applied to $\varphi((1_i, \mathbf{z}))$ and $\varphi((0_i, \mathbf{z}))$. So, for instance, applying (2.23) to these functions with pivotal element e_j, an obvious modification of the notation (2.15), yields

$$\varphi((1_i, \mathbf{z})) = z_j \varphi((1_i, 1_j, \mathbf{z})) + \overline{z}_j \varphi((1_i, 0_j, \mathbf{z})),$$
$$\varphi((0_i, \mathbf{z})) = z_j \varphi((0_i, 1_j, \mathbf{z})) + \overline{z}_j \varphi((0_i, 0_j, \mathbf{z})).$$

Inserting these decompositions into (2.23) yields

$$\varphi(\mathbf{z}) = z_i z_j \varphi((1_i, 1_j, \mathbf{z})) + z_i \overline{z}_j \varphi((1_i, 0_j, \mathbf{z}))]$$
$$+ \overline{z}_i z_j \varphi((0_i, 1_j, \mathbf{z})) + \overline{z}_i \overline{z}_j \varphi((0_i, 0_j, \mathbf{z}))].$$

(In this representation of $\varphi(\mathbf{z})$, all terms are disjoint.)   Proceeding in this way till all the variables z_i in $\mathbf{z}$ are fixed by 0 or 1, yields the following representation of $\varphi(\mathbf{z})$:

$$\varphi(\mathbf{z}) = \sum_{\mathbf{x} \in V_n} \varphi(\mathbf{x}) \prod_{k=1}^{n} z_k^{x_k} \overline{z}_k^{\overline{x}_k}, \tag{2.24}$$

where $\mathbf{x} = (x_1, x_2, ..., x_n)$. In view of its construction, all the terms in (2.24) are disjoint. Hence, this $\varphi(\mathbf{z})$ has structure (2.12) so that it is a *disjoint sum form*. Principally, generating the form (2.24) requires a complete enumeration of all of the 2^n possible vectors $\mathbf{x} \in V_n$ and knowledge of the corresponding values $\varphi(\mathbf{x})$ of the structure function. Hence, with increasing order n of the system, the computational effort for determining $\varphi(\mathbf{z})$ grows exponentially fast. If the z_i are independent, then (2.24) yields the system availability

$$p_s = \sum_{\mathbf{x} \in V_n} \varphi(\mathbf{x}) \prod_{k=1}^{n} p_k^{x_k} \overline{p}_k^{\overline{x}_k}.$$

By taking the expected value on both sides of (2.23) one obtains a *decomposition formula with pivotal element e_i* for the system availability:

$$p_s = p_i h((1_i, \mathbf{p})) + \overline{p}_i h((0_i, \mathbf{p})); \quad i = 1, 2, ..., n, \tag{2.25}$$

with

$$h((r_i, \mathbf{p})) = \mathbb{E}(\varphi((r_i, \mathbf{z}))), \quad r = 0 \text{ or } 1. \tag{2.26}$$

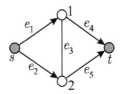

Figure 2.3: Bridge structure

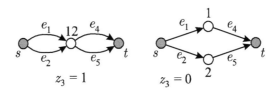

Figure 2.4: Decomposition of the bridge structure

Example 2.7 *Consider a system with the reliability block diagram given by Figure 2.3. (Note that edge e_3 is undirected.) This is the bridge structure, a standard example in reliability textbooks. To derive its structure function, we apply formula (2.23) with pivotal element e_3. State vector $(1_3, \mathbf{z})$ implies that there is always a connection between nodes 1 and 2 in the bridge structure. Hence, the reliability block diagram belonging to $\varphi((1_3, \mathbf{z}))$ arises from the bridge structure by fusing nodes 1 and 2 to the node 12. When doing this, edge e_3 becomes a loop, which will be removed (Figure 2.4 left). The resulting structure is a series-parallel system, where the two parallel systems consist of* *elements e_1, e_2 and e_4, e_5, respectively. State vector $(0_3, \mathbf{z})$ implies that edge e_3 is not available. Hence, the reliability block diagram belonging to $\varphi((0_3, \mathbf{z}))$ arises from the bridge structure simply by removing edge e_3. The resulting structure is a parallel-series system, where the two series systems consist of elements e_1, e_4 and e_2, e_5, respectively (Figure 2.4 right). Thus,*

$$\varphi((1_3, \mathbf{z})) = (1 - \overline{z}_1 \overline{z}_2)(1 - \overline{z}_4 \overline{z}_5),$$
$$\varphi((0_3, \mathbf{z})) = 1 - (1 - z_1 z_4)(1 - z_2 z_5).$$

Inserting these results into (2.23) with $i = 3$ and $n = 5$, simplifying as far as possible, yields

$$\varphi(\mathbf{z}) = z_1 z_4 + z_2 z_5 + z_1 z_3 z_5 + z_2 z_3 z_4 - z_1 z_2 z_3 z_4 - z_1 z_2 z_3 z_5 \qquad (2.27)$$
$$- z_1 z_2 z_4 z_5 - z_1 z_3 z_4 z_5 - z_2 z_3 z_4 z_5 + 2 z_1 z_2 z_3 z_4 z_5.$$

This representation is called the linear form *of the structure function.*

Generally, the *linear form* of the structure function of a monotone system is given by

$$\varphi(\mathbf{z}) = \sum_{i=1}^{n} a_i z_i + \sum_{\substack{i,j=1 \\ i<j}}^{n} a_{ij} z_i z_j + \sum_{\substack{i,j,k=1 \\ i<j<k}}^{n} a_{ijk} z_i z_j z_k + \cdots + a_{12\ldots n} z_1 z_2 \cdots z_n, \quad (2.28)$$

where the coefficients a_i, a_{ij}, ..., $a_{12\ldots n}$ are integers. The advantage of having the structure function given by a disjoint sum form or by the linear form results from formula (2.2): The system availability is equal to the sum of the expected values of the terms in these forms. In particular, if the elements operate independently, i.e. the z_i are independent, one obtains the system availability by replacing in the linear form the z_i with the corresponding element availabilities $p_i = \Pr(z_i = 1)$. This gives the system availability $p_s = h(\mathbf{p})$ in form of a polynomial, the so-called *reliability (availability) polynomial* of the system. If in addition $p_i = p$, $i = 1, 2, ..., n$, the reliability polynomial has the form

$$h(p) = d_1 p + d_2 p^2 + \cdots + d_n p^n$$

with integers d_i satisfying $d_1 + d_2 + \cdots + d_n = 1$. Thus, from (2.27), for identical elements the reliability polynomial of the bridge structure is

$$h(p) = 2p^2 + 2p^3 - 5p^4 + 2p^5.$$

Example 2.8 *The structure function of the 2-out-of-3 system is given by Table 2.1. In the corresponding disjoint sum form (2.24), only the state vectors with $k = 5, 6, 7, 8$ generate nondisappearing terms:*

$$\varphi(\mathbf{z}) = z_1^1 \overline{z}_1^0 z_2^1 \overline{z}_2^0 z_3^0 \overline{z}_3^1 + z_1^1 \overline{z}_1^0 z_2^0 \overline{z}_2^1 z_3^1 \overline{z}_3^0 + z_1^0 \overline{z}_1^1 z_2^1 \overline{z}_2^0 z_3^1 \overline{z}_3^0 + z_1^1 \overline{z}_1^0 z_2^1 \overline{z}_2^0 z_3^1 \overline{z}_3^0$$
$$= z_1 z_2 (1 - z_3) + z_1 (1 - z_2) z_3 + (1 - z_1) z_2 z_3 + z_1 z_2 z_3$$

k	1	2	3	4	5	6	7	8
$\mathbf{x}$	$0,0,0$	$1,0,0$	$0,1,0$	$0,0,1$	$1,1,0$	$1,0,1$	$0,1,1$	$1,1,1$
$\varphi(\mathbf{x})$	0	0	0	0	1	1	1	1

Table 2.1: Structure function of the 2-out-of-3 system

Hence, the linear form of φ is

$$\varphi(\mathbf{z}) = z_1 z_2 + z_1 z_3 + z_2 z_3 - 2 z_1 z_2 z_3.$$

This form is logically equivalent to (2.14) with $k = 2$ and $n = 3$. The corresponding reliability polynomial in case of identical elements is

$$h(p) = 3p^2 - 2p^3.$$

2.2.2 Path-and-Cut Representation of Structure Functions

Path-and-Cut Vectors (Sets)   A structure function partitions the set of all state vectors V_n into two disjoint subsets $V_n^{(1)}$ and $V_n^{(0)}$ in the following way:

$$V_n^{(1)} = \{\mathbf{z},\ \varphi(\mathbf{z}) = 1\} \quad \text{and} \quad V_n^{(0)} = \{\mathbf{z},\ \varphi(\mathbf{z}) = 0\}.$$

Thus, the sets $V_n^{(1)}$ and $V_n^{(0)}$ contain all those state vectors for which the system is available or not available, respectively. The vectors in $V_n^{(1)}$ are called *path vectors* and the vectors in $V_n^{(0)}$ are called *cut vectors*. To path/cut vectors $\mathbf{z} = (z_1, z_2, ..., z_n)$ sets of indices are assigned as follows:

$$P = \{j,\ z_j = 1\}, \quad C = \{k,\ z_k = 0\}.$$

If $\mathbf{z}$ is a path vector, then P is called the *path set* belonging to P, and if $\mathbf{z}$ is a cut vector, then C is called the *cut set* belonging to $\mathbf{z}$. The concepts of path/cut vectors and path/cut sets are obviously equivalent. Hence, we frequently will only use the simplified terminology *paths* and *cuts*. Sometimes it is more convenient to identify the indices specified by path sets and cut sets with their corresponding elements. In view of (2.16), vector $\mathbf{1} = (1, 1, ..., 1)$ is always a path vector and vector $\mathbf{0} = (0, 0, ..., 0)$ is always a cut vector. For this reason, $\mathbf{1}$ is said to be the *trivial path vector* and $\mathbf{0}$ is said to be the *trivial cut vector*.

Note that by definition the elements with indices outside a path set are not available, whereas the elements with indices outside a cut set are available.

A series system has only one path vector (the trivial one), and a parallel system has only one cut vector (the trivial one). According to Table 2.1, a 2-out-of-3 system has path vectors $(1, 1, 0)$, $(1, 0, 1)$, $(0, 1, 1)$, $(1, 1, 1)$, and cut vectors $(0, 0, 0)$, $(1, 0, 0)$, $(0, 1, 0)$, $(0, 0, 1)$.

A structure function defines the paths and cuts of a system. Conversely, from the paths or cuts of a system, its structure function can be generated. However, for the latter purpose, it is more efficient to only use the so-called minimal path sets and minimal cut sets, respectively. To introduce these concepts, we first need to introduce a partial order in the state space V_n. Let

$$\mathbf{x} = (x_1, x_2, ..., x_n) \text{ and } \mathbf{z} = (z_1, z_2, ..., z_n) \in V_n.$$

If $x_i \leq z_i$, $i = 1, 2, ..., n$, then we write $\mathbf{x} \leq \mathbf{z}$. If $\mathbf{x} \leq \mathbf{z}$ and there exists at least one i with property that $x_i < z_i$, then we write $\mathbf{x} < \mathbf{z}$. In the latter case, vector $\mathbf{x}$ is called *smaller than vector* $\mathbf{z}$, or, equivalently, *vector* $\mathbf{z}$ *is greater than vector* $\mathbf{x}$. The relation "$<$" is a *partial order* in V_n, since, for instance, if $\mathbf{x} = (1, 0, 1)$ and $\mathbf{z} = (0, 1, 1)$, there is neither $\mathbf{x} < \mathbf{z}$ nor $\mathbf{x} > \mathbf{z}$.

Definition 2.9 *A path vector* $\mathbf{z}$ *is called* minimal *if for all* $\mathbf{x} \in V_n$ *with* $\mathbf{x} < \mathbf{z}$, $\varphi(\mathbf{x}) = 0$. *A cut vector* $\mathbf{z}$ *is called* minimal *if for all* $\mathbf{x} \in V_n$ *with* $\mathbf{x} > \mathbf{z}$, $\varphi(\mathbf{x}) = 1$. *To minimal path (cut) vectors there belong* minimal path (cut) sets.

A failure of only one element operating under a minimal path vector induces a system failure, whereas the repair of only one element being unavailable under a minimal cut vector will render the system available. Thus, "minimal" in connection with minimal path (cut) vectors refers to the number of elements being available (not available).

The minimal path (cut) sets of the bridge structure (Figure 2.3) are

$$P_1 = \{1, 4\}, \ P_2 = \{2, 5\}, \ P_3 = \{1, 3, 5\}, \ P_4 = \{2, 3, 4\}, \qquad (2.29)$$
$$C_1 = \{1, 2\}, \ C_2 = \{4, 5\}, \ C_3 = \{1, 3, 5\}, \ C_4 = \{2, 3, 4\}.$$

We see from this example that minimal path (minimal cut) sets (vectors) may coincide, and that neither minimal path sets nor minimal cut sets need to be disjoint.

Minimal Path Representation of the Structure Function   Let system S have the minimal path sets

$$P_1, P_2, ..., P_a.$$

Then, for any vector $\mathbf{z} = (z_1, z_2, ..., z_n) \in \mathbf{V}_n$, the conjunction (product)

$$A_j(\mathbf{z}) = \bigwedge_{i \in P_j} z_i = \prod_{i \in P_j} z_i \qquad (2.30)$$

is called the j th *minimal path series structure*, $j = 1, 2, ..., a$. (Note that $A_j(\mathbf{z})$ is the structure function of the series connection of all elements e_i with $i \in P_j$.) If $A_j(\mathbf{z}) = 1$, then $\mathbf{z}$ must be a path vector of S. Hence, the availability of any minimal path series structure implies the availability of S. Thus, S is available iff at least one of its a minimal path series structures is available. Therefore, the structure function of S can be represented as

$$\varphi(\mathbf{z}) = \bigvee_{i=1}^{a} A_j(\mathbf{z}) = \bigvee_{i=1}^{a} \bigwedge_{i \in P_j} z_i \tag{2.31}$$

or

$$\varphi(\mathbf{z}) = \prod_{i=1}^{a} \coprod_{i \in P_j} z_i = \max_{1 \le j \le a} \min_{i \in P_j} z_i. \tag{2.32}$$

By (2.10), representation (2.31) of φ is a *disjunctive normal form*.

Minimal Cut Representation of a Structure Function   Let the system S have the minimal cut sets

$$C_1, C_2, ..., C_b.$$

Then, for any vector $\mathbf{z} = (z_1, z_2, ..., z_n) \in V_n$, the disjunction

$$B_k(\mathbf{z}) = \bigvee_{i \in C_k} z_i = \coprod_{i \in C_k} z_i \tag{2.33}$$

is called the k th *minimal cut parallel structure*, $k = 1, 2, ..., b$. (Note that $B_k(\mathbf{z})$ is the structure function of the parallel connection of all elements e_i with $i \in C_k$.) If $B_k(\mathbf{z}) = 0$, then $\mathbf{z}$ must be a cut vector of S. Hence, the nonavailability of any minimal cut parallel structure implies the nonavailability of S. Thus, S is not available iff at least one out of its b minimal cut parallel structures is not available. Therefore, the structure function of S can be represented as

$$\varphi(\mathbf{z}) = \bigwedge_{k=1}^{b} B_k(\mathbf{z}) = \bigwedge_{k=1}^{b} \bigvee_{i \in C_k} z_i \tag{2.34}$$

or

$$\varphi(\mathbf{z}) = \prod_{k=1}^{b} \coprod_{i \in C_k} z_i = \min_{1 \le k \le b} \max_{i \in C_k} z_i. \tag{2.35}$$

By (2.11), representation (2.34) of φ is a *conjunctive normal form*. The representations (2.31) and (2.34) of the structure function imply an interesting corollary (see Figures 2.5 and 2.6):

Corollary 2.10 *The reliability theoretic structure of every coherent system can be represented by a parallel connection of a series systems or by a series connection of b parallel systems.*

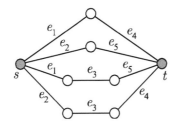

Figure 2.5: Minimal path representation of the bridge structure

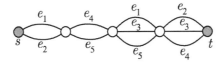

Figure 2.6: Minimal cut representation of the bridge structure

Example 2.11 *Let us consider the reliability block diagram given by Figure 2.7 with entrance node s and exit node t. The minimal path sets are*

$$P_1 = \{1,6\}, \ P_2 = \{1,5,7\}, \ P_3 = \{2,3,6\}, \ P_4 = \{2,4,7\}, \qquad (2.36)$$
$$P_5 = \{2,3,5,7\}.$$

For determining the structure function φ, we will apply the pivotal decomposition with pivotal element e_4, $\mathbf{z} = (z_1, z_2, ..., z_7) \in V_7$:

$$\varphi(\mathbf{z}) = z_4 \varphi((1_4, \mathbf{z})) + \overline{z}_4 \varphi((0_4, \mathbf{z})). \qquad (2.37)$$

Figure 2.8 shows the reliability block diagrams belonging to $\varphi((0_4, \mathbf{z}))$ and $\varphi((1_4, \mathbf{z}))$. The Boolean function $\varphi((1_4, \mathbf{z}))$ is determined by pivotal decomposition with pivotal element e_7:

$$\varphi((1_4, \mathbf{z})) = z_7 \varphi((1_4, 1_7, \mathbf{z})) + \overline{z}_7 \varphi((1_4, 0_7, \mathbf{z})).$$

The respective minimal path sets for $\varphi((1_4, 1_7, \mathbf{z}))$ and $\varphi((1_4, 0_7, \mathbf{z}))$ are $\{2\}$, $\{1,5\}$, $\{1,6\}$, and $\{1,6\}$, $\{2,3,6\}$. Hence,

$$\varphi((1_4, \mathbf{z})) = z_7 \left(z_2 \vee z_1 z_5 \vee z_1 z_6 \right) + \overline{z}_7 (z_1 z_6 \vee z_2 z_3 z_6),$$

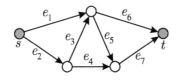

Figure 2.7: Directed example graph

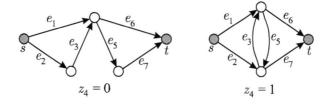

$$z_4 = 0 \qquad\qquad\qquad z_4 = 1$$

Figure 2.8: Pivotal decomposition of the directed example graph

or, equivalently,

$$\varphi((1_4, \mathbf{z})) = z_7 \left(z_2 + z_1 z_5 - z_1 z_2 z_5 + z_1 z_6 - z_1 z_2 z_6 - z_1 z_5 z_6 + z_1 z_2 z_5 z_6\right)$$
$$+ \overline{z}_7 (z_1 z_6 + z_2 z_3 z_6 - z_1 z_2 z_3 z_6).$$

$\varphi((0_4, \mathbf{z}))$ *is the structure function of a series-parallel-system:*

$$\varphi((0_4, \mathbf{z})) = (z_1 \vee z_2 z_3)(z_6 \vee z_5 z_7)$$
$$= (z_1 + z_2 z_3 - z_1 z_2 z_3)(z_6 + z_5 z_7 - z_5 z_6 z_7).$$

With these results, φ is given by (2.37). This representation of φ easily yields the reliability polynomial of the system in case of identical, independent elements:

$$h(p) = p^2 + 3p^3 - p^4 - 6p^5 + 5p^6 - p^7. \tag{2.38}$$

The graph of $h(p)$, $0 \le p \le 1$, has the typical S-form (Figure 2.9).

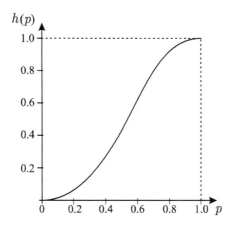

Figure 2.9: Availability graph of the extended bridge structure

Inversion of the Minimal Paths By applying the de Morgan rules (2.9), the conjunctive normal form (2.34) for φ can be transformed into a disjunctive normal form for $1 - \varphi$ in the variables $\bar{z}_i = 1 - z_i$:

$$1 - \varphi(\mathbf{z}) = \bigvee_{k=1}^{b} \bigwedge_{i \in C_k} \bar{z}_i \qquad (2.39)$$

This formula can be used to determine the minimal cuts of a system from its minimal paths. Let us illustrate the approach by the bridge structure (Figure 2.3). Its minimal path sets are given by (2.29) so that the corresponding disjunctive normal form is

$$\varphi(\mathbf{z}) = z_1 z_4 \vee z_2 z_5 \vee z_1 z_3 z_5 \vee z_2 z_3 z_4.$$

By applying the de Morgan rule,

$$1 - \varphi(\mathbf{z}) = (\bar{z}_1 \vee \bar{z}_4)(\bar{z}_2 \vee \bar{z}_5)(\bar{z}_1 \vee \bar{z}_3 \vee \bar{z}_5)(\bar{z}_2 \vee \bar{z}_3 \vee \bar{z}_4).$$

By making use of the distributive and absorption properties (2.4), this representation of $1 - \varphi(\mathbf{z})$ is seen to be logically equivalent to

$$1 - \varphi(\mathbf{z}) = \bar{z}_1 \bar{z}_2 \vee \bar{z}_4 \bar{z}_5 \vee \bar{z}_1 \bar{z}_3 \bar{z}_5 \vee \bar{z}_2 \bar{z}_3 \bar{z}_4.$$

But this is exactly the representation (2.39). Hence, the minimal cut sets are $C_1 = \{1, 2\}$, $C_2 = \{4, 5\}$, $C_3 = \{1, 3, 5\}$, $C_4 = \{2, 3, 4\}$.

The principle "inversion of minimal paths into minimal cuts" (or vice versa) has been dealt with in numerous papers to obtain computationally advantageous algorithms, see e.g. [252, 207, 153].

2.2.3   Generation of Disjoint Sum Forms

As mentioned before, the advantage of having a structure function given as a disjoint sum form is that the system availability is equal to the sum of the expected values of the terms in such a form. Moreover, if the elements operate independently and the terms have structure

$$\prod_{i_k \in \mathbf{U}} z_{i_k} \prod_{j_k \in \overline{\mathbf{U}}} \bar{z}_{j_k}, \qquad (2.40)$$

where $\mathbf{U}$ and $\overline{\mathbf{U}}$ are disjoint subsets of $\{1, 2, ..., n\}$, then the terms have the expected values $(\bar{p}_i = 1 - p_i)$

$$\prod_{i_k \in \mathbf{U}} p_{i_k} \prod_{j_k \in \overline{\mathbf{U}}} \bar{p}_{j_k}. \qquad (2.41)$$

As *Ball and Provan* [13], we will call terms having structure (2.40) *simple products*. Note that (2.41) is the probability that the simple product (2.40) assumes value 1.

In what follows, we next will generate representations of φ as sums of disjoint simple products based on its minimal path representation (2.31). (We could as well start with the minimal cut representation (2.34) for $1 - \varphi$.) From (2.4),

$$A_1 \vee A_2 = A_1 + \overline{A}_1 A_2.$$

By applying properties (2.5) and (2.9),

$$A_1 \vee A_2 \vee A_3 = (A_1 + \overline{A}_1 A_2) \vee A_3 = A_1 + \overline{A}_1 A_2 + \overline{A_1 + \overline{A}_1 A_2}\, A_3$$

$$= A_1 + \overline{A}_1 A_2 + \overline{A}_1 \overline{\overline{A}_1 A_2}\, A_3 = A_1 + \overline{A}_1 A_2 + \overline{A}_1 \overline{A}_2 A_3.$$

By induction,

$$\varphi = \bigvee_{i=1}^{a} A_j = \sum_{k=1}^{a} G_k, \tag{2.42}$$

where the disjoint Boolean functions G_k are defined by $G_1 = A_1$ and

$$G_1 = A_1, \quad G_k = \overline{A}_1 \overline{A}_2 \cdots \overline{A}_{k-1} A_k, \quad k = 2, 3, ..., a. \tag{2.43}$$

Most of the known algorithms dealing with the representation of φ as a sum of disjoint simple products are derived from an algorithm developed by *Abraham* [1]. The motivation for constructing ever new algorithms is to reduce the complexity of the arising structure functions φ as far as possible, in particular with regard to the number of disjoint terms involved. This number determines the *length* of the Boolean function φ. Algorithms which yield the shortest possible structure function exist only in special cases.

The Abraham Algorithm   The Abraham algorithm starts with the disjunctive normal form (2.31) of the structure function and generates sets of disjoint simple products M_k so that

$$G_1 = A_1, \quad G_k = \sum_{D \in M_k} D, \quad k = 2, 3, ..., a. \tag{2.44}$$

Thus, each G_k as given by (2.43) is replaced with a logically equivalent disjoint sum form. The sets M_k are successively obtained from sets $M_{0,k}$, $M_{1,k}$, ..., $M_{k-1,k}$ with $M_{0,k} = \{A_k\}$, $M_{k-1,k} = M_k$. To describe the procedure, let A be a product of some (nonnegated) z_i, B be a simple product, and $C(A, B)$ be the set of all those z_i, which are factors in A, but not in B. Then,

  1) If z_i exists in A and $\overline{z}_i$ in B, then A and B are disjoint.
  2) If A and B are not disjoint and $C(A, B) = \varnothing$, then $A \vee B = A$.
  3) If A and B are not disjoint and $C(A, B) = (C_1, C_2, ..., C_c) \neq \varnothing$, then

$$A \vee B = A + \overline{C}_1 B + C_1 \overline{C}_2 B + C_1 C_2 \cdots C_{c-1} \overline{C}_c B.$$

To obtain $M_{j,k}$ from $M_{j-1,k}$, $j = 1, 2, \cdots, k-1$, let $A = A_j$ and $B \in M_{j-1,k}$. If a z_i exists in A_j and $\overline{z}_i$ in B, then B is also element of B. If A_j and

B are not disjoint and $C(A_j, B) = \emptyset$, then B is dropped. If A_j and B are not disjoint and $C(A_j, B) = (C_1, C_2, ..., C_c) \neq \emptyset$, then the simple products $\overline{C}_1 B$, $C_1 \overline{C}_2 B$, ..., $C_1 C_2 \cdots C_{c-1} \overline{C}_c B$ are added to $M_{j,k}$. With M_1, M_2, ..., M_a constructed in this way, the Abraham disjoint sum for φ is given by (2.42) and (2.44). The $M_{j,k}$ have property

$$\bigvee_{i=1}^{j} A_i \vee A_k = \bigvee_{i=1}^{j} A_i + \sum_{D \in M_{j,k}} D,$$

$j = 1, 2, ..., k-1$, $M_{0,k} = \{A_k\}$, $M_{k-1,k} = M_k$.

Let $M_\varphi = \{M_1 \cup M_2 \cup ... \cup M_a\}$ be the set of all simple products D the sum of which generates an Abraham disjoint sum form of φ, i.e.

$$\varphi = \sum_{D \in M_\varphi} D. \tag{2.45}$$

Then Abraham's algorithm can be formalized as follows:

Algorithm
0. Order the A_j increasingly according to the number of their factors.
1. Initialize $M_\varphi = \{A_1\}$, $k = 2$.
2. Initialize $M_{0,k} = \{A_k\}$, $j = 1$.
3. Initialize $M_{j,k} = \emptyset$.
4. Select a $B \in M_{j-1,k}$.
4.1 If A_j and B are disjoint, add B to $M_{j,k}$ and take the next B.
4.2 Determine $C(A_j, B)$.
4.3 If $C(A_j, B) = \emptyset$, remove this B and take the next B.
4.4 If $C(A_j, B) = (C_1, C_2, ..., C_c)$, expand $M_{j,k}$ by the simple products $\overline{C}_1 B$, $C_1 \overline{C}_2 B$, ..., $C_1 C_2 \cdots C_{c-1} \overline{C}_c B$.
4.5 Go to 4 and proceed with another $B \in M_{j-1,k}$.
5. If all the $B \in M_{j-1,k}$ have been processed, then,
5.1 if $j < k-1$, then go to 3 with $j \longleftarrow j+1$,
5.2 if $j = k-1$, expand M_φ by $M_{k-1,k}$.
6. For any $k = 2, 3, ..., a$,
6.1 if $k < a$, then $k \longleftarrow k+1$ and go to 2,
6.2 if $k = a$, M_φ is complete, and φ is given by (2.42). **Stop.**

Applying the Abraham algorithm to the bridge structure (Figure 2.3) with $A_1 = z_1 z_4$, $A_2 = z_2 z_5$, $A_3 = z_1 z_3 z_5$, $A_4 = z_2 z_3 z_4$ yields the disjoint sum form

$$\varphi = z_1 z_4 + \overline{z}_1 z_2 z_5 + z_1 z_2 \overline{z}_4 z_5 + z_1 \overline{z}_2 z_3 \overline{z}_4 z_5 + \overline{z}_1 z_2 z_3 z_4 \overline{z}_5.$$

(Compare to the corresponding linear form (2.27), which contains 10 terms.) Here we consider a somewhat more complicated example.

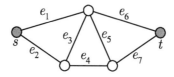

Figure 2.10: Extended bridge structure

Example 2.12 *Figure 2.10 shows the reliability block diagram of a coherent system with 7 elements (*extended bridge structure*). The edges e_1, e_2, e_4, e_6, and e_7 have a "natural order", whereas edges e_3 and e_5, contrary to Figure 2.7, are truly undirected. To generate a disjoint sum form for its structure function φ, the minimal path-series structures of this system are used in the order*

$$A_1 = z_2 z_7, \quad A_2 = z_1 z_4 z_6, \quad A_3 = z_1 z_3 z_7, \quad A_4 = z_2 z_5 z_6,$$
$$A_5 = z_1 z_3 z_5 z_6, \quad A_6 = z_1 z_4 z_5 z_7, \quad A_7 = z_2 z_3 z_4 z_6.$$

k = 2 : $M_{0,2} = \{A_2\} = \{z_1 z_4 z_6\}$.
$C(A_1, A_2) = \{z_2, z_7\}, \quad M_1 = M_{1,2} = \{z_1 \overline{z}_2 z_4 z_6, \ z_1 z_2 z_4 z_6 \overline{z}_7\}$

k = 3 : $M_{0,3} = \{A_3\} = \{z_1 z_3 z_7\}$
$C(A_1, A_3) = \{z_2\}, \quad M_{1,3} = \{B = z_1 \overline{z}_2 z_3 z_7\}$
$C(A_2, B) = \{z_4, z_6\}, \quad M_3 = M_{2,3} = \{z_1 \overline{z}_2 z_3 \overline{z}_4 z_7, \ z_1 \overline{z}_2 z_3 z_4 \overline{z}_6 z_7\}$

k = 4 : $M_{0,4} = \{A_4\} = \{z_2 z_5 z_6\}$
$C(A_1, A_4) = \{z_7\}, \quad M_{1,4} = \{B = z_2 z_5 z_6 \overline{z}_7\}$
$C(A_2, B) = \{z_1, z_4\}, \quad M_{2,4} = \{B_1 = \overline{z}_1 z_2 z_5 z_6 \overline{z}_7, \ B_2 = z_1 z_2 \overline{z}_4 z_5 z_6 \overline{z}_7\}$
A_3 is disjoint both with B_1 and B_2. Hence,
$M_4 = M_{3,4} = M_{2,4} = \{\overline{z}_1 z_2 z_5 z_6 \overline{z}_7, \ z_1 z_2 \overline{z}_4 z_5 z_6 \overline{z}_7\}$.

k = 5 : $M_{0,5} = \{A_5\} = \{z_1 z_3 z_5 z_6\}$
$C(A_1, A_5) = \{z_2, z_7\}, \quad M_{1,5} = \{B_1 = z_1 \overline{z}_2 z_3 z_5 z_6, \ B_2 = z_1 z_2 z_3 z_5 z_6 \overline{z}_7\}$
$C(A_2, B_1) = \{z_4\}, \quad C(A_2, B_2) = \{z_4\}$
$M_{2,5} = \{B_1 = z_1 \overline{z}_2 z_3 \overline{z}_4 z_5 z_6, \ B_2 = z_1 z_2 z_3 \overline{z}_4 z_5 z_6 \overline{z}_7\}$
$C(A_3, B_1) = \{z_7\}, \quad A_3$ disjoint with B_2. Hence,
$M_{3,5} = \{B_1 = z_1 \overline{z}_2 z_3 \overline{z}_4 z_5 z_6 \overline{z}_7, \ B_2 = z_1 z_2 z_3 \overline{z}_4 z_5 z_6 \overline{z}_7\}$
A_4 disjoint with B_1, $C(A_4, B_2) = \oslash$. Hence,
$M_5 = M_{4,5} = \{z_1 \overline{z}_2 z_3 \overline{z}_4 z_5 z_6 \overline{z}_7\}$

k = 6 : $M_{0,6} = \{A_6\} = \{z_1 z_4 z_5 z_7\}$
$C(A_1, A_6) = \{z_2\}, \quad M_{1,6} = \{z_1 \overline{z}_2 z_4 z_5 z_7\}$
$C(A_2, B) = \{z_6\}, \quad M_{2,6} = \{z_1 \overline{z}_2 z_4 z_5 \overline{z}_6 z_7\}$
$C(A_3, B) = \{z_3\}, \quad M_{3,6} = \{z_1 \overline{z}_2 \overline{z}_3 z_4 z_5 \overline{z}_6 z_7\}$
A_4 disjoint with B. Hence, $M_{4,6} = M_{3,6} = \{z_1 \overline{z}_2 \overline{z}_3 z_4 z_5 \overline{z}_6 z_7\}$
A_5 disjoint with B. Hence, $M_6 = M_{5,6} = M_{4,6} = \{z_1 \overline{z}_2 \overline{z}_3 z_4 z_5 \overline{z}_6 z_7\}$

$\mathbf{k} = 7 : M_{0,7} = \{A_7\} = \{z_2 z_3 z_4 z_6\}$
$C(A_1, A_7) = \{z_7\}, \quad M_{1,7} = \{z_2 z_3 z_4 z_6 \overline{z}_7\}$
$C(A_2, B) = \{z_1\}, \quad M_{2,7} = \{\overline{z}_1 z_2 z_3 z_4 z_6 \overline{z}_7\}$
A_3 *disjoint with* B. *Hence,* $M_{3,7} = M_{2,7} = \{\overline{z}_1 z_2 z_3 z_4 z_6 \overline{z}_7\}$
$C(A_4, B) = \{z_5\}, \quad M_{4,7} = \{\overline{z}_1 z_2 z_3 z_4 \overline{z}_5 z_6 \overline{z}_7\}$
A_5 *disjoint with* B. *Hence,* $M_{5,7} = M_{4,7} = \{\overline{z}_1 z_2 z_3 z_4 \overline{z}_5 z_6 \overline{z}_7\}$.
A_6 *disjoint with* B. *Hence,* $M_7 = M_{6,7} = M_{5,7} = \{\overline{z}_1 z_2 z_3 z_4 \overline{z}_5 z_6 \overline{z}_7\}$

Therefore, a disjoint sum form for φ *in terms of simple products is*

$$\varphi = z_2 z_7 + z_1 \overline{z}_2 z_4 z_6 + z_1 z_2 z_4 z_6 \overline{z}_7 + z_1 \overline{z}_2 z_3 \overline{z}_4 z_7 + z_1 \overline{z}_2 z_3 z_4 \overline{z}_6 z_7$$
$$+ \overline{z}_1 z_2 z_5 z_6 \overline{z}_7 + z_1 z_2 \overline{z}_4 z_5 z_6 \overline{z}_7 + z_1 \overline{z}_2 z_3 \overline{z}_4 z_5 z_6 \overline{z}_7 \tag{2.46}$$
$$+ z_1 \overline{z}_2 \overline{z}_3 z_4 z_5 \overline{z}_6 z_7 + \overline{z}_1 z_2 z_3 z_4 \overline{z}_5 z_6.$$

Short Disjoint Sums via the Abraham Algorithm Ordering the minimal path-series structures increasingly according to the number of their factors (*Abraham order of the* A_j) is generally a good start for achieving a short disjoint sum form for φ, since in this case the sets $C(A_j, A_k)$, $j < k$, tend to have small cardinalities. But this is only a partial order in the set $\mathbf{A} = \{A_1, A_2, ..., A_a\}$, since usually some of the A_j will have the same number of factors. Hence, let $\mathbf{A}_{n_i}$ be that subset of $\mathbf{A}$ the elements of which have exactly n_i factors, $1 \leq n_1 < n_2 < \cdots < n_s \leq n$. As Spross (correct spelling: Sproß) in [283], see also [45], demonstrated, suitably ordering the products within the $\mathbf{A}_{n_i}$ may also substantially reduce the cardinality of M_φ. To introduce one of the orders proposed in this paper, let $K_{n_r}(z_l)$ be the number of those A_j, which contain z_l as a factor and which have less than n_r factors, and let

$$N_r(A_j) = \sum_{l, \, l \in P_j} K_{n_r}(z_l), \quad r = 2, 3, ..., s, \tag{2.47}$$

$$N_1(A_j) = N_2(A_j).$$

Sproß order of the $\mathbf{A}_{n_r}$ If $A_i, A_j \in \mathbf{A}_{n_r}$, then A_i is placed before A_j if

$$N_r(A_i) > N_r(A_j), \quad r = 1, 2, ..., r.$$

The $N_r(A_j)$ may be interpreted as *weights* assigned to the A_j determining their position in $\mathbf{A}_{n_r}$. The paper [45] also proposes some other well-motivated orders of the $\mathbf{A}_{n_r}$. Actually, to establish a partial order of the A_j, weights defined analogously to (2.47) [35] or the ones of *Balan* and *Traldi* [9] can be assigned to all $A_1, A_2, ...$, irrespective of the number of their factors. This may lead to even shorter disjoint sums than applying Abraham's order. These and other orders within the Abraham algorithm make sure that the following two guidlines, aimed at constructing short disjoint sum forms, are taking into account:

1. The simple products $B \in M_{j-1,k}$ should be disjoint with as many as possible of the products $A_{j+1}, A_{j+2}, ..., A_{k-1}$.

2. The simple products $B \in M_{j-1,k}$ should have as many as possible common variables with those $A_{j+1}, A_{j+2}, ..., A_{k-1}$, which are not disjoint with B. (Then small cardinalities of the sets $C(A_{j+1}, B)$, $B \in M_{j,k}$, can be expected.)

But also the order of the elements within $C(A_j, B)$, $B \in M_{j-1,k}$, strongly influences the cardinality d of M_φ. The guidelines 1 and 2 motivate the following order of $C(A_j, B) = \{C_1, C_2, ..., C_c\}$, $B \in M_{j-1,k}$, published in [43]:

Spross order of the C$(\cdot, \cdot)$ Let m_i be the number of those $A_{j+1}, A_{j+2}, ...,$ A_{k-1} in which C_i exists and which are not disjoint with B. If $m_1 \geq m_2 \geq \cdots \geq m_c$, then, when doing Step 4.4 in the Abraham algorithm, use $C(A_j, B)$, $B \in M_{j-1,k}$, in the form $\{C_{m_1}, C_{m_2}, ..., C_{m_c}\}$.

Removing Unnecessary Products and Variables Apart from "preprocessing" the minimal path-series-structures A_j, eliminating superfluous A_j and variables z_i in the process of applying the Abraham algorithm will lead to shorter disjoint sum forms for φ, see [43]. To specify these A_j and z_i, let $C(A_j, A_k) = (C_1, C_2, ..., C_c)$, $j < k$, and $(Y_1, Y_2, ..., Y_y)$ be the set of those variables z, which exist both in A_j and A_k. Then,

$$A_j = \prod_{r=1}^{c} C_r \prod_{s=1}^{y} Y_s.$$

In view of $\overline{Y}_s A_k = 0$ and the de Morgan rules (2.9),

$$\overline{A}_j \, A_k = \left[\left(\bigvee_{r=1}^{c} \overline{C}_r\right) \vee \left(\bigvee_{s=1}^{y} \overline{Y}_s\right)\right] A_k = \left(\bigvee_{r=1}^{c} \overline{C}_r\right) A_k.$$

Hence, the variables $Y_1, Y_2, ..., Y_s$ have no influence on $\overline{A}_j \, A_k$ and can, therefore, be deleted from A_j. Thus, letting

$$\overline{A}_{j,k} \, A_k = \left(\bigvee_{r=1}^{c} \overline{C}_r\right) A_k$$

yields a less complex representation for G_k, which is logically equivalent to (2.43):

$$G_k = \overline{A}_{1,k} \, \overline{A}_{2,k} \cdots \overline{A}_{k-1,k} \, A_k. \tag{2.48}$$

This representation had already been used by *Rai* and *Agarwal* [252]. In addition, for the representation of G_k unnecessary A_j can be identified as follows: If there is an integer l so that

$$\overline{A}_{j,k} \, \overline{A}_{l,k} = \overline{A}_{l,k}, \quad j \neq l, \ 1 \leq j, l \leq k-1, \tag{2.49}$$

then G_k is not affected if in (2.48) $\overline{A}_{j,k}$ is removed (i.e. replaced by "1"). In this case, the minimal path-series structure A_j can be omitted when generating G_k. By the de Morgan rules, condition (2.49) is equivalent to

$$A_{j,k} \vee A_{l,k} = A_{l,k}, \quad j \neq l, \ 1 \leq j, l \leq k - 1. \tag{2.50}$$

Note that the relationship (2.50) can be true, but $A_j \vee A_l = A_l$ is never true for some $j \neq l$, $1 \leq j, l \leq a$, since the A_i correspond to minimal path sets. Removing from (2.47) all $\overline{A}_{j,k}$ with property (2.50) yields a reduced, but logically equivalent representation of G_k as defined by (2.43):

$$G_k = \overline{A}_{i_1,k} \, \overline{A}_{i_2,k} \, \cdots \overline{A}_{i_h,k} \, A_k, \quad 1 \leq h \leq k - 1. \tag{2.51}$$

Subproduct Inversion It is surely an interesting exercise to explore the potentialities of the Abraham algorithm for generating possibly short disjoint sums of simple products of structure functions. But the method of subproduct inversion, invented by *Heidtmann* [154], see e.g. also [155], is significantly superior to all the other numerous modifications of the Abraham algorithm with regard to the length of disjoint sums for φ. The factors of a *subproduct* of an underlying product π of Boolean variables or Boolean functions are a subset of the factors of π. Heidtmann essentially follows the Abraham approach, but the output is a representation of the structure function by a sum of disjoint mixproducts. A *mixproduct* is a product of uncomplemented variables z_i with one or more complemented products of some other z_j. Of course, as with simple products, all variables z_i involved in a mixproduct are different. Hence, mixproducts contain simple products as special cases. The "theoretical fundament" of the Heidtmann algorithm is the following obvious fact:

Let $\pi = z_{i_1} z_{i_2} \cdots z_{i_k}$ be a subproduct of the product Π. Then the products Π and Π' are disjoint if Π' contains the factor $\overline{\pi} = \overline{z_{i_1} z_{i_2} \cdots z_{i_k}}$.

Note that if the elements operate independently with availabilities p_i, then, given $\Pi = \pi_1 \pi_2$ with $\pi_1 = z_{i_1} z_{i_2} \cdots z_{i_k}$ and $\pi_2 = \overline{z_{j_1} z_{j_2} \cdots z_{j_l}}$,

$$\mathbb{E}(\Pi) = \Pr(\Pi = 1) = p_{i_1} p_{i_2} \cdots p_{i_k} (1 - p_{j_1} p_{j_2} \cdots p_{j_l}).$$

Applying subproduct inversion and proceeding otherwise analogous to Example 2.12 would yield the structure function belonging to the reliability block diagram of Figure 2.13 as a sum of disjoint simple products and mixproducts:

$$\varphi = z_2 z_7 + \overline{z_2 z_7} z_1 z_4 z_6 + \overline{z_4 z_6} z_1 \overline{z}_2 z_3 z_7 + \overline{z_1 z_4} z_2 z_5 z_6 \overline{z}_7 \tag{2.52}$$
$$+ z_1 \overline{z}_2 z_3 \overline{z}_4 z_5 z_6 \overline{z}_7 + z_1 \overline{z}_2 \overline{z}_3 z_4 z_5 \overline{z}_6 z_7 + \overline{z}_1 z_2 z_3 z_4 \overline{z}_5 z_6 \overline{z}_7.$$

Thus, subproduct inversion reduces the number of disjoint terms from 10 to 7 (compare formulas (2.52) and (2.46)). For instance, in step **k** = **3** of Example 2.12 the contribution of set $M_3 = \{z_1 \overline{z}_2 z_3 \overline{z}_4 z_7, \ z_1 \overline{z}_2 z_3 z_4 \overline{z}_6 z_7\}$ to φ is replaced with the single term $\overline{z_4 z_6} z_1 \overline{z}_2 z_3 z_7$:

$$\overline{z_4 z_6} z_1 \overline{z}_2 z_3 z_7 = z_1 \overline{z}_2 z_3 \overline{z}_4 z_7 + z_1 \overline{z}_2 z_3 z_4 \overline{z}_6 z_7.$$

In the recent literature, representation of the structure function by sums of simple products is called the (*method of*) *single variable inversion* (SVI), and representation of the structure function by sums of (disjoint) mixproducts is called the (*method of*) *multiple variable inversion* (MVI). The algorithm of *Luo* and *Trivedi* [211] uses both SVI and MVI. For a survey of some papers on the (*method of*) *sum of disjoint products* (SDP), including both SVI and MVI up till 1994, see [253]. For a comparison of three early papers, see [205]. For noncoherent structure functions and application to artificial intelligence, see *Anrig and Beichelt* [6].

Abraham's Example Network   In view of the attention Abraham's example network has found in the literature, we cannot ignore it. Neither Abraham's order of the A_j nor the Spross orders have been tailored to minimize the lengths of disjoint sums of φ for example systems, but to serve as general guidelines for constructing short disjoint sum forms of structure functions of any coherent systems. Abraham's network (Figure 2.11), which is actually due to *Fratta* and *Montanari* [128], symbolizes the reliability block diagram of a system consisting of 12 elements with nodes s and t as entrance and exit node, respectively. This system has 24 minimal path vectors or 24 minimal path-series-structures A_j, respectively. *Abraham* [1] generated a disjoint sum form with 71 simple terms, *Beichelt* and *Spross* [43] (Abraham's algorithm with the Spross order of the $\mathbf{A}_{n_r}$ and elimination of redundant variables and A_j) yielded 63 simple terms. *Locks* [208] generated 61 terms with another approach. *Beichelt* and *Spross* [45] complemented their 1987 paper with the Spross order of the $\mathbf{C}(\cdot,\cdot)$ and generated a disjoint sum form with 55 simple terms. Because of two typographical errors, their list of the disjoint simple terms only contains 53 terms, and in this typewritten list with 336 entries one entry is wrong, see *Spross* [283] and *Beichelt* [35] for the complete and correct list. In view of this, *Balan* and *Traldi* [9] qualify the disjoint sum analysis done in [45] as "mistaken". The fact is, however, that up till now the Spross approach to SDP-SVI belongs to the top ones: by following the paper [45], changing only the position of one product, *Balan* and *Traldi* generate a disjoint sum form with 54 simple products, and with their own order they manage to reduce the number of simple terms to 53, see also [206]. Moreover,

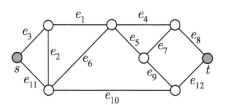

Figure 2.11: Abraham's example network

Traldi [296] showed that an SVI disjoint sum for the Abraham problem must have at least 54 simple products if Abraham's order is applied. *Heidtmann* [154], by applying his MVI-method, generated a disjoint sum form with 41 mixproducts. *Lian-Chang Zhao* and *Jun-Chen Xu* [314] essentially use Heidtmann's MVI-method and principles published in [43] as well as in [13] to obtain a disjoint sum form for the Abraham example with 36 mixproducts. *Traldi* [296] showed that an MVI disjoint sum for the Abraham problem must have at least 35 mixproducts if Abraham's order is applied.

2.2.4 Inclusion–Exclusion, Formations, and Signature

Inclusion–Exclusion From the representation $\varphi = A_1 \vee A_2 \vee \cdots \vee A_a$ the linear form of the structure function can be generated by applying the principle of inclusion–exclusion, which is well-known from set theory. One starts with writing the disjunction $A_1 \vee A_2$ in terms of arithmetic operations according to formula (2.5): $A_1 \vee A_2 = A_1 + A_2 - A_1 A_2$. Now (2.5) is applied to $A_1 \vee A_2 \vee A_3$ with $x = A_1 \vee A_2$ and $y = A_3$. This yields

$$A_1 \vee A_2 \vee A_3 = A_1 + A_2 + A_3 - A_1 A_2 - A_1 A_3 - A_2 A_3 + A_1 A_2 A_3.$$

Continuing in this way, more exactly, by induction, one obtains

$$\varphi = \sum_{k=1}^{a} (-1)^{k-1} T_k \quad \text{with } T_k = \sum_{1 \le i_1 < i_2 < \cdots < i_k \le a} A_{i_1} A_{i_2} \cdots A_{i_k}. \tag{2.53}$$

This representation is called the *inclusion–exclusion-form* of φ. In T_k, the sum runs over all subsets of $\{1, 2, ..., a\}$ with cardinality k. Hence, T_k comprises exactly $\binom{a}{k}$ terms. In view of this, the inclusion–exclusion form has

$$\binom{a}{1} + \binom{a}{2} + \cdots + \binom{a}{a} = 2^a - 1$$

terms. Thus, with increasing a, the computation time of the inclusion–exclusion form grows exponentially fast. Since determining the number of minimal paths a also grows exponentially fast with increasing system complexity, application of (2.53) is actually a doubly exponential algorithm for determining φ. Thus, its practical application is limited to "sufficiently small" systems. The linear form of the structure function we get from (2.53) by making use of the idempotency property of Boolean variables. By taking the expected value on both sides of (2.53),

$$p_s = \mathbb{E}(\varphi) = \sum_{k=1}^{a} (-1)^{k-1} \mathbb{E}(T_k). \tag{2.54}$$

Formations A closer look at the inclusion–exclusion-formula (2.53) leads to a more efficient way for deriving the structure function of a binary coherent system based on its minimal path (cut) sets. For illustration, let us first consider an example.

Example 2.13 *The bridge structure given by Figure 2.3 has the minimal path sets*

$$P_1 = \{1,4\}, \ \ P_2 = \{2,5\}, \ \ P_3 = \{1,3,5\}, \ \ P_4 = \{2,3,4\}.$$

Hence, formula (2.53) yields (replacing every $z_i z_i$ with z_i):

$$\varphi(\mathbf{z}) = z_1 z_4 + z_2 z_5 + z_1 z_3 z_5 + z_2 z_3 z_4$$

$$-z_1 z_2 z_4 z_5 - z_1 z_3 z_4 z_5 - z_1 z_2 z_3 z_4 - z_1 z_2 z_3 z_5 - z_2 z_3 z_4 z_5$$

$$-z_1 z_2 z_3 z_4 z_5 + z_1 z_2 z_3 z_4 z_5 + z_1 z_2 z_3 z_4 z_5 + z_1 z_2 z_3 z_4 z_5 + z_1 z_2 z_3 z_4 z_5$$

$$-z_1 z_2 z_3 z_4 z_5.$$

Two pairs of terms in this representation of $\varphi(\mathbf{z})$ cancel each other so that in the end we obtain the linear form (2.27).

This example shows that some of the terms in the inclusion–exclusion form (2.53) can be identical, possibly apart from the sign. As first shown by *Satyanarayana* and *Prabhakar* [270], identifying those terms before determining the structure function via the inclusion–exclusion principle may substantially reduce the computational effort for obtaining φ, in particular for large systems.

Firstly we note that every term $A_{i_1} A_{i_2} \cdots A_{i_k}$ in (2.53) is the product of those indicator variables z_i, whose indices are in the union of the corresponding minimal path sets $P_{i_1} \cup P_{i_2} \cup \cdots \cup P_{i_k}$. Hence,

$$A_{i_1} A_{i_2} \cdots A_{i_k} = 1 \ \text{ if } \ z_i = 1 \text{ for all } i \in P_{i_1} \cup P_{i_2} \cup \cdots \cup P_{i_k}.$$

In this sense, the union $P_{i_1} \cup P_{i_2} \cup \cdots \cup P_{i_k}$ *generates* the product $A_{i_1} A_{i_2} \cdots A_{i_k}$. But different unions of minimal path sets may generate one and the same product. For instance, as we have seen in Example 2.13, $P_2 \cup P_3 = \{1, 2, \cdots, 5\}$ generates the product $z_1 z_2 z_3 z_4 z_5$, and so does $P_1 \cup P_2 \cup P_3$. According to (2.53), if the number of minimal path sets in the union is even, then the sign of the arising product term is "+", otherwise it is "−". This situation motivates the following terminology/notation:

A *formation* F is the union of some minimal path sets. A *k-representation* of F is its union of k minimal path sets. A k-representation is called *odd (even)* if k is odd (even). If there are r_o (r_e) odd (even) representations of F, the difference $d = r_o - r_e$ is called the *signed domination* of F. Let the formations of a coherent system be given by $F_1, F_2, ..., F_m$ with respective signed dominations $d_1, d_2, ..., d_m$. Then the inclusion–exclusion-formula is logically equivalent to

$$\varphi = \sum_{j=1}^{m} d_j \prod_{i \in F_j} z_i, \tag{2.55}$$

which is the linear form of the structure function. If the elements operate independently, then (2.55) yields the system availability

$$p_s = \sum_{j=1}^{m} d_j \prod_{i \in F_j} p_i, \tag{2.56}$$

In particular, for identical, independent elements with joint availability p,

$$p_s = h(p) = \sum_{j=1}^{m} d_j\, p^{n_j}, \qquad (2.57)$$

where $n_j = |F_j|$ denotes the cardinality of F_j. We will write this formula more efficiently as

$$p_s = h(p) = \sum_{i=1}^{n} d_i\, p^i, \qquad (2.58)$$

letting $d_i = 0$ if there is no formation with cardinality i. Otherwise d_i is the sum of all signed dominations of those formations, which have cardinality i. Hence, in this special case, formations need only be distinguished with regard to their cardinality, i.e. two formations are different iff they have different cardinalities. Then the system has at most n formations, and their signed dominations are again denoted as d_i.

i	Formation	r_o	r_e	d_i	i	Formation	r_o	r_e	d_i
1	$\{1,6\}$	1	0	1	11	$\{2,3,4,5,7\}$	0	1	-1
2	$\{1,5,7\}$	1	0	1	12	$\{2,3,4,6,7\}$	0	1	-1
3	$\{2,3,6\}$	1	0	1	13	$\{2,3,5,6,7\}$	0	1	-1
4	$\{2,4,7\}$	1	0	1	14	$\{1,2,3,4,5,7\}$	1	0	1
5	$\{2,3,5,7\}$	1	0	1	15	$\{1,2,3,4,6,7\}$	1	0	1
6	$\{1,2,3,6\}$	0	1	-1	16	$\{1,2,3,5,6,7\}$	4	3	1
7	$\{1,5,6,7\}$	0	1	-1	17	$\{1,2,4,5,6,7\}$	1	0	1
8	$\{1,2,3,5,7\}$	0	1	-1	18	$\{2,3,4,5,6,7\}$	1	0	1
9	$\{1,2,4,5,7\}$	0	1	-1	19	$\{1,2,3,4,5,6,7\}$	3	4	-1
10	$\{1,2,4,6,7\}$	0	1	-1					

Table 2.2: Formation analysis for example 2.14

Example 2.14 *The minimal path sets of a coherent system are*

$$P_1 = \{1,6\}\,, P_2 = \{1,5,7\}\,, P_3 = \{2,3,6\}\,, P_4 = \{2,4,7\}\,, P_5 = \{2,3,5,7\}$$

(Example 2.11). Table 2.2 shows its formations, the number of odd and even representations, and signed dominations. Thus, the linear form only consists of 19 terms, whereas the inclusion–exclusion formula generates $2^5 - 1 = 31$ terms. Table 2.2 yields again the reliability polynomial (2.38) for independent elements with joint availability p, derived there by pivotal decomposition. In this special case, there are 6 formations F_1, F_2,...,F_6 with respective cardinalities 2, 3, 4, 5, 6, 7 and signed dominations 1, 3,-1, -6, 5, -1.

Signature Closely related to the signed domination is the *signature* of a coherent system, which was introduced by F. J. Samaniego [265], see also [183, 62]. We will consider this concept for coherent systems with n independently operating elements having iid lifetimes $L_1, L_2, ..., L_n$. The system starts operating at time $t = 0$. If an element fails, it will not be repaired or be replaced by another one. Let σ_i be the probability that the ith element failure causes the system to fail. Then, if L_s denotes the system lifetime,

$$\sigma_i = \Pr(L_s = L_{k_i}), \ i = 1, 2, ..., n,$$

where $(L_{k_1}, L_{k_2}, ..., L_{k_n})$, $L_{k_1} < L_{k_2} < \cdots < L_{k_n}$, is the ordered sequence of the L_i. The vector $\sigma = (\sigma_1, \sigma_2, ..., \sigma_n)$ is called the *signature* of the system. It has the property

$$\sum_{i=1}^{n} \sigma_i = 1. \tag{2.59}$$

Obviously, a series system has signature $(1, 0, ..., 0)$, and the signature of a parallel system is $(0, 0, ..., 0, 1)$. The order of the failures of the elements is given by a permutation $(e_{k_1}, e_{k_2}, ..., e_{k_n})$ of the set $(e_1, e_2, ..., e_n)$. Under our iid lifetime assumption, every permutation $(e_{k_1}, e_{k_2}, ..., e_{k_n})$ has the same probability to occur. This gives a principally elementary way to exactly determine the probabilities σ_i: The failure of element e_{k_i} causes a system failure iff e_{k_i} is the last element in a minimal cut, which was still operating. Hence, we have to count all those permutations $(e_{k_1}, e_{k_2}, ..., e_{k_n})$ in which this happens, and divide the result by $n!$ – the total number of permutations.

Example 2.15 *Let us illustrate the approach by the bridge structure (Example 2.7). Its minimal cut sets are $C_1 = \{1, 2\}$, $C_2 = \{4, 5\}$, $C_3 = \{1, 3, 5\}$, $C_4 = \{2, 3, 4\}$. Obviously, the first failing element cannot cause a system failure: $\sigma_1 = 0$, neither can the 5th failure (the system fails at latest at the 4th element failure): $\sigma_5 = 0$. The second element failure causes a system failure if the permutations have structure*

$$(e_1, e_2, e_{k_3}, e_{k_4}, e_{k_5}), \ (e_2, e_1, e_{k_3}, e_{k_4}, e_{k_5}),$$
$$(e_4, e_5, e_{k_3}, e_{k_4}, e_{k_5}), \ (e_5, e_4, e_{k_3}, e_{k_4}, e_{k_5}).$$

There is a total of $4 \cdot 3! = 24$ of such permutations. Hence, $\sigma_2 = 24/5! = 0.2$. The third element failure causes the system to fail if the permutations have one of the following structures:

$$(e_1, e_{k_2}, e_2, e_{k_4}, e_{k_5}), \ (e_2, e_{k_2}, e_1, e_{k_4}, e_{k_5}),$$
$$(e_{k_1}, e_1, e_2, e_{k_4}, e_{k_5}), \ (e_{k_1}, e_2, e_1, e_{k_4}, e_{k_5}),$$
$$(e_4, e_{k_2}, e_5, e_{k_4}, e_{k_5}), \ (e_5, e_{k_2}, e_4, e_{k_4}, e_{k_5}),$$
$$(e_{k_1}, e_4, e_5, e_{k_4}, e_{k_5}), \ (e_{k_1}, e_5, e_4, e_{k_4}, e_{k_5}),$$

or if the permutations start with elements e_1, e_3, e_5 (in any order) or with elements e_2, e_3, e_5 (in any order). There is a total of $8 \cdot 3! + 2 \cdot 3! \cdot 2 = 72$ of such permutations. Hence, $\sigma_3 = 72/5! = 0.6$. Analogously or by making use of (2.59) and $\sigma_1 = \sigma_5 = 0$, one gets $\sigma_4 = 0.2$. Thus, the signature of the bridge structure is $\sigma = (0, 0.2, 0.6, 0.2, 0)$.

Let $p = p(t) = \Pr(L_i > t)$, $i = 1, 2, ..., n$, be the joint availability of the elements at any fixed point in time t. Then, on condition that the jth failing element causes the system to fail, the availability of the system (at time t) is

$$h(p\,|\,j) = \sum_{k=0}^{j-1} \binom{n}{k} p^{n-k}(1-p)^k.$$

Hence, by the formula of the total probability, the unconditional system availability $p_s = h(p) = \Pr(L_s > t)$ is

$$h(p) = \sum_{j=1}^{n} \sigma_j \sum_{k=0}^{j-1} \binom{n}{k} p^{n-k}(1-p)^k. \tag{2.60}$$

By changing the order of summation and subsequent index transformation,

$$h(p) = \sum_{k=1}^{n} \left(\sum_{j=n-k+1}^{n} \sigma_j \right) \binom{n}{k} p^k (1-p)^{n-k}.$$

Representing $(1-p)^{n-k}$ as a binomial series, another index transformation, and changing the order of summation once more yield

$$h(p) = \sum_{i=1}^{n} \left\{ \sum_{k=1}^{i} (-1)^{i-k} \left(\sum_{j=n-k+1}^{n} \sigma_j \right) \binom{n}{k} \binom{n-k}{i-k} \right\} p^i.$$

Comparing the coefficients of p^i in this formula and (2.58) gives the desired relationship between domination and signature of a coherent system:

$$d_i = \sum_{k=1}^{i} (-1)^{i-k} \left(\sum_{j=n-k+1}^{n} \sigma_j \right) \binom{n}{k} \binom{n-k}{i-k}; \quad i = 1, 2, ..., n. \tag{2.61}$$

Example 2.16 Let us again consider the reliability block diagram of Figure 2.7. By example 2.14, its vector of signed dominations is

$$d_1 = 0, \; d_2 = 1, \; d_3 = 3, \; d_4 = -1, \; d_5 = -6, \; d_6 = 5, \; d_7 = -1,$$

where d_i is the factor of p^i in the corresponding reliability polynomial (2.38):

$$h(p) = p^2 + 3p^3 - p^4 - 6p^5 + 5p^6 - p^7.$$

In this case, $\sigma_1 = \sigma_7 = 0$, since the system will always survive the failure of only one element, and it will fail at latest at the failure of the 6th element (the latter can only happen if e_2 and e_6 are the last elements to fail). In view of $\sigma_7 = 0$, for $k = 1$ the first term in (2.61) is 0 for all i. For $i = 2, 3, ..., 6$, the respective equations (2.61) are

$$+1 = 21\sigma_6$$
$$+3 = -70\sigma_6 + 35\sigma_5$$
$$-1 = 105\sigma_6 - 105\sigma_5 + 35\sigma_4$$
$$-6 = -84\sigma_6 + 126\sigma_5 - 84\sigma_4 + 21\sigma_3$$
$$+5 = 35\sigma_6 - 70\sigma_5 + 70\sigma_4 - 35\sigma_3 + 7\sigma_2$$

The solution is

$$\sigma_2 = 70/735, \quad \sigma_3 = 224/735, \quad \sigma_4 = 273/735, \quad \sigma_5 = 133/735, \quad \sigma_6 = 35/735.$$

In Chapter 3 we will return to the signature of a coherent system and consider some applications.

2.2.5 Bounds on the System Availability

Since the computation time of the exact value of the system availability p_s generally increases exponentially fast with increasing system complexity, bounds on p_s, the calculation of which requires less effort or can even be done in polynomial time, are of particular interest.

1. Inclusion–Exclusion Bounds   The inclusion–exclusion representation (2.53) of the system availability immediately yields the following upper and lower bounds on p_s:

$$p_s \leq \mathbb{E}(T_1)$$
$$p_s \geq \mathbb{E}(T_1) - \mathbb{E}(T_2)$$
$$p_s \leq \mathbb{E}(T_1) - \mathbb{E}(T_2) + \mathbb{E}(T_3)$$
$$p_s \geq \mathbb{E}(T_1) - \mathbb{E}(T_2) + \mathbb{E}(T_3) - \mathbb{E}(T_4)$$

$$\vdots$$

These upper/lower bounds obviously converge to p_s, but not necessarily monotone. Computationally, the construction of these bounds can be coupled with the determination of the minimal paths. Then only such a number of minimal paths needs to be generated as necessary for achieving the desired degree of accuracy.

Since the system availability $p_s = \mathbb{E}(A_1 \vee A_2 \vee \cdots \vee A_a)$ is the probability that at least one of the A_i is equal to 1, the *Bonferroni inequality* (see e.g. [120]) is applicable to p_s :

$$\mathbb{E}(T_1) - \mathbb{E}(T_2) \leq p_s \leq \mathbb{E}(T_1).$$

These bounds are given by the first two inclusion–exclusion inequalities above. Numerous modifications/extensions of the Bonferroni inequality for p_s have been proposed. Sometimes all of them, including all possible inclusion–exclusion bounds, are called *Bonferroni (type) inequalities*. (For more information, see Section 6.4.)

2. Disjoint Sum Bounds By formula (2.42), the structure function can be represented as

$$\varphi = \sum_{k=1}^{a} G_k, \quad G_k = \overline{A}_1 \overline{A}_2 \cdots \overline{A}_{k-1} A_k.$$

Similarly, by formula (2.39), $1 - \varphi$ can be represented as

$$1 - \varphi = \sum_{j=1}^{b} H_j, \quad H_j = \overline{D}_1 \overline{D}_2 \cdots \overline{D}_{j-1} D_j,$$

where $D_j = \bigwedge_{i \in C_j} \overline{z}_i$; $j = 1, 2, ..., b$. Hence, for any integers g and h with $1 \leq g \leq a$ and $1 \leq h \leq b$, lower bounds $l(g)$ and upper bounds $u(h)$ for p_s are

$$l(g) = \sum_{k=1}^{g} \mathbb{E}(G_k) \leq p_s \leq 1 - \sum_{j=1}^{h} \mathbb{E}(H_j) = u(h). \qquad (2.62)$$

Simple as these bounds may be, they yield excellent results even if only a very small percentage of all minimal paths/cuts is used for their calculation. This will be illustrated by the following example; more have been analysed by *Beichelt and Spross* [44] and *Beichelt* [35].

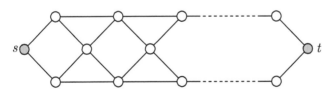

Figure 2.12: Variable example reliability block diagram

Example 2.17 *The reliability block diagram of a system is given by Figure 2.12. Its entrance node $s = 1$ is fixed, whereas the exit node t is variable and is equal to the number of nodes in the diagram. The corresponding number of elements (edges) is $n = 5 + 7(t-1)$, which are assumed to operate independently with common availability $p = 0.8$. In this way, we are able to study the behavior of the bounds (2.62) in dependence on the complexity of the system. The*

t	a	b
4	3	4
7	21	18
10	151	56
13	1081	148
16	7739	356
19	55405	806

Table 2.3: Number of minimal paths and cuts

expected values $\mathbb{E}(G_k)$ *and* $\mathbb{E}(H_j)$ *have been determined by representing* G_k *and* H_j *as disjoint sums. Table 2.3 shows the number of minimal paths/cuts for t ranging from 7 to 19, and Table 2.4 gives the lower and upper bounds in case of a node number of* $t = 19$, *i.e. for a system with 131 elements. Only 43 out of 55 405 minimal paths (0.08%) together with only 15 out of 806 minimal cuts (1.86%) yield sufficiently sharp bounds, namely* $0.9871 \le p_s \le 0.9975$. *A better numerical result than the one achieved scarcely has practical significance due to the fact that the underlying availability of the elements cannot be provided with absolute accuracy.*

g	$l(g)$	h	$u(h)$
5	0.6453	2	0.9980
10	0.8069	4	0.9978
20	0.9040	8	0.9976
40	0.9765	12	0.9976
43	0.9871	15	0.9975

Table 2.4: Lower and upper bounds for t=19

3. Elementary bounds From (2.17),

$$E\left(\prod_{i=1}^{n} z_i\right) \le p_s \le E\left(\coprod_{i=1}^{n} z_i\right).$$

In particular, for independent z_i, i.e. independently operating elements,

$$\prod_{i=1}^{n} p_i \le p_s \le 1 - \prod_{i=1}^{n}(1 - p_i). \tag{2.63}$$

4. Min-Max Bounds Sharper bounds than (2.63) can be derived from the min/max representations (2.32) and (2.35) of the structure function: For any $j = 1, 2, ..., a$ and any $k = 1, 2, ..., b$,

$$\min_{i \in P_j} z_i \le \varphi(\mathbf{z}) \le \max_{i \in C_k} z_i.$$

Hence, by taking the expected values,

$$\mathbb{E}\left(\min_{i\in P_j} z_i\right) \le p_s \le \mathbb{E}\left(\max_{i\in C_k} z_i\right).$$

Since these inequalities are true for any P_j and C_k,

$$\max_{1\le j\le a} \mathbb{E}\left(\min_{i\in P_j} z_i\right) \le p_s \le \min_{1\le k\le b} \mathbb{E}\left(\max_{i\in C_k} z_i\right).$$

In particular, for independent z_i, letting $\mathbf{p} = (p_1, p_2, ..., p_n)$,

$$l_4(\mathbf{p}) = \max_{1\le j\le a}\left(\prod_{i\in P_j} p_i\right) \le p_s \le \min_{1\le k\le b}\left(\coprod_{i\in C_k} p_i\right) = u_4(\mathbf{p}). \qquad (2.64)$$

The bounds (2.63) and (2.64) even hold if the random indicator variables $z_1, z_2, ..., z_n$ are statistically dependent. More exactly, they are allowed to be associated, see Definition 3.7.

5. Esary-Proschan Bounds   For independent z_i,

$$l_5(\mathbf{p}) = \prod_{k=1}^{b}\coprod_{i\in C_k} p_i \le p_s \le \coprod_{j=1}^{a}\prod_{i\in P_j} p_i = u_5(\mathbf{p}). \qquad (2.65)$$

Formally one obtains these bounds by replacing the z_i in (2.32) and (2.35), respectively, with p_i [116].

The lower (upper) bounds in (2.64) are not uniformly better or worse than the ones in (2.65). Hence, it makes sense to combine these bounds:

$$\max(l_4(\mathbf{p}), l_5(\mathbf{p})) \le p_s \le \min(u_4(\mathbf{p}), u_5(\mathbf{p})).$$

6. Fu-Koutras Bounds   Numerical examples indicate that the lower (upper) bounds (2.65) are good approximations to p_s if the p_i are high (low). To narrow the difference between lower and upper bounds, in particular for the practically important case of systems with highly reliable elements, the construction of more sophisticated bounds is necessary. Along the line of the bounds (2.64) and (2.65) are the following ones proposed by *Fu* and *Koutras* [130] for systems with independently operating elements.

In what follows, let $\emptyset$ be the empty set and

$$P = \{1, 2, ..., a\}, \quad C = \{1, 2, ..., b\}.$$

a) If the minimal path sets $\{P_1, P_2, ..., P_a\}$ are given, let $K_1' = \emptyset$ and

$$K_j' = \{i; \; P_i \cap P_j \ne \emptyset, \; 1 \le i < j\}, \; j = 2, 3, ..., a.$$

For every non empty K'_j, let K_j be a subset of $P\backslash P_j$ satisfying

$$K_j \cap P_i \neq \emptyset \quad \text{for every} \quad i \in K'_j.$$

Then, setting $K_j = \emptyset$ if $K'_j = \emptyset$ and $q_i = 1 - p_i$, a lower bound for p_s is

$$l_6(\mathbf{p}) = \prod_{j=1}^{a} \left(\prod_{i \in P_j} p_i \prod_{i \in K_j} q_i \right). \tag{2.66}$$

b) Analogously, if the minimal cut sets $C_1, C_2, ..., C_b$ are given, let $L'_1 = \emptyset$ and

$$L'_j = \{i; \ C_i \cap C_j \neq \emptyset, \ 1 \le i < j\}, \ j = 2, 3, ..., b.$$

For every non empty L'_j, let L_j be any subset of $C \backslash C_j$ satisfying

$$L_j \cap C_i \neq \emptyset \quad \text{for every} \quad i \in L'_j.$$

Then, putting $L_j = \emptyset$ if $L'_j = \emptyset$, an upper bound for p_s is

$$u_6(\mathbf{p}) = \prod_{j=1}^{b} \left(1 - \prod_{i \in L_j} p_i \prod_{i \in C_j} q_i \right). \tag{2.67}$$

The bounds (2.66) and (2.67) depend on the order of the minimal path and cut sets, respectively. The following example shows that generally there is some freedom in the selection of the K_j and L_j. The optimal order of the minimal path and cut sets as well as the optimal selection of the K_j and L_j depend on $\mathbf{p}$. ("Optimal" refers to maximizing $l_6(\mathbf{p})$ and minimizing $u_6(\mathbf{p})$, respectively.) For equal element availability, sets K and L with smallest possible cardinality are best.

Example 2.18 *Let us return to the system that has the reliability block diagram given by Figure 2.7. Its minimal path sets are*

$$P_1 = \{1, 6\}, \ P_2 = \{1, 5, 7\}, \ P_3 = \{2, 3, 6\}, \ P_4 = \{2, 4, 7\},$$
$$P_5 = \{2, 3, 5, 7\}.$$

Hence,

$$K'_1 = \emptyset, \ K'_2 = \{1\}, K'_3 = \{1\}, K'_4 = \{2, 3\}, \ K'_5 = \{2, 3, 4\},$$
$$K_1 = \emptyset, \ K_2 = \{6\}, \ K_3 = \{1\}, \ K_4 = \{1, 3\}, \ K_5 = \{1, 4, 6\}.$$

Assuming $p_i = p$, $i = 1, 2, ..., 7$, the lower bound (2.66) becomes

$$l_6(p) = 1 - \left(1 - p^2\right) \left(1 - p^3 q\right)^2 \left(1 - p^3 q^2\right) \left(1 - p^4 q^3\right).$$

p	l_4	l_5	l_6	p_s	u_4	u_5	u_6
.05	.00250	.00000	.00286	.00287	.09750	.00288	.00908
.10	.01000	.00019	.01265	.01284	.19000	.01307	.03318
.15	.02250	.00171	.03077	.03166	.27750	.03285	.06866
.85	.72250	.94267	.77483	.94581	.97750	.99238	.94651
.90	.81000	.97618	.83799	.97687	.99000	.99870	.97770
.95	.90250	.99451	.91088	.99455	.99750	.99995	.99456

Table 2.5: Comparison of bounds 4, 5, and 6

The minimal cuts sets are

$$C_1 = \{1, 2\}, \ C_2 = \{6, 7\}, \ C_3 = \{1, 3, 4\}, \ C_4 = \{1, 3, 7\},$$
$$C_5 = \{2, 5, 6\}, \ C_6 = \{4, 5, 6\}.$$

Hence,

$$L_1' = \emptyset, L_2' = \emptyset, L_3' = \{1\}, L_4' = \{1, 2, 3\}, L_5' = \{1, 2\}, L_6' = \{2, 3, 5\}.$$
$$L_1 = \emptyset, L_2 = \emptyset, L_3 = \{2\}, L_4 = \{2, 4, 6\}, L_5 = \{1, 7\}, L_6 = \{2, 3, 7\}.$$

Assuming $p_i = p$, $i = 1, 2, ..., 7$, the upper bound (2.67) becomes

$$u_6(p) = \left(1 - q^2\right)^2 \left(1 - pq^3\right) \left(1 - p^2 q^3\right) \left(1 - p^3 q^3\right)^2.$$

Table 2.5 shows the bounds 4, 5, and 6 together with the exact system avail-ability p_s given by (2.38) for some low and high values of p. (The zeros before the dots have been omitted.) The table supports the recommendation of Fu and Koutras to use their lower bounds l_6 and the upper Esary-Proschan bounds u_5 in case of low element availabilities and the lower Esary-Proschan bounds l_5 and the upper Fu-Koutras bounds u_6 in case of high element availabilities. In our example, the corresponding intervals are very tight.

An improvement of the *Fu-Koutras* bounds was obtained by *Koutras, Papastavridis, and Petakos* [186].

2.2.6   Importance Criteria

Introduction

Generally, elements (subsystems) of a system have different degrees of influ-ence on its failure behavior and its availability. For the reliability engineer, this influence is interesting with regard to the following questions:

1) What is the contribution of an element to the system availability?

2) How does the system availability vary depending on the availability of an element?

3) What maximum increase in system availability can be achieved if the availability of an element is increased?

4) What element most likely causes a system failure?

5) What elements most likely contribute to a system failure?

The significance of the influence of elements with regard to one or more of these questions is quantitatively measured by so-called *importance criteria*. Knowledge of these criteria is a key information for the design and possible upgrading of systems from the reliability point of view, as well as for their efficient maintenance. For instance, if the availability of a system has to be increased by upgrading its elements, then ordering the elements with regard to importance criteria referred to in questions 1 to 3 is crucial. On the other hand, if the order of the elements with regard to importance criteria referred to in questions 4 and/or 5 is known, then this will facilitate the search for the cause of a system failure. Of course, importance criteria can also be combined with economic aspects.

Assumption   In this section, if the opposite is not explicitly stated, elements operate independently.

Birnbaum-Importance

The first, simplest, but nevertheless crucial importance measure up to now is due to *Birnbaum* [51].

Definition 2.19 *The* Birnbaum-importance (B-importance) $I_B(i, \mathbf{p})$ *of element e_i is defined as*

$$I_B(i, \mathbf{p}) = \frac{\partial h(\mathbf{p})}{\partial p_i}; \quad i = 1, 2, ..., n. \tag{2.68}$$

Therefore, the total differential of $h(\mathbf{p})$ can be written as

$$dh(\mathbf{p}) = \sum_{i=1}^{n} I_B(i, \mathbf{p}) dp_i . \tag{2.69}$$

Thus, the increment $dh(\mathbf{p})$ of the system availability is a weighted sum of the increments of the element availabilities dp_i, where the weights are the corresponding B-importances. As a corollary: The larger the B-importance of an element, the larger is its influence on the system availability. This fact is even better illustrated by the following representation of the B-importance: Partially differentiating the pivotal decomposition formula (2.23) with regard to p_i yields

$$I_B(i, \mathbf{p}) = h((1_i, \mathbf{p})) - h((0_i, \mathbf{p})); \quad i = 1, 2, ..., n. \tag{2.70}$$

Thus, $I_B(i, \mathbf{p})$ is the maximally possible increase in system availability due to element e_i. Equivalent to (2.70) is

$$I_B(i, \mathbf{p}) = \Pr\left(\varphi((1_i, \mathbf{z})) - \varphi((0_i, \mathbf{z})) = 1\right); \quad i = 1, 2, ..., n. \tag{2.71}$$

Note that $\varphi((1_i, \mathbf{z})) - \varphi((0_i, \mathbf{z})) = 1$ implies that on condition $z_i = 0$ the system is down, and that the repair of element e_i renders the system operating. This leads to the following definition:

Definition 2.20 $\mathbf{z}$ *is a* critical path vector *for element* e_i *if*

$$\varphi((0_i, \mathbf{z})) = 0, \quad \varphi((1_i, \mathbf{z})) = 1.$$

Hence, the B-importance $I_B(i, \mathbf{p})$ of element e_i can be interpreted as the probability that a critical path vector for e_i is operating.

The total number $n(i)$ of critical path vectors for e_i is formally given by

$$n(i) = \sum_{\mathbf{z} \in V_n} [\varphi((1_i, \mathbf{z})) - \varphi((0_i, \mathbf{z}))]. \tag{2.72}$$

Example 2.21 *a) For a series system,* $h(\mathbf{p}) = p_1 p_2 \cdots p_n$. *Hence, letting* $p_0 = p_{n+1} = 1$,

$$I_B(i, \mathbf{p}) = p_1 p_2 \cdots p_{i-1} p_{i+1} \cdots p_n; \quad i = 1, 2, ..., n.$$

If $p_1 \leq p_2 \leq \cdots \leq p_n$, *then*

$$I_B(1, \mathbf{p}) \geq I_B(2, \mathbf{p}) \geq \cdots \geq I_B(n, \mathbf{p}).$$

Thus, the element with the smallest availability has the highest B-importance. ("A chain is as weak as its weakest link.")

b) For a parallel system, $h(\mathbf{p}) = 1 - \bar{p}_1 \bar{p}_2 \cdots \bar{p}_n$. *Hence, letting* $p_0 = p_{n+1} = 0$,

$$I_B(1, \mathbf{p}) = \bar{p}_1 \bar{p}_2 \cdots \bar{p}_{i-1} \bar{p}_{i+1} \cdots \bar{p}_n; \quad i = 1, 2, ..., n.$$

Thus, if $p_1 \leq p_2 \leq \cdots \leq p_n$,

$$I_B(1, \mathbf{p}) \leq I_B(2, \mathbf{p}) \leq \cdots \leq I_B(n, \mathbf{p}).$$

c) The reliability block diagram of a system consisting of 3 elements is given by Figure 2.13. It has the availability $h(\mathbf{p}) = p_1(p_2 + p_3 - p_2 p_3)$. *Hence,*

$$I_B(1, \mathbf{p}) = p_2 + p_3 - p_2 p_3, \quad I_B(2, \mathbf{p}) = p_1(1 - p_3), \quad I_B(3, \mathbf{p}) = p_1(1 - p_2).$$

Depending on the numerical values of the p_i, *every order of the* $I_B(1, \mathbf{p})$, $I_B(2, \mathbf{p})$, $I_B(3, \mathbf{p})$ *is possible.*

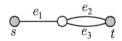

Figure 2.13: System with 3 elements

Since the sum of the B-importances of the elements generally does not add up to 1, a *standardised B-importance* $I_{B,SD}(i, \mathbf{p})$ may be a more useful importance measure:

$$I_{B,SD}(i, \mathbf{p}) = \frac{I_B(i, \mathbf{p})}{\sum_{k=1}^{n} I_B(k, \mathbf{p})}, \quad i = 1, 2, ..., n. \tag{2.73}$$

The B-importance depends heavily on the availabilities of the elements. If there is no or only little information on these availabilities, an importance measure based only on the structure of the system is interesting. The *structural Birnbaum-importance (structural B-importance)* $I_{SB}(i)$ of element e_i, $i = 1, 2, ..., n$, is defined as

$$I_{SB}(i) = I_{SB}(i, \mathbf{p}) \text{ with } \mathbf{p} = \left(\frac{1}{2}, \frac{1}{2}, \cdots, \frac{1}{2} \right).$$

Since all elements have the same availability $1/2$, every state vector $\mathbf{z} \in V_n$ has the same probability to operate the system, namely $(1/2)^n$. Hence, $I_{SB}(i)$ can be obtained by the classical definition of the probability:

$$I_{SB}(i) = \frac{n(i)}{2^{n-1}}, \quad i = 1, 2, ..., n,$$

where $n(i)$ is given by (2.72). (Note that there are 2^{n-1} state vectors $\mathbf{z}$ with the property that the state of e_i is fixed.)

In the literature, importances that depend on element availabilities, are frequently called *reliability (availability) importances*. But in case of $p = p_i$, $i = 1, 2, ..., n$, the corresponding availability importance may give even more information on the respective structural importance than the one obtained from the artificial assumption $p_i = 1/2$. For instance, if the p_i are known, then it might be a better idea to get information on the structural importance of elements when letting $p = \bar{p} = \frac{1}{n} \sum_{i=1}^{n} p_i$.

Example 2.22 *a) A series (parallel) system has only one critical path vector for every e_i. Hence, $I_{SB}(i) = 2^{-(n-1)}$.*

b) For the system with reliability block diagram given by Figure 2.13, the critical path vectors for e_1 are (0,1,1),(0,1,0), (0,0,1). Both for e_2 and e_3 there is only the critical path vector (1,0,0). Hence, $I_{SB}(1) = 3/4$ and $I_{SB}(2) = I_{SB}(3) = 1/4$. This result is quite intuitive, since a failure of e_1 always induces a system failure, whereas this need not be the case if e_2 or e_3 fail.

c) In a k-out-of-n-system, there are $\binom{n-1}{k-1}$ critical path vectors for every element e_i. Hence,

$$I_{SB}(i) = \binom{n-1}{k-1} 2^{-(n-1)}; \quad i = 1, 2, ..., n.$$

Availability Improvement Potential

Strictly speaking, the B-importance of elements only answers question 3 stated in the introduction. This is because $I_B(i, \mathbf{p})$ is the maximally possible increase in system availability due to e_i (namely, if the availability of e_i is increased from 0 to 1). Such an increase is, however, a hypothetical parameter, since normally an element comes with a positive availability, which will never reach level 1. Another disadvantage, $I_B(i, \mathbf{p})$ does not depend on p_i. Hence, an element can have a high B-importance, but if its availability is next to 1, then it may be uninteresting to the reliability engineer, since as a highly reliable element it will scarcely be the cause of a system failure. Instead, the actual changes in system availability are of interest, which are induced by changing the availability of an element. A standard situation is the following one: The availability of a system has to be increased. How should limited finances be spent on increasing the availabilities of its elements so that the increase in system availability becomes maximal? Helpful for solving this problem and in answering the questions 1 and 2 in the introduction are the following two modified B-importance criteria, also proposed by *Birnbaum* [51]:

$$I_0(i, \mathbf{p}) = h(\mathbf{p}) - h((0_i, \mathbf{p})), \tag{2.74}$$
$$I_1(i, \mathbf{p}) = h((1_i, \mathbf{p})) - h(\mathbf{p}).$$

$I_0(i, \mathbf{p})$ is the increase in system availability if element e_i with availability p_i is installed in the system, and $I_1(i, \mathbf{p})$ is the maximally possible availability gain if the availability of e_i is increased, starting from its present level p_i. By applying the pivotal decomposition formula (2.25) one gets the following relationships with the B-importance:

$$I_0(i, \mathbf{p}) = p_i I_B(i, \mathbf{p}), \quad I_1(i, \mathbf{p}) = (1 - p_i) I_B(i, \mathbf{p}). \tag{2.75}$$

Hence,

$$I_0(i, \mathbf{p}) = \frac{p_i}{1 - p_i} I_1(i, \mathbf{p}), \quad I_1(i, \mathbf{p}) = \frac{1 - p_i}{p_i} I_0(i, \mathbf{p}).$$

Thus, $I_0(i, \mathbf{p})$ $(I_1(i, \mathbf{p}))$ increases (decreases) linearly with increasing p_i, $0 \leq p_i \leq 1$. Analogously to (2.73), standardised versions of I_0 and I_1 can be defined.

Let us now explicitly consider the time-dependency of the p_i and $p_s = h(\mathbf{p})$ in systems without the renewal of failed elements: $p_i(x) = \overline{F}_i(x)$, $h(\mathbf{p}(x)) = \overline{F}_s(x)$, $f_s(x) = dF_s(x)/dx$. Then, from (2.69) and (2.75),

$$dF_s(x) = \sum_{i=1}^{n} I_0(i, \mathbf{p}(x)) d(\ln \overline{F}_i(x)).$$

Since $-d(\ln \overline{F}_i(x)) = \lambda_i(x) dx$,

$$f_s(x) = \sum_{i=1}^{n} I_0(i, \mathbf{p}(x)) \lambda_i(x).$$

Thus, the density of the system lifetime is the weighted sum of the failure rates of its elements, where the weights are the importance measures $I_0(i, \mathbf{p}(x))$ [33].

Importance measures, requiring the same input as I_B, I_0, and I_1, are the ratios

$$I_2 = \frac{h((1_i, \mathbf{p}))}{h(\mathbf{p})}, \quad I_3 = \frac{h((0_i, \mathbf{p}))}{h(\mathbf{p})}, \quad I_4 = \frac{1 - h(\mathbf{p})}{1 - h((1_i, \mathbf{p}))}, \quad I_5 = \frac{1 - h((0_i, \mathbf{p}))}{1 - h(\mathbf{p})}.$$

These ratios relate the effect of the availability (unavailability) of an element to the availability (unavailability) of the system.

Vesely–Fussel Importance

When a system fails, then it is possible that all elements of two or more minimal cut sets have failed. This implies that after the repair or replacement of a failed element the system may not necessarily become available. For instance, if $\{1, 2, 3\}$ and $\{2, 3, 4\}$ are the minimal cut sets of a system and if the elements fail in the order e_1, e_2, e_4, e_3, then the failure of e_3 has caused the system failure. In this case, the repair of only e_1 or only e_4 will not enable the system to continue its work. That means, an element e_i may not have caused the system to fail, but it contributes to the system failure. This happens when all elements of at least one of those minimal cut sets which contain e_i fail. The probability of this event is

$$h_i(\mathbf{p}) = \Pr\left(\bigwedge_{j=1}^{n_i} \bigvee_{k \in C_{j,i}} z_k = 0\right) = 1 - E\left(\bigwedge_{j=1}^{n_i} \bigvee_{k \in C_{j,i}} z_k\right),$$

where $\{C_{1,i}, C_{2,i}, \cdots, C_{n_i,i}\}$ is the set of all those minimal cut sets which contain e_i.

Definition 2.23 *The Vesely–Fussel (VF-) importance $I_{VF}(i)$ of element e_i is defined as the conditional probability that e_i contributes to a system failure given that the system has failed, i.e.*

$$I_{VF}(i) = \frac{h_i(\mathbf{p})}{1 - h(\mathbf{p})}.$$

Hence, the Vesely–Fussel importance refers to question 5 of the introduction.

Example 2.24 *a) For a series system, each element defines a minimal cut set so that $n_i = 1$, $i = 1, 2, ..., n$. Hence,*

$$I_{VF}(i) = \frac{1 - p_i}{1 - p_1 p_2 \cdots p_n}.$$

b) For a parallel system, there is only one minimal cut containing all n elements: $C_1 = \{1, 2, ..., n\} = C_{1,i}$ for all $i = 1, 2, ..., n$. Hence, in case of a system failure, every element contributes to the failure so that

$$I_{VF}(i) = 1, \quad i = 1, 2, ..., n.$$

More exactly, this is true since

$$h_i(\mathbf{p}) = \Pr\left(\bigvee_{k \in C_{1,i}} z_k = 0\right) = 1 - h(\mathbf{p}).$$

c) The system with the reliability block diagram (2.13) has the two minimal cut sets $C_1 = \{1\}$ and $C_2 = \{2, 3\}$. Hence, $C_{1,1} = C_1$ with $n_1 = 1$ and $C_{1,2} = C_{1,3} = C_2$ with $n_2 = n_3 = 1$. This implies

$$h_1(\mathbf{p}) = \Pr(z_1 = 0) = 1 - p_1, \quad h_2(\mathbf{p}) = h_3(\mathbf{p}) = (1 - p_2)(1 - p_3).$$

Therefore, with $h(\mathbf{p}) = p_1 [p_2 + p_3 - p_2 p_3]$,

$$I_1(1) = \frac{1 - p_1}{1 - h(\mathbf{p})}, \quad I_2(2) = I_{VF}(3) = \frac{(1 - p_2)(1 - p_3)}{1 - h(\mathbf{p})}.$$

A structural VF-importance $I_{SFV}(i)$ is defined analogously to the structural B-importance by letting $p_i = 1/2$ for all $i = 1, 2, ..., n$. For instance, the elements of the system with the reliability block diagram given by Figure (2.13) have the structural VF-importances

$$I_{SVF}(1) = 0.8, \quad I_{SVF}(2) = I_{SVF}(3) = 0.4.$$

Butler's Structural Cut Importance

Butler [76] *argued that the artificial choice of $p_i = 1/2$, $i = 1, 2, ..., n$, may at least quantitatively give inaccurate information on the structural importances of elements within a system, even if the rank order of the elements based on structural importances obtained from the assumption $p = p_i$, $i = 1, 2, ..., n$, is the same for all p, $0 \leq p \leq 1$. To derive an alternative concept of structural importance, the inclusion–exclusion approach can be exploited similar to Section 2.2.4, but applied to formula (2.39):*

$$p_s = 1 - \sum_{k=1}^{b}(-1)^{k-1}E(T_k), \quad T_k = \sum_{1 \leq j_1 < j_2 \cdots < j_k \leq b} D_{j_1} D_{j_2} \cdots D_{j_k}, \quad (2.76)$$

where

$$D_j = \prod_{i \in C_j} \overline{z}_i, \quad j = 1, 2, ..., b.$$

There holds $D_{j_1} D_{j_2} \cdots D_{j_k} = 1$ iff $z_i = 0$ for all i with $i \in C_{j_1} \cup C_{j_2} \cup \cdots \cup C_{j_k}$. Let $n_{k,r}^{(i)}$ be the number of all those unions of exactly k minimal cut sets, so that the corresponding term $D_{j_1} D_{j_2} \cdots D_{j_k}$ is the product of $\overline{z}_i$ and $r - 1$ factors $\overline{z}_j$ different from $\overline{z}_i$, and

$$w_r^{(i)} = \sum_{k=1}^{b}(-1)^{k-1}n_{k,r}^{(i)}; \quad r = 1, 2, ..., n. \quad (2.77)$$

Definition 2.25 *Element e_i is structurally more (Butler) important than element e_j if vector $\mathbf{w}^{(i)} = \left(w_1^{(i)}, w_2^{(i)}, ..., w_n^{(i)} \right)$ is lexicographically larger than vector $\mathbf{w}^{(j)} = \left(w_1^{(j)}, w_2^{(j)}, ..., w_n^{(j)} \right)$, $i, j = 1, 2, ..., n$, i.e. if $w_r^{(i)} > w_r^{(j)}$ for all $r = 1, 2, ..., m$ with $m = \min(r, \; w_r^{(i)} \neq w_r^{(j)})$, $i \neq j$. In this case, we write $I_{Bu}(i) > I_{Bu}(j)$.*

To motivate Butler's structural importance ranking, note that under the assumption all elements operate independently and have the same availability p, all terms in (2.76) with r factors $\bar{z}_k$ contribute the same *absolute* amount to the system availability. This absolute amount is the larger, the smaller r. The total weight of the terms with r factors in (2.77) on condition that one of them is $\bar{z}_i$ is given by $w_r^{(i)}$. Under this assumption, the system availability (2.76) can be represented as

$$p_s = 1 - \sum_{r=1}^{n} w_r^{(i)} (1-p)^r - h_{-e_i}(p), \quad i = 1, 2, ..., n, \tag{2.78}$$

where function $h_{-e_i}(p)$ does not depend on the availability $p_i = p$ of e_i. For p close to 1, the influence of e_i on the system availability will be dominated by the first term of the sum in (2.78). Hence, Butler's importance ranking is particularly valid for systems with high element availabilities. (The individual availabilities may, however, differ.) Moreover, under our assumption $p_i = p$ there is a close relationship between Butler's and Birnbaum's importance ranking: From (2.78), taking into account that the Birnbaum-importance of e_i is the partial derivative of $p_s = h(\mathbf{p})$ with regard to p_i,

$$I_B(i, p) = \sum_{r=1}^{n} w_r^{(i)} (1-p)^{r-1}. \tag{2.79}$$

Thus, for p being sufficiently large, Birnbaum's importance ranking with $p_i = p$, $i = 1, 2, ..., n$, and Butler's structural importance ranking of elements coincide.

Butler's approach can also be applied if instead of the minimal cuts the minimal paths are given. Then $n_{k,r}^{(i)}$ denotes the number of all those unions of exactly k minimal path sets, so that in (2.53) the corresponding term $A_{j_1} A_{j_2} \cdots A_{j_k}$ is the product of z_i and $r - 1$ factors z_j different from z_i. Defining weights $w_r^{(i)}$ again by (2.77) yields

$$p_s = \sum_{r=1}^{n} w_r^{(i)} p^r + h_{-e_i}(p), \quad I_B(i, p) = \sum_{r=1}^{n} w_r^{(i)} p^{r-1}.$$

Thus, in this case Butler's structural (path) ranking and Birnbaum's importance ranking with $p_i = p$, $i = 1, 2, ..., n$, coincide for p being sufficiently small.

Example 2.26 *The minimal cut sets of the system considered in Example 2.11 (Figure 2.7) are* $C_1 = \{1,2\}$, $C_2 = \{6,7\}$, $C_3 = \{1,3,4\}$, $C_4 = \{1,3,7\}$, $C_5 = \{2,5,6\}$, $C_6 = \{4,5,6\}$. *The corresponding weight vectors* $\mathbf{w}^{(i)}$, $i = 1, 2, ..., 7$, *are*

$$\mathbf{w}^{(1)} = (0,\ 1,\ 2,-5,\ 2,\ 2,-1)$$
$$\mathbf{w}^{(2)} = (0,\ 1,\ 1,-6,\ 4,\ 3,-1)$$
$$\mathbf{w}^{(3)} = (0,\ 0,\ 2,-4,\ 1,\ 2,-1)$$
$$\mathbf{w}^{(4)} = (0,\ 0,\ 2,-4,\ 1,\ 2,-1)$$
$$\mathbf{w}^{(5)} = (0,\ 0,\ 2,-4,\ 2,\ 2,-1)$$
$$\mathbf{w}^{(6)} = (0,\ 1,\ 2,-6,\ 3,\ 2,-1)$$
$$\mathbf{w}^{(7)} = (0,\ 1,\ 1,-6,\ 4,\ 1,-1)$$

Thus, Butler's structural importance ranking of the elements is

$$I_{Bu}(1) > I_{Bu}(6) > I_{Bu}(2) > I_{Bu}(7) > I_{Bu}(5) > I_{Bu}(4) = I_{Bu}(3). \quad (2.80)$$

From (2.79), the structural Birnbaum-importances order the elements in the same way:

$$I_{SB}(1) = 0.5469,\ I_{SB}(6) = 0.4844,\ I_{SB}(2) = 0.3281,$$
$$I_{SB}(7) = 0.1805,\ I_{SB}(5) = 0.1719,\ I_{SB}(4) = I_{SB}(3) = 0.1094.$$

It is interesting to compare Butler's ranking with the one generated by the structural Vessely–Fussel importances, since both rankings are based on the minimal cuts.

$$I_{SVF}(1) = I_{SVF}(6) = 0.6197,\ I_{SVF}(2) = I_{SVF}(7) = 0.5634,$$
$$I_{SVF}(4) = 0.3944,\ I_{SVF}(5) = I_{SVF}(3) = 0.3380. \quad (2.81)$$

As seen in this example, a disadvantage of Butler's importance ranking is that it is a qualitative one; no quantitative information on element importance is given.

Meng's Importance Ranking

Meng [218] *proposed another importance ranking concept, which is related to the Birnbaum-importance. It can be considered to be a mixture between the availability and the structural importance approach. As with Butler's importance ranking, no numerical importance measures for the elements are provided.*

Definition 2.27 *Let* $\mathbf{p} = (p_1, p_2, \cdots, p_n)$. *Element* e_i *is said to be more (Meng–) important than element* e_j *if for all* $r \in [0, 1]$

$$I_B(i, (r_j, \mathbf{p})) > I_B(j, (r_i, \mathbf{p})). \quad (2.82)$$

In this case we write $I_M(i) > I_M(j)$.

(Remember that the (reliability) Birnbaum-importance $I_B(i, \mathbf{p})$) does not depend on p_i, and that according to notation (2.15) the i th component of vector $(r_i, \mathbf{p})$, $i = 1, 2, ..., n$, is r.) To intuitively interpret this definition: Given $\mathbf{p}$, the validity of (2.82) for all $r \in [0, 1]$ is due to the system structure. But for other vectors $\mathbf{p}$, (2.82) may not be true.

By making use of the pivotal decomposition formula (2.23), it can be shown that

$$I_M(i) > I_M(j) \quad \text{iff} \quad h((r_i, s_j, \mathbf{p})) > h((s_i, r_j, \mathbf{p})) \qquad (2.83)$$

for all pairs (r, s) with $0 \leq s < r \leq 1$. (The notation $(r_i, s_j, \mathbf{p})$ means that the i th component of vector $\mathbf{p}$ is r and its j th component is s.)

Example 2.28 *For the sake of comparison, we consider the same system as in Example 2.26. For independent elements, the availability of this system can easily be obtained from Example 2.11:*

$$\begin{aligned}
h(\mathbf{p}) = {} & p_1 p_6 + p_1 p_5 p_7 + p_2 p_3 p_6 + p_2 p_4 p_7 - p_1 p_2 p_3 p_6 \\
& - p_1 p_5 p_6 p_7 + p_2 p_3 p_5 p_7 - p_1 p_2 p_3 p_5 p_7 - p_1 p_2 p_4 p_5 p_7 \\
& - p_1 p_2 p_4 p_6 p_7 - p_2 p_3 p_4 p_5 p_7 - p_2 p_3 p_4 p_6 p_7 - p_2 p_3 p_5 p_6 p_7 \\
& + p_1 p_2 p_3 p_4 p_5 p_7 + p_1 p_2 p_3 p_4 p_6 p_7 + p_1 p_2 p_3 p_5 p_6 p_7 \\
& + p_1 p_2 p_4 p_5 p_6 p_7 + p_2 p_3 p_4 p_5 p_6 p_7 - p_1 p_2 p_3 p_4 p_5 p_6 p_7.
\end{aligned}$$

In what follows, the assumption $p_i = p$, $i = 1, 2, ..., n$, is made. Then Meng's ranking is essentially based on the system structure. Motivated by Butler's ranking (2.80), we next order e_1 and e_6. The system availability in dependence on p_1, p_6, and p is

$$\begin{aligned}
h(p_1, p_6, p) = {} & (p^2 - 2p^4 + p^5)p_1 + (p^2 - 2p^4 + p^5)p_6 \\
& (1 - 2p^2 - p^3 + 3p^4 - p^5)p_1 p_6 + 2p^3 + p^4 - p^5.
\end{aligned}$$

Hence, $I_B(1, (r_6, p)) = I_B(6, (r_1, p))$ *so that* $I_M(1) = I_M(6)$. *Based on* $h(p_2, p_6, p)$ *one obtains* $I_M(6) > I_M(2)$, *and finally*

$$I_M(1) = I_M(6) > I_M(2) = I_M(7) > I_M(5) = I_M(4) = I_M(3).$$

Note that the structural Vesely–Fussel ranking (2.81) for this example is:

$$\begin{aligned}
I_{SVF}(1) = I_{SVF}(6) > {} & I_{SVF}(2) = I_{SVF}(7) > I_{SVF}(4) \\
& > I_{SVF}(5) = I_{SVF}(3).
\end{aligned}$$

In another paper, *Meng* [219] proposed a *criticality importance measure* and the corresponding structural importance. He showed that this structural importance as well as the structural Vessely–Fussel importance can be represented by the cardinalities of certain sets, see also [220, 221, 222].

In what follows, some time-dependent importance measures will be discussed, always on condition that failed elements are not repaired/replaced.

Barlow–Proschan Importance

We now consider time-dependent element availabilities $p_i = p_i(x) = \overline{F}_i(x)$ for systems without repair or replacement of failed elements. (The latter assumption does not matter when assuming constant availabilities p_i.) The densities $f_i(x) = F_i'(x)$, $i = 1, 2, ..., n$, are assumed to exist. Let $\mathbf{z}(x) = (z_1(x), z_2(x), ..., z_n(x))$ be the vector of the states of the elements at time x. Then the probability that the system fails at time x if e_i fails at time x is

$$\Pr(\varphi((1_i, \mathbf{z}(x)) - \varphi((0_i, \mathbf{z}(x)) = 1) = h((1_i, \mathbf{p}(x)) - h((0_i, \mathbf{p}(x)).$$

Hence, the probability that e_i causes a system failure in the time interval $(x, x + \Delta x]$ is

$$[h((1_i, \mathbf{p}(x)) - h((0_i, \mathbf{p}(x))] \, f_i(x)\Delta x + o(\Delta x).$$

Therefore and in view of (2.70), the probability that e_i causes the system to fail in $(0, t]$ is

$$\int_0^t I_B(i, \mathbf{p}(x)) \, f_i(x)dx.$$

Definition 2.29 *The* Barlow–Proschan importance (BP-importance) $I_{BP}(i, t, \mathbf{p})$ *of element* e_i *is defined as*

$$I_{BP}(i, t, \mathbf{p}) = \frac{\int_0^t I_B(i, \mathbf{p}(x)) \, f_i(x)dx}{\sum_{k=1}^n \int_0^t I_B(k, \mathbf{p}(x)) \, f_k(x)dx}. \tag{2.84}$$

Thus, given the system fails in $(0, t]$, $I_{BP}(i, t, \mathbf{p})$ is the probability that the failure of e_i has triggered this system failure. Hence, the BP-importances refer to question 4 of the introduction. For any $t > 0$ they have properties

$$0 \le I_{BP}(i, t, \mathbf{p}) \le 1, \quad \sum_{i=1}^n I_{BP}(i, t, \mathbf{p}) = 1.$$

It follows

$$\lim_{t \to \infty} \sum_{i=1}^n \int_0^t I_B(i, \mathbf{p}(x)) \, f_i(x)dx = \lim_{t \to \infty} [1 - h(\mathbf{p}(t))] = 1. \tag{2.85}$$

Hence, letting $t \to \infty$, $I_{BP}(i, t, \mathbf{p})$ becomes

$$I_{BP}(i, \mathbf{p}) = \int_0^\infty I_B(i, \mathbf{p}(x)) \, f_i(x)dx. \tag{2.86}$$

$I_{BP}(i, \mathbf{p})$ is simply the probability that the system failure, occurring anytime in $(0, \infty)$, has been triggered by the failure of e_i.

Example 2.30 *Let us again consider the system with the reliability block diagram given by Figure 2.13. The lifetimes of its elements are assumed to have exponential distributions:* $p_i(x) = \overline{F}_i(x) = e^{-\lambda_i x}$, $x \geq 0$, $i = 1, 2, 3$. *Then,*

$$h(\mathbf{p}(x)) = e^{-\rho_1 x} + e^{-\rho_2 x} - e^{-\rho_3 x}$$

with $\rho_1 = \lambda_1 + \lambda_2$, $\rho_2 = \lambda_1 + \lambda_3$, $\rho_3 = \lambda_1 + \lambda_2 + \lambda_3$, *and*

$$I_{BP}(1, t, \mathbf{p}) = \frac{\lambda_1}{1 - h(\mathbf{p}(t))} \left[\frac{1 - e^{-\rho_1 t}}{\rho_1} + \frac{1 - e^{-\rho_2 t}}{\rho_2} - \frac{1 - e^{-\rho_3 t}}{\rho_3} \right],$$

$$I_{BP}(2, t, \mathbf{p}) = \frac{\lambda_2}{1 - h(\mathbf{p}(t))} \left[\frac{1 - e^{-\rho_1 t}}{\rho_1} - \frac{1 - e^{-\rho_3 t}}{\rho_3} \right],$$

$$I_{BP}(3, t, \mathbf{p}) = \frac{\lambda_3}{1 - h(\mathbf{p}(t))} \left[\frac{1 - e^{-\rho_2 t}}{\rho_2} - \frac{1 - e^{-\rho_3 t}}{\rho_3} \right].$$

For $t \to \infty$,

$$I_{BP}(1) = \lambda_1 \left[\rho_1^{-1} + \rho_2^{-1} - \rho_3^{-1} \right],$$

$$I_{BP}(2) = \lambda_2 \left[\rho_1^{-1} - \rho_3^{-1} \right],$$

$$I_{BP}(3) = \lambda_3 \left[\rho_2^{-1} - \rho_3^{-1} \right].$$

In particular, for $\lambda = \lambda_1 = \lambda_2 = \lambda_3$,

$$I_{BP}(1) = 2/3, \ I_{BP}(2) = I_{BP}(3) = 1/6. \tag{2.87}$$

The *structural BP-importance* $I_{SBP}(i)$ we get from (2.86) by letting there be $\overline{F}_i(t) = p$ for all $i = 1, 2, ..., n$. Then,

$$I_{SBP}(i) = \int_0^1 I_B(i, \mathbf{p}) \, dp \tag{2.88}$$

with $\mathbf{p} = (p, p, ..., p)$. Thus, the structural BP-importance is the expected value of the B-importance on condition that the joint element availability is a random variable, which has a uniform distribution over the interval $[0, 1]$. An equivalent representation of $I_{SBP}(i)$ is

$$I_{SBP}(i) = \frac{1}{n} \sum_{r=1}^{n} \binom{n-1}{r-1}^{-1} n_r(i), \tag{2.89}$$

where $n_r(i)$ is the number of critical path vectors for element e_i of order r, i.e. exactly r of their n components have value 1.

Example 2.31 *With regard to Figure 2.13, the critical path vectors for* e_1 *are* $(1, 2), (1, 3)$ *and* $(1, 2, 3)$, *the respective critical path vectors for* e_2 *and* e_3 *are* $(1, 2)$ *and* $(1, 3)$. *Hence,* $n_1(1) = 0$, $n_2(2) = 2$, *and* $n_3(1) = 1$, *so that by (2.88),*

$$I_{SBP}(1) = \frac{2}{3}, \ I_{SBP}(2) = I_{SBP}(3) = \frac{1}{6}.$$

These structural importances coincide with the corresponding "stationary importances" $I_{BP}(i, \mathbf{p})$ *given by (2.87).*

Natvig Importance

Natvig [241, 242] considered that element most important, the failure of which reduces the remaining system lifetime most.

Definition 2.32 *Let $L_{r,i}$ be the remaining system lifetime from the time point at which element e_i fails. Then the* Natvig-importance *of e_i is defined as*

$$I_N(i) = \frac{E(L_{r,i})}{\sum_{j=1}^{n} E(L_{r,j})} \tag{2.90}$$

with

$$E(L_{r,i}) = -\int_0^\infty I_B(i, \mathbf{p}(x))\overline{F}_i(x)\ln \overline{F}_i(x)dx.$$

For extensions of the Natvig importance see [240].

Bergman Importance

In case of time-dependent element availabilities, *Bergman* [48] proposed a powerful generalization of the transition from the B-importance $I_B(i, \mathbf{p})$ to $I_0(i, \mathbf{p}) = h(\mathbf{p}) - h((0_i, \mathbf{p}))$ (formula (2.74)). Given the survival function $\overline{F}_i$ of element e_i, he was interested in the increase of the expected system lifetime if $\overline{F}_i$ is replaced with the survival function $\overline{G}_i$. Let $\mathbf{p} = (\overline{F}_1, \overline{F}_2, ..., \overline{F}_n)$, i.e. no repair or replacement of failed elements. By making use of formula (1.2) with L as the system lifetime, and letting $(\overline{G}_i, \mathbf{p}) = (p_1, p_2, ..., p_{i-1}, \overline{G}_i, p_{i+1}, ..., p_n)$, this increase is

$$\Delta_{Be}(i, \overline{\mathbf{G}}, \mathbf{p}) = \int_0^\infty h((\overline{G}_i, \mathbf{p}(x)))dx - \int_0^\infty h(\mathbf{p}(x))dx.$$

$\overline{G}_i$ is assumed to be such that $\Delta_{Be}(i, \overline{\mathbf{G}}, \mathbf{p}) \geq 0$. The pivotal decomposition formula (2.23), applied to both integrals with pivotal element e_i and taking into account (2.70), yields

$$\Delta_{Be}(i, \overline{\mathbf{G}}, \mathbf{p}) = \int_0^\infty \left[(\overline{G}_i(x) - \overline{F}_i(x))I_B(i, \mathbf{p}(x))\right] dx, \quad i = 1, 2, ..., n. \tag{2.91}$$

Thus, a sufficient condition for achieving the desired nonnegative increase $\Delta_{Be}(i, \overline{\mathbf{G}}, \mathbf{p})$ is that L_{G_i} is larger than L_{F_i} with respect to the *usual stochastic order*, i.e that $\overline{G}_i(x) \geq \overline{F}_i(x)$ for all $x \geq 0$.

Definition 2.33 *With regard to a vector $\overline{\mathbf{G}} = (\overline{G}_1, \overline{G}_2, ..., \overline{G}_n)$ of survival functions, element e_i is said to have the* Bergman-importance

$$I_{Be}(i, \overline{\mathbf{G}}, \mathbf{p}) = \frac{\Delta_{Be}(i, \overline{\mathbf{G}}, \mathbf{p})}{\sum_{k=1}^{n} \Delta_{Be}(k, \overline{\mathbf{G}}, \mathbf{p})}, \quad i = 1, 2, ..., n. \tag{2.92}$$

Example 2.34 *The elements of the system with the reliability block diagram given by Figure 2.13 have availabilities $p_i(x) = \overline{F}(x)$ with $\overline{F}(x) = e^{-\lambda x}$, $i = 1, 2, 3$. Then, by Example 2.21 c),*

$$I_B(1, \mathbf{p}(x)) = e^{-\lambda x}(2 - e^{-\lambda x}),$$
$$I_B(2, \mathbf{p}(x)) = I_B(3, \mathbf{p}(x)) = e^{-\lambda x}(1 - e^{-\lambda x}).$$

If the corresponding replacement availabilities are

$$\overline{G}_i(x) = e^{-\mu x}, \quad x \geq 0; \quad 0 < \mu < \lambda, \quad i = 1, 2, 3, \tag{2.93}$$

then

$$\Delta_{Be}(1, \overline{\mathbf{G}}, \mathbf{p}) = \frac{2}{\lambda + \mu} - \frac{1}{2\lambda + \mu} - \frac{4}{6\lambda},$$

$$\Delta_{Be}(2, \overline{\mathbf{G}}, \mathbf{p}) = \Delta_{Be}(3, \overline{\mathbf{G}}, \mathbf{p}) = \frac{1}{\lambda + \mu} - \frac{1}{2\lambda + \mu} - \frac{1}{6\lambda}.$$

Hence,

$$I_{Be}(1, \overline{\mathbf{G}}, \mathbf{p}) = \frac{10\lambda^2 - 6\lambda\mu - 4\mu^2}{18\lambda^2 - 12\lambda\mu - 6\mu^2},$$

$$I_{Be}(2, \overline{\mathbf{G}}, \mathbf{p}) = I_{Be}(3, \overline{\mathbf{G}}, \mathbf{p}) = \frac{4\lambda^2 - 3\lambda\mu - \mu^2}{18\lambda^2 - 12\lambda\mu - 6\mu^2}.$$

In view of assumption (2.93), $I_{Be}(1, \overline{\mathbf{G}}, \mathbf{p}) > I_{Be}(2, \overline{\mathbf{G}}, \mathbf{p})$. In particular, for $\mu = \lambda/2$, i.e. elements are replaced with ones having double the lifetime,

$$I_{Be}(1, \overline{\mathbf{G}}, \mathbf{p}) = 0.5714, \quad I_{Be}(2, \overline{\mathbf{G}}, \mathbf{p}) = I_{Be}(3, \overline{\mathbf{G}}, \mathbf{p}) = 0.2143. \tag{2.94}$$

Xie's Yield Importance

Bergman's importance concept is based on a yield (increase) in the expected system lifetime. He already suggested that a more general yield concept could serve as a basis for a more general importance measure. To construct such a measure for nonrepairable elements, *Xie* [312] introduced a nondecreasing function $Y(x)$ as a function of the system lifetime $L = x$ with the properties $Y(+0) = 0$ and $Y(\infty) < \infty$. $Y(x)$ is called the *yield function* and, in case of its existence, the corresponding first derivative $y(x) = dY(x)/dx$ is the *yield rate*. Since $F_s(x) = P(L \leq x) = 1 - h(\mathbf{p}(x))$, the *expected yield* during the random lifetime L of the system is

$$\mathbb{E}(Y(L)) = \int_0^\infty Y(x)dF_s(x) = \int_0^\infty h(\mathbf{p}(x))y(x)dx.$$

Analogously to the derivation of (2.91), the expected yield increase (assumed to be nonnegative) if replacing $\overline{F}_i$ with $\overline{G}_i$ is seen to be

$$\Delta_X(i, Y, \overline{\mathbf{G}}, \mathbf{p}) = \int_0^\infty \left[(\overline{G}_i(x) - \overline{F}_i(x))y(x)I_B(i, \mathbf{p}(x)) \right] dx, \quad i = 1, 2, ..., n. \tag{2.95}$$

Definition 2.35 Xie's lifetime yield importance *for element e_i is defined as*

$$I_X(i, Y, \overline{\mathbf{G}}, \mathbf{p}) = \frac{\Delta_X(i, Y, \overline{\mathbf{G}}, \mathbf{p})}{\sum_{k=1}^{n} \Delta_X(k, Y, \overline{\mathbf{G}}, \mathbf{p})}, \quad i = 1, 2, ..., n. \tag{2.96}$$

It underlines the significance of Xie's approach that some key importance criteria, although motivated in quite another way, are formally special cases of (2.96):

1. For the linear yield function $Y(x) = x$, Xie's yield importance (2.96) becomes the Bergman importance (2.92).

2. The Barlow–Proschan importance (2.86) arises from Xie's yield importance by letting $y(x) = f_i(x)/(\overline{G}_i(x) - \overline{F}_i(x))$, $x > 0$:

$$\Delta_i^Y = \int_0^\infty \left[(\overline{G}_i(x) - \overline{F}_i(x)) y(x) I_B(i, \mathbf{p}(x)) \right] dx$$

$$= \int_0^\infty I_B(i, \mathbf{p}(x)) \, f_i(x) dx = I_{BP}(i, \mathbf{p}).$$

The more general Barlow–Proschan importance (2.84) we get from (2.96) by letting

$$y(x) = \begin{cases} f_i(x)/(\overline{G}_i(x) - \overline{F}_i(x)), & 0 < x \le t \\ 0, & t < x \end{cases}.$$

3. The Natvig importance (2.90) is seen to be a special case of (2.96) if letting

$$y(x) = \frac{\overline{F}_i(x) \ln \overline{F}_i(x)}{\overline{G}_i(x) - \overline{F}_i(x)}, \quad x > 0.$$

For a detailed discussion of Xie's yield importance see Xie and Bergman (1991).

Availability Yield Importance

Instead of relating a yield to the lifetime of a system, it may frequently be a more adequate approach to assign a yield to the system availability (reliability); again on condition that failed components are not repaired. Let $Y = Y(z)$ be a real-valued nondecreasing function on $[0, 1]$ with properties $Y(0) = 0$, $Y(1) < \infty$, and

$$I_A(S) = \int_0^\infty Y(h(\mathbf{p}(x))) dx < \infty.$$

(Note that the function $\widetilde{Y}(x) = Y(h(\mathbf{p}(\mathbf{x})))$ is decreasing in x and tends to 0 as $x \to \infty$.) $I_A(S)$ is the *total availability yield* due to the work of the system to its failure. Analogously to the derivation of (2.91), the increase in $I_A(S)$ if $\overline{F}_i$ is replaced with $\overline{G}_i$ is seen to be

$$\Delta_A(i, Y, \overline{\mathbf{G}}, \mathbf{p}) = \int_0^\infty \left[(\overline{G}_i(x) - \overline{F}_i(x)) I_{B,Y}(i, \mathbf{p}(x)) \right] dx, \quad i = 1, 2, ..., n,$$

where the *Birnbaum yield importance* $I_{B,Y}(i,\mathbf{p})$ of element e_i is defined as

$$I_{B,Y}(i,\mathbf{p}) = Y(h((1_i,\mathbf{p}))) - Y(h((0_i,\mathbf{p}))), \quad i = 1, 2, ..., n.$$

$I_{B,Y}(i,\mathbf{p})$ can be interpreted as the economical or any other profit arising from installing the absolutely reliable element e_i in the system. (For a somewhat related idea see [266].)

Definition 2.36 *With regard to a vector of survival functions*

$$\overline{\mathbf{G}} = (\overline{G}_1, \ \overline{G}_2, ...\overline{G}_n),$$

the availability yield importance *for element e_i if $\overline{F}_i$ is replaced with $\overline{G}_i$ is defined as*

$$I_A(i,Y,\overline{\mathbf{G}},\mathbf{p}) = \frac{\Delta_A(i,Y,\overline{\mathbf{G}},\mathbf{p})}{\sum_{k=1}^{n}\Delta_A(k,Y,\overline{\mathbf{G}},\mathbf{p})}, \quad i = 1, 2, ..., n. \qquad (2.97)$$

As with some importance measures dealt with before, if only an importance ranking of the components is required, then the calculation of the $\Delta_A(i,Y,\overline{\mathbf{G}},\mathbf{p})$, $i = 1, 2, ..., n$, is sufficient.

Xie's lifetime yield importance (2.96) is formally seen to be a special case of the availability yield importance when replacing in (2.97) $\widetilde{Y}(x) = Y(h(\mathbf{p}(x))$ with $\widetilde{Y}(x) = y(x)h(\mathbf{p}(x))$. (But $\widetilde{Y}(x) = y(x)h(\mathbf{p}(x))$ will usually not be a decreasing function in x on the whole half axis $[0,\infty)$.)

Example 2.37 *Under otherwise the same assumptions as in example 2.33, let $Y(y) = y^2$, $0 \le y \le 1$. Since $h(\mathbf{p}) = p_1(p_2 + p_3 - p_2 p_3)$,*

$$h((1_1,\mathbf{p})) = 2p - p^2, \quad h((0_1,\mathbf{p})) = 0,$$
$$h((1_2,\mathbf{p})) = h((1_3,\mathbf{p})) = p, \quad h((0_2,\mathbf{p})) = h((0_3,\mathbf{p})) = p^2,$$
$$I_{B,Y}(1,\mathbf{p}(x)) = 4e^{-2\lambda x} - 4e^{-3\lambda x} + e^{-4\lambda x},$$
$$I_{B,Y}(2,\mathbf{p}(x)) = I_{B,Y}(3,\mathbf{p}(x)) = e^{-2\lambda x} - e^{-4\lambda x}.$$

Hence,

$$\Delta_A(1,Y,\overline{\mathbf{G}},\mathbf{p}) = \frac{4}{2\lambda+\mu} - \frac{4}{3\lambda+\mu} + \frac{1}{4\lambda+\mu} - \frac{8}{15\lambda},$$
$$\Delta_A(2,Y,\overline{\mathbf{G}},\mathbf{p}) = \Delta_A(3,Y,\overline{\mathbf{G}},\mathbf{p}) = \frac{1}{2\lambda+\mu} - \frac{1}{4\lambda+\mu} - \frac{2}{15\lambda}.$$

In particular, for $\mu = \lambda/2$,

$$I_A(1,Y,\overline{\mathbf{G}},\mathbf{p}) = 0.6216, \quad I_A(2,Y,\overline{\mathbf{G}},\mathbf{p}) = I_A(3,Y,\overline{\mathbf{G}},\mathbf{p}) = 0.1892.$$

As expected, if emphasis is on the availability yield rather than the lifetime yield,

$$I_A(1,Y,\overline{\mathbf{G}},\mathbf{p}) = 0.6216 > I_{Be}(1,\overline{\mathbf{G}},\mathbf{p}) = 0.5714$$

see (2.94).

Remark 2.38 *The total availability yield due to the work of the system in the finite interval* $[0, t]$ *is*

$$I_A(S, t) = \int_0^t Y(h(\mathbf{p}(x)))dx.$$

$I_A(S, t)$ *can serve as the basis for defining importance measures* $I_A(i, Y, \mathbf{G}, \mathbf{p}, t)$ *for the elements analogously to (2.97).*

2.2.7   Modular Decomposition

The technological structures of coherent systems frequently show more or less closed subsystems. If the knowledge of the state of such a subsystem (available or not) provides the same information as the knowledge of the states of all components of the subsystem with regard to the state of the system, then such a subsystem is called a *module* of the system. For the reliability analysis of a system with modules the following approach is a simple and efficient way to reduce the computational effort:

1. All modules are identified and their mutual reliability-theoretic connection within the system is established. This process generates a *modular decomposition* of the system.

2. The reliability block diagram of the system is simplified by reducing the modules to simple systems (elements).

3. The availability of the modules is determined by any suitable method dealt with in the previous sections, and, based on the simplified reliability block diagram of the system, its availability is calculated.

Note that the modules, which generate a modular decomposition of a system, comprise disjoint sets of elements, the union of which is the set of all elements of the system.

The modular decomposition of a system is not necessarily unique, since a module in its turn may consist of submodules. Hence, the three steps for determining the system availability, depending on the complexity of the modules, can be applied to modules and their submodules as well. Thus, the modular decomposition approach to the reliability analysis of a complex system may include modular decomposition on several levels.

Every system consisting of n elements has $(n + 1)$ *trivial modules*, namely the system itself and all of its elements can be considered to be modules. If only trivial modules exist, then modular decomposition simply replicates the system.

For large, complicated structured systems with nontrivial modules, modular decomposition may considerably reduce the time for their numerical reliability analysis. On the other hand, if statistically dependent elements can be comprised in suitably chosen modules in such a way that the modules operate independently, then even more computational advantages can be expected.

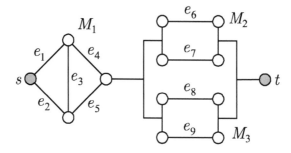

Figure 2.14: System with modules

Example 2.39 *We will formalize the generation of a modular decomposition of a coherent system by a system S with 9 elements. Figure 2.14 shows its reliability block diagram. The 3 modules M_1, M_2, and M_3 are clearly visible. Module M_1 is the bridge structure given by Figure 2.3. Hence, by formula (2.27), its structure function is*

$$u = u(z_1, z_2, z_3, z_4, z_5) = z_1 z_4 + z_2 z_5 + z_1 z_3 z_5 + z_2 z_3 z_4 - z_1 z_2 z_3 z_4$$
$$- z_1 z_2 z_3 z_5 - z_1 z_2 z_4 z_5 - z_1 z_3 z_4 z_5 - z_2 z_3 z_4 z_5 + 2 z_1 z_2 z_3 z_4 z_5.$$

The modules M_2 and M_3 are parallel systems with the respective structure functions

$$v = v(z_6, z_7) = 1 - \overline{z}_6 \overline{z}_7,$$
$$w = w(z_8, z_9) = 1 - \overline{z}_9 \overline{z}_9.$$

This modular decomposition simplifies the reliability block diagram of the system with 9 elements to the one of a system consisting of only 3 elements, which is given by Figure 2.13. Thus, the structure function $\varphi(\mathbf{z})$ with $\mathbf{z} = (z_1, ..., z_9)$ can be rewritten in terms of the indicator variables $u, v,$ and w of the modules as follows:

$$\varphi(\mathbf{z}) = \psi(u, v, w) = uv + uw - uvw.$$

The Boolean function ψ is called the organizing *structure function of the (original) system.*

2.3   Exercises

Exercise 2.1 *a) Determine the structure function of the bridge structure (Figure 2.3) by generating its disjoint sum form (2.24).*
b) Assuming $p_i = p$, $i = 1, 2, ..., 5$, show that its availability function $h(p)$ has an S-form.

Exercise 2.2 *a) Determine the structure function of the extended bridge structure (Figure 2.10) by pivotal decomposition with pivotal element e_4.*
b) Determine the Birnbaum-importance and the structural Birnbaum-importance of element e_4 on condition $p_i = p$, $i = 1, 2, ..., 7$.

Exercise 2.3 *a) Determine the structure function of a system with reliability block diagram Figure 2.7 by pivotal decomposition with pivotal element e_5.*
b) Determine the Birnbaum-importances of elements e_6 and e_7 on condition $p_i = p$, $i = 1, 2, ..., 7$, and give an intuitive explanation why they differ.

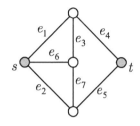

Figure 2.15: Reliability block diagram for Exercise 2.4

Exercise 2.4 *Figure 2.15 shows the reliability block diagram of a system.*
a) Determine its structure function by pivotal decomposition in such a way that you can principally make use of the structure function of the bridge structure Figure 2.3.
b) Find all minimal path- and cut sets of this system and construct the minimal path- and minimal cut representation of its reliability block diagram.

Exercise 2.5 *Find the minimal path- and the minimal cut representation of the reliability block diagram of a 2-out-of-3 system (see also Exercise 1.12).*

Exercise 2.6 *Show that the minimal path (cut) sets of a coherent system S are the minimal cut (path) sets of the corresponding dual system S_d.*

Exercise 2.7 *a) Show that you get a cut set of a coherent system S if you take out an element from each minimal path set of S.*
b) Check this with Figure 2.15.

Exercise 2.8 *Determine a disjoint sum form for the structure function of a system with reliability block diagram Figure 2.7 by applying the Abraham algorithm.*

Exercise 2.9 *a) Determine the bounds (2.64) and (2.65) for the availability of the bridge structure (Figure 2.3) on condition $p_i = p$, $i = 1, 2, ..., 5$.*
b) For what values of p are the differences between the corresponding upper and lower bounds smallest?
c) Determine the same bounds on condition $p_1 = 0.4$, $p_2 = 0.6$, $p_3 = 0.8$, $p_4 = p_5 = p$, $0 < p < 1$.

Exercise 2.10 *a) Determine the bounds (2.64) and (2.65) for the availability of the extended bridge structure given by Figure 2.10 on condition $p_i = p$, $i = 1, 2, ..., 7$.*
b) Under the same condition, determine the lower and upper bounds (2.66) and (2.67) for the availability of the extended bridge structure.

Exercise 2.11 *Verify the weight vectors for Butler's importance ranking $\mathbf{w}^{(i)}$, $i = 1, 2, ..., 7$, given in Example 2.26.*

Exercise 2.12 *Determine Butler's importance ranking for the elements of the bridge structure (Figure 2.3).*

Exercise 2.13 *For the bridge structure Figure 2.3 let vector $\mathbf{p}(t)$ and replacement vector $\mathbf{G}(t)$ be given by*

$$\mathbf{p}(t) = (e^{-2t}, \; e^{-2t}, \; e^{-2t}, \; e^{-2t}, \; e^{-2t}),$$
$$\mathbf{G}(t) = (e^{-t}, \; e^{-t}, \; e^{-t}, \; e^{-2t}, \; e^{-2t}).$$

a) Determine Xie's yield-importances of all 5 elements if $y(t) = e^t$.
b) From Xie's yield-importances, derive the BP-importances (2.86) of all elements.

Exercise 2.14 *Prove that*

$$I_M(i) > I_M(j) \quad \text{iff} \quad h((r_i, s_j, \mathbf{p})) > h((s_i, r_j, \mathbf{p}))$$

for all pairs (r, s) with $0 \le s < r \le 1$, where $I_M(i)$ is defined by Definition 2.27. (The notation $(r_i, s_j, \mathbf{p})$ means that the ith component of vector $\mathbf{p}$ is r and its jth component is s.).

Exercise 2.15 *Show that the structural BP-importance $I_{SBP}(i)$ as defined by formula (2.88) can be equivalently represented by formula (2.89).*

Exercise 2.16 *Consider a system with independent elements, the reliability block diagram of which is given by Figure 2.14. The availabilities of the elements are $p_i = 0.7$, $i = 1, 2, 3, 4, 5$; $p_6 = p_7 = 0.8$; and $p_8 = p_9 = 0.9$. 1) Determine the bounds (2.63), (2.64), and (2.65) on the system availability p_s without modular decomposition. 2) Determine the same bounds after having done the same modular decomposition as in example 2.39. 3) Determine the same bounds after having done a modular decomposition which only uses the bridge structure and the subsystem consisting of the elements $e_6, e_7, ..., e_9$ as modules.*

Chapter 3

Lifetime of Coherent Systems

3.1 Independent Element Lifetimes

In case of independent element lifetimes and no repair of failed elements, the survival (availability) function of the system is fully determined by the vector of the survival functions of the elements $\mathbf{p}(t) = (\overline{F}_1(t), \overline{F}_2(t), ..., \overline{F}_n(t))$ and the availability function of the system $h(\cdot)$:

$$\overline{F}_s(t) = 1 - F_s(t) = h(\mathbf{p}(t)).$$

An already classic problem related to $F_s(t)$ is the IFR *lifetime closure problem* for coherent systems: If the $F_i(t)$, $i = 1, 2, ..., n$, are IFR (Definition 1.1), is $F_s(t)$ IFR as well? *Esary and Proschan* [117] gave a counter example: The lifetime distribution of a parallel system consisting of two independent elements with different exponential lifetime distributions is not IFR. (The exponential distribution is both IFR and DFR.) Actually, the failure rate of such a system increases up to a certain time point t_0, and after t_0 it decreases. Later *Birnbaum et al.* [50] showed that the lifetime closure problem for coherent systems with independent elements has a positive answer for the IFRA-distribution (Definition 1.2): If the $F_i(t)$, $i = 1, 2, ..., n$, are IFRA, then $F_s(t)$ is IFRA.

However, there are cases where $F_i(t)$, $i = 1, 2, ..., n$, being IFR implies $F_s(t)$ being IFR, too. *Samaniego* [265] gave a "simple" condition: $F_s(t)$ is IFR if the elements have iid IFR lifetimes and the function

$$g(x) = \frac{\sum_{i=0}^{n-1}(n-i)\sigma_{i+1}\binom{n}{i}x^i}{\sum_{i=0}^{n-1}\left(\sum_{j=i+1}^{n}\sigma_j\right)\binom{n}{i}x^i}$$

is increasing in x, $x > 0$, where $(\sigma_1, \sigma_2, ..., \sigma_n)$ is the signature of the system (Section 2.2.4).

For instance, the bridge structure has signature $\sigma = (0,\ 0.2,\ 0.6,\ 0.2,\ 0)$ (Example 2.15). Hence, the corresponding function $g(x)$ is

$$g(x) = \frac{4x + 18x^2 + 4x^3}{1 + 5x + 8x^2 + 2x^3}.$$

Since $dg(x)/dx > 0$ for all $x > 0$, $g(x)$ is strictly increasing. Thus, if $F(t) = F_i(t)$, $i = 1, 2, ..., n$, is IFR, then $F_s(t)$ is IFR, too. But this is not the case in the following more complex example.

Example 3.1 *The system with the reliability block diagram given by Figure 2.7 has signature (Example 2.16)*

$$\sigma = \left(0,\ \frac{70}{735},\ \frac{224}{735},\ \frac{273}{735},\ \frac{133}{735},\ \frac{35}{735},\ 0\right).$$

Hence,

$$g(x) = \frac{4x + 32x^2 + 52x^3 + 19x^4 + 2x^5}{1 + 7x + 19x^2 + 21x^3 + 8x^4 + x^5}.$$

In $[0, \infty)$, this function strictly increases from $x = 0$ to $x_m \approx 3.84$ from its minimum value $g(0) = 0$ to its maximum value $g(x_m) \approx 2.27$ and decreases then for $x \to \infty$ to $g(\infty) = 2$. Hence, $F(t)$ being IFR does not imply that $F_s(t)$ is IFR as well.

By (2.60), for i.i.d. element lifetimes with distribution function $F(t)$, the system has the survival function

$$\overline{F}_s(t) = \Pr(L_s > t) = \sum_{j=1}^{n} \sigma_j \sum_{k=0}^{j-1} \binom{n}{k} \overline{F}(t)^{n-k} F(t)^k. \tag{3.1}$$

This relationship implies three corollaries:

Corollary 3.2 *The probability density of the system lifetime L_s is*

$$f_s(t) = -\frac{d}{dt}\overline{F}_s(t) = \sum_{k=1}^{n} k\sigma_k \binom{n}{k} \overline{F}(t)^{n-k} F(t)^{k-1} f(t),$$

where $f(t) = dF(t)/dt$ is the density of the lifetimes of the elements.

Corollary 3.3 *The system failure rate $\lambda_s(t) = f_s(t)/\overline{F}_s(t)$ is*

$$\lambda_s(t) = \frac{\sum_{k=1}^{n} k\sigma_k \binom{n}{k} \overline{F}(t)^{n-k+1} F(t)^{k-1}}{\sum_{j=1}^{n} \sigma_j \sum_{k=0}^{j-1} \binom{n}{k} \overline{F}(t)^{n-k} F(t)^k} \lambda(t), \tag{3.2}$$

where $\lambda(t)$ denotes the failure rate of the elements.

Corollary 3.4 *Let S_1 and S_2 be two coherent systems with n elements each, respective lifetimes L_{s_1} and L_{s_2}, and respective signatures $\sigma_1 = (\sigma_{1,1}, \sigma_{1,2}, ..., \sigma_{1,n})$ and $\sigma_2 = (\sigma_{2,1}, \sigma_{2,2}, ..., \sigma_{2,n})$. Then L_{s_2} is stochastically greater than L_{s_1} with regard to the usual stochastic order, i.e. it is $\Pr(L_{s_1} \geq t) \leq \Pr(L_{s_2} \geq t)$ for all $t \geq 0$ if*

$$\sum_{j=k+1}^{n} \sigma_{1,j} \leq \sum_{j=k+1}^{n} \sigma_{2,j} \quad for \ k = 0, 1, ..., n-1.$$

Note that (3.1) is equivalent to

$$\overline{F}_s(t) = \sum_{k=0}^{n-1} \left(\sum_{j=k+1}^{n} \sigma_j \right) \binom{n}{k} \overline{F}(t)^{n-k} F(t)^k.$$

For other useful properties and applications of signatures in reliability engineering see the monograph [267].

3.2   Dependent Element Lifetimes

3.2.1   Introduction

The assumption of independently operating elements is convenient for the mathematical analysis of coherent systems, since, at least for complex systems, explicit and tractable formulas for the exact system availability and other criteria can in most cases only be derived in case of statistically independent element lifetimes $L_1, L_2, ..., L_n$. Hence, the assumption of independent L_i has been a favorite one. Moreover, analytical results obtained under the assumption of independence may serve as bounds or approximations in case of dependence. On the other hand, in reality, elements (subsystems) are more or less dependent with regard to their failure behavior. They are usually subject to the same environmental conditions, such as temperature, humidity, and pressure. Dependency of their lifetimes also arises if in the process of their construction the same material and the same technology has been used. Available elements may have to share extra work load caused by failures of other elements. This will usually generate a strong dependence between their lifetimes. *Common failure causes* like natural disasters, power cuts et al. may lead to failures of different elements within the same short time interval or even at the same time point. Dependence is also generated if systems and their elements are operated and/or maintained by the same people.

While the concept of independence of random variables does not allow for subclassifications, dependence between random variables can be classified according to characteristic properties of the respective dependency. *Lehmann* [199] is generally recognized as the first one who summarized, analyzed and extended concepts of statistical dependence between two random variables (bivariate dependence).

3.2.2   Positive Quadrant Dependence

In problems of reliability and maintenance, the most encountered dependence property is *positive dependence*. Intuitively, this property means that a large (small) value of X in a joint observation (X, Y) increases the probability that Y has a large (small) value as well. (Analogously, there is a *negative dependence* between X and Y if a large (small) value of X increases the probability that Y has a small (large) value.) It is well-known that positive (negative) *linear* dependence between random variables X and Y exists if their correlation coefficient $\rho(X, Y)$ is positive (negative).

To be able to give more general characterizations of positive dependence, let (X, Y) be a bivariate random vector with marginal distributions $F_X(x) = \Pr(X \leq x)$, $F_Y(y) = \Pr(Y \leq y)$, and joint distribution (survival) function

$$F(x, y) = \Pr(X \leq x, Y \leq y), \quad \overline{F}(x, y) = \Pr(X > x, Y > y).$$

Note that

$$\overline{F}(x, y) = 1 - F_X(x) - F_Y(y) + F(x, y). \tag{3.3}$$

Definition 3.5 *Two random variables X and Y are said to be* positively quadrant dependent *(PQD) if for all $x, y \in \mathbf{R}^2$,*

$$F(x, y) \geq F_X(x) F_Y(y), \ \text{or equivalently,} \ \overline{F}(x, y) \geq \overline{F}_X(x) \overline{F}_Y(y). \tag{3.4}$$

Thus, in case of **PQD** with unknown joint distribution of (X, Y), its marginal distributions yield lower bounds for the joint distribution characteristics $F(x, y)$ and $\overline{F}(x, y)$. Or, with regard to the intuitive interpretation of PQD: Conditions (3.4) imply that the joint probabilities of the random variables X and Y being both small (large) are greater or equal than in case of their independence.

The inequalities (3.4) are equivalent to

$$\Pr(X \leq x \,|\, Y \leq y) \geq \Pr(X \leq x) = \Pr(X \leq x \,|\, Y \leq +\infty),$$
$$\Pr(X > x \,|\, Y > y) \geq \Pr(X > x) = \Pr(X > x \,|\, Y > -\infty),$$

respectively. This gives rise to stronger conditions than (3.4) [117].

Definition 3.6 1. *X is said to be* left tail decreasing *(LTD) in Y if $\Pr(X \leq x \,|\, Y \leq y)$ is a nonincreasing function in y for all x.*
2. *X is said to be* right tail increasing *(RTI) in Y if $\Pr(X > x \,|\, Y > y)$ is a nondecreasing function of y for all x.*

Obviously, in definition 3.6 the random variables X and Y can be exchanged. The following dependence concept was introduced particularly for applications in reliability by *Esary et. al.* [118]:

Definition 3.7 *Two random variables X and Y are said to be* associated *if for all nondecreasing functions U and V for which the covariance $Cov(U(X), V(Y))$ exists,*

$$Cov(U(X), V(Y)) \geq 0. \tag{3.5}$$

An advantage of this dependence concept is that it can readily be extended to the multivariate case with

$$\mathbf{X} = (X_1, X_2, ..., X_{n_x}) \text{ and } \mathbf{Y} = (Y_1, Y_2, ..., Y_{n_y})$$

for X and Y, respectively. Letting $U = V \equiv 1$ implies that associated random variables have a nonnegative correlation coefficient: $\rho(X, Y) \geq 0$. Moreover, definition 3.7 implies that independent random variables are associated. Usually it is difficult or even impossible to verify condition (3.5) for all functions U and V with the given properties. But the easier to verify conditions LTD and RTI may be helpful: They are sufficient for X and Y to be associated. As an application in reliability, the bounds (2.63) and (2.64) on the system availability p_s hold for associated random indicator variables $z_1, z_2, ..., z_n$ [19].

Another important dependence concept in reliability implying PQD is the following one [173]:

Definition 3.8 *A function $T(x, y)$ is said to be* totally positive of order 2 *(TP_2) on $\mathbf{R}^2$ if for all $x_1 < x_2$ and $y_1 < y_2$*

$$\begin{vmatrix} T(x_1, y_1) & T(x_1, y_2) \\ T(x_2, y_1) & T(x_2, y_2) \end{vmatrix} \geq 0,$$

i.e. if

$$T(x_1, y_1) \, T(x_2, y_2) \geq T(x_1, y_2) \, T(x_2, y_1). \tag{3.6}$$

If, in particular, the joint density $f(x, y)$ of a random vector (X, Y) is TP_2, then the corresponding joint distribution function $F(x, y)$ and the joint survival function $\overline{F}(x, y)$ are TP_2 as well. From definition 3.8 it is obvious how totally positive functions $T(x_1, x_2, ..., x_n)$ of order n are defined.

Note that if the density $f(x)$ of a random variable X is PD_2, then the function $T(x, y) = f(x - y)$ is TP_2, $x, y \in \mathbf{R}$.

There is the following chain of implications for the above dependence concepts:

3.2.3   Bivariate PQD Probability Distributions

According to *Lai and Xie* [191], a random vector (X, Y) with $X \geq 0$, $Y \geq 0$, and continuous marginal distribution functions $F_X(x)$, $F_Y(y)$ with densities $f_X(x)$, $f_Y(y)$ has a PQD distribution if its joint distribution function can be represented as

$$F(x, y) = F_X(x)F_Y(y) + H(x, y), \tag{3.7}$$

where $H(x, y)$ is any nonnegative, twice partially differentiable function satisfying the two conditions

1) $\lim\limits_{x \to \infty} H(x, \infty) = \lim\limits_{y \to \infty} H(\infty, y) = 0, \quad H(x, 0) = H(0, y) = 0,$

2) $\dfrac{\partial^2 H(x, y)}{\partial x \partial y} + f_X(x)f_Y(y) \geq 0.$

These conditions allow the construction of PQD bivariate distributions, which suit the need of special applications, or they are used for verifying that the following or other bivariate distributions are PQD. Note that, by (3.3), condition (3.7) is equivalent to

$$\overline{F}(x, y) = \overline{F}_X(x)\overline{F}_Y(y) + H(x, y).$$

Bivariate Exponential Distributions

1. Let us consider a 2-element system, which is subject to three independent types of shocks [216]: A type i-shock occurs at a random time T_i with $\Pr(T_i > t) = e^{-\lambda_i t}$ and destroys element e_i, $i = 1, 2$. A shock, causing a common cause failure at a random time T_c with $\Pr(T_c > t) = e^{-\lambda_c t}$, destroys both elements. Hence, given that there are no other failure causes, the lifetimes of the elements are

$$L_1 = \min(T_1, T_c), \quad L_2 = \min(T_2, T_c)$$

with independent T_1, T_2, and T_c. Thus, L_1 and L_2 are clearly dependent, and their joint survival probability is

$$\overline{F}(t_1, t_2) = \Pr(L_1 > t_1, L_2 > t_2) = e^{-\lambda_1 t_1 - \lambda_2 t_2 - \lambda_3 \max(t_1, t_2)}. \tag{3.8}$$

It can be easily shown that

$$\Pr(L_1 > t_1 + x, L_2 > t_2 + x \,|\, L_1 > x, L_2 > x) = \Pr(L_1 > t_1, L_2 > t_2).$$

This equation implies that the joint survival probability of two elements, both of age x, is the same as the one of a pair of new elements. In terms of $\overline{F}$, for $t_i \geq 0$, $x \geq 0$, the equation can be written as

$$\overline{F}(t_1 + x, t_2 + x) = \overline{F}(t_1, t_2)\overline{F}(x, x). \tag{3.9}$$

This is the bivariate analog to (1.12). Moreover, $\overline{F}$ given by (3.8) is the only function, which satisfies (3.9).

This bivariate distribution can be generalized to an n-dimensional distribution if a system of order n is hit by independent shocks in the following way: A type i-shock occurs at a random time T_i with $\Pr(T_i > t) = e^{-\lambda_i t}$ and destroys element e_i at arrival, $i = 1, 2, ..., n$. A common cause failure shock occurs at a random time T_c with $\Pr(T_c > t) = e^{-\lambda_c t}$ and destroys all n elements. Then the joint survival probability of all elements is

$$\overline{F}(t_1, t_2, ..., t_n) = e^{-\lambda_1 t_1 - \lambda_2 t_2 - \cdots - \lambda_n t_n - \lambda_c \max(t_1, t_2, ..., t_n)}. \tag{3.10}$$

The marginals of this distribution are

$$\overline{F}_i(t) = e^{-(\lambda_i + \lambda_c)t}, \quad i = 1, 2, ..., n.$$

A further generalization is possible if shocks may destroy any subset of the elements.

2. *Gumbel* [147] discussed a bivariate exponential distribution with joint survival probability

$$\overline{F}(t_1, t_2) = e^{-\lambda_1 t_1 - \lambda_2 t_2 - \lambda t_1 t_2}, \quad \lambda_i > 0, \ i = 1, 2, \ 0 \leq \lambda \leq 1. \tag{3.11}$$

3. A third special bivariate exponential distribution is given by the joint survival probability

$$\overline{F}(t_1, t_2) = (e^{+\lambda_1 t_1} + e^{+\lambda_2 t_2} - 1)^{-1}, \quad \lambda_i > 0, \ t_i \geq 0, \ i = 1, 2. \tag{3.12}$$

The marginals of the exponential-type distributions 2 and 3 are

$$\overline{F}_i(t) = e^{-\lambda_i t}, \quad i = 1, 2.$$

The joint distribution function $F(x, y)$ of the respective random vectors (L_1, L_2) one gets from (3.3).

Morgenstern–Farlie Distribution

$$F(t_1, t_2) = F_1(t_1) F_2(t_2) \left[1 + \alpha \overline{F}_1(t_1) \overline{F}_2(t_2) \right], \ 0 < \alpha \leq 1.$$

This bivariate distribution was first proposed by *Morgenstern* [228], analyzed by *Gumbel* [147] for exponential marginals F_i, $i = 1, 2$, and extended by *Farlie* [119]. This distribution provides an easy option to construct a bivariate distribution with given marginals.

Bivariate Normal Distribution

Let $L_i = N(\mu_i, \sigma_i^2)$, $i = 1, 2$, and $\rho = \rho(L_1, L_2)$ be the correlation coefficient between L_1 and L_2. Then the random vector (L_1, L_2) has a *bivariate normal distribution* if it has for $t_i \in (-\infty, +\infty)$, $i = 1, 2$, the joint density

$$f(t_1, t_2) = \frac{1}{2\pi\sigma_1\sigma_2\sqrt{1-\rho^2}} \exp\left\{ -\frac{1}{2(1-\rho^2)} \left[\frac{(t_1-\mu_1)^2}{\sigma_1^2} \right.\right.$$
$$\left.\left. -2\rho\frac{(t_1-\mu_1)(t_2-\mu_2)}{\sigma_1\sigma_2} + \frac{(t_2-\mu_2)^2}{\sigma_2^2} \right] \right\}.$$

Then L_1 and L_2 are associated and, hence, PQD if $\rho \geq 0$ [19].

For many other bivariate and multivariate joint probability distributions, we refer to the monographs [164, 163, 243] (in the context of copulas). [192] and [123] are up to date presentations of the subject with applications to reliability and maintenance. However, in the area of modeling dependence, there is still a big gap between the state of the art of the theory and its application in engineering reliability.

Example 3.9 *Let us consider the 3-element system with reliability block diagram Figure 2.13. The joint survival probability of the elements is assumed to be*

$$\overline{F}(t_1, t_2, t_3) = e^{-\lambda_1 t_1 - \lambda_2 t_2 - \lambda_3 t_3 - \lambda_c \max(t_1, t_2, t_3)}, \quad t_i \geq 0, \; \lambda_i, \lambda_c \geq 0.$$

Let $A_i = \;'L_i > t\;'$, $i = 1, 2, 3$. The system availability at time t is

$$h(t, \overline{F}) = \Pr(A_1 A_2) + \Pr(A_1 A_3) - \Pr(A_1 A_2 A_3)$$
$$= \overline{F}(t, t, 0) + \overline{F}(t, 0, t) - \overline{F}(t, t, t).$$

Hence,

$$h(t, \overline{F}) = e^{-(\lambda_1 + \lambda_c)t} \left[e^{-\lambda_2 t} + e^{-\lambda_3 t} - e^{-(\lambda_2 + \lambda_3)t} \right].$$

For determining the Birnbaum-importances of the elements, we need to apply their representations (2.70) as a function of time: With obvious modification of the notation (2.26),

$$I_B(i, t) = h(t, (1_i, \overline{F})) - h(t, (0_i, \overline{F})), \quad i = 1, 2, 3.$$

If e_1 has availability 0, i.e. $\lambda_1 = \infty$, then the system is down: $h(t, (0_1, \overline{F})) = 0$. But if element e_1 is absolutely reliable ($\lambda_1 = 0$), then the system availability is equal to the availabilty of the parallel system consisting of e_2 and e_3. Hence,

$$I_B(1, t) = e^{-\lambda_c t} \left[e^{-\lambda_2 t} + e^{-\lambda_3 t} - e^{-(\lambda_2 + \lambda_3)t} \right].$$

Analogously,

$$I_B(2, t) = e^{-(\lambda_1 + \lambda_c)t} \left[1 - e^{-\lambda_3 t} \right],$$
$$I_B(3, t) = e^{-(\lambda_1 + \lambda_c)t} \left[1 - e^{-\lambda_2 t} \right].$$

For $\lambda_c \to \infty$, the importances of the elements tend to 0, but their importance ranking does not change. In particular, the ratios $I_B(i, t)/I_B(1, t)$, $i = 1, 2$, do not depend on λ_c.

3.3   Exercises

Exercise 3.1 *The reliability block diagram of a system with 5 elements is given by Figure 3.1. The lifetimes of its elements are independent and identically distributed with the IFR-distribution function $F(t)$. Check whether the system lifetime X_s has an IFR-distribution as well.*

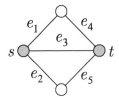

Figure 3.1: Reliability block diagram for Exercise 3.1

Exercise 3.2 *Prove formula (3.3).*

Exercise 3.3 *The joint survival probability $\overline{F}(x,y)$ of a random vector (L_1, L_2) is given by (3.12).*
a) By verifying formula (3.4) show that this distribution is PQD.
b) Let L_1 and L_2 be the random lifetimes of a 2-element parallel system. Determine the Birnbaum-importances of its elements via (2.70).
c) Let L_1 and L_2 be the random lifetimes of a 2-element series system. Determine the Birnbaum-importances of its elements via (2.70).

Exercise 3.4 *Let X and Y be random $(0,1)$-variables and $Cov(X,Y) \geq 0$. Show that X and Y are associated.*

Exercise 3.5 *Verify that Gumbel's bivariate exponential distribution given by (3.11) has structure (3.7).*

Exercise 3.6 *Prove: If the joint density $f(x,y)$ of a random vector (X,Y) is TP_2, then the corresponding joint distribution (survival) function $F(x,y)$ $(\overline{F}(x,y))$ is TP_2 as well.*

Exercise 3.7 *Show: If the random vector (L_1, L_2) has the joint survival function (3.8), then*

$$\Pr(L_1 > t_1 + x, L_2 > t_2 + x \mid L_1 > x, L_2 > x) = \Pr(L_1 > t_1, L_2 > t_2).$$

Exercise 3.8 *Figure 2.13 shows the reliability block diagram of a system with 3 elements. The vector of their lifetimes (L_1, L_2, L_3) has the joint survival function*

$$\overline{F}(t_1, t_2, t_3) = e^{-t_1 - 2t_2 - 3t_3 - \max(t_1, t_2, t_3)}.$$

a) Determine the survival probability of this system in the interval $[0, 2]$.
b) Determine the availability potentials $I_1(i, \mathbf{p})$, $i = 1, 2, 3$.

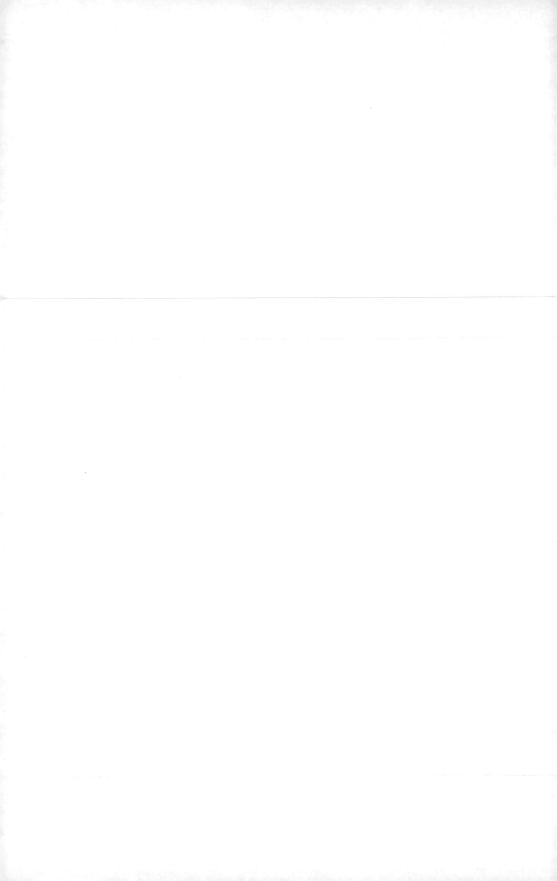

Part II

Network Reliability

Chapter 4

Modeling Network Reliability with Graphs

4.1 Introduction to Network Reliability

Technical systems such as communication networks, power grids, traffic systems, or command and control systems have a *network-like structure*. Let us consider a computer network as one typical example. The topology of a computer network is characterized by two main components, namely computers (server, router, terminals) and transmission lines between computers (copper cable, fiber-optic lines, wireless channels). In a mathematical model, a so-called *graph*, these two types of network components are represented by *vertices* (or *nodes*) and *edges*. One way to specify a network topology or a graph is to give a list of vertices and edges together with an incidence function that indicates the end vertices of each edge. However, there is a more convenient way to visualize the structure of a network by *drawing* the corresponding graph. We represent vertices by small circles and edges as lines connecting

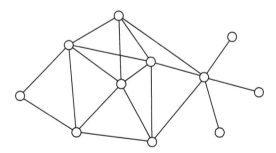

Figure 4.1: A graph

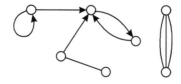

Figure 4.2: A graph with directed edges, multiple edges, and loops

those two circles that correspond to their end vertices. Figure 4.1 shows an example.

A graph may have directed edges that can be traversed in only one direction or multiple edges connecting the same pair of vertices. A loop is an edge for which the end vertices coincide. All these possibilities are illustrated in Figure 4.2. The next section of this chapter is devoted to a more precise foundation of graph theoretic tools that are indispensable in network reliability analysis.

Besides its topology a network is characterized by a great diversity of parameters assigned to edges or vertices. An edge may have a length, a capacity (a data transfer rate), costs, and an availability (reliability, i.e. the probability of operating properly during a given period of time). A vertex might have geographic coordinates (a location), costs, and a reliability. The network itself has been designed in order to support a communication process. The description of the communication process leads to a new class of parameters that describe dynamic properties such as packet delay, throughput, or packet loss. An even closer look at the inner structure of the edges and vertices of the network is required if we have to evaluate the network performance or the quality of service. In this case, traffic parameters, protocol properties, and server characteristics have to be incorporated in network analysis. Here we restrict our attention to questions of network reliability.

The first basic observation in network reliability analysis is the fact that vertices and edges of a network are subject to random failure. Failures of vertices and edges may be caused by natural disasters such as thunderstorms, by human errors, or technical defects, for instance as a result of overstressing. We assume that the failure distributions of the components (vertices and edges) are known. In general, we presume only the knowledge of the probability that a component operates within a given period of time. These data may be derived from statistical observations, stress tests, or manufacturer quality certificates.

Usually reliability is defined as the probability that a component or system is operating within a certain period of time. When do we consider a network as being operating? The minimum requirement for a successful communication from one vertex in a network to another one is the existence of a path of operating edges and vertices connecting the two terminal vertices in question. Thus the first question in network reliability analysis is the determination of the probability that there exists at least one operating path between two given terminal vertices. We can generalize this concept to an arbitrary set of terminal vertices of a graph and ask for the probability that all these terminal vertices are connected by operating paths. In the following sections we will introduce further generalizations and alternative reliability measures.

Once we have introduced a suitable reliability measure the question of determining the reliability emerges naturally. Unfortunately, most of the reliability problems considered here turn out to be computationally intractable, i.e. (loosely spoken) there exist only algorithms that require an exponentially with the size of the network increasing computation time. Consequently, in order to analyze large networks we are interested in fast approximative methods or in developing bounds for the desired reliability measure. Finally, for huge networks where all analytic methods fail, simulation may help. Often, even when the computation of the reliability is computationally hard, we can find interesting classes of networks that can be analyzed efficiently. One such class consists of so-called series-parallel networks that can be easily reduced.

The paper on reliable relays by *Moore* and *Shannon* [227] can be considered as the pioneering work in this field. An enormous increase of publications in network reliability started around 1980. We can mention here only some milestones: *Satyanarayana* and his colleagues [269], [272] developed factoring and reduction methods for network reliability calculations. *Valiant* [300] showed that many important reliability problems are #P-complete, i.e. computational intractable. The first two monographs on network reliability by *Colbourn* [92] and *Shier* [278] deal with combinatorial and algebraic aspects of network reliability, respectively.

4.2 Foundations of Graph Theory

This section provides a short introduction to basic concepts of graph theory. We refer the interested reader to the textbooks [64], [306], and [144] for a more thorough treatment of this field.

4.2.1 Undirected Graphs

The mathematical model for the representation of network structures is a graph. In the following, we use the notation

$$\binom{X}{2} = \{\{x, y\} : x, y \in X, x \neq y\}$$

for the set of all two-element subsets of a set X.

Definition 4.1 *An* undirected graph $G = (V, E)$ *consists of two sets, a vertex set V and an edge set E and an incidence function $\phi : E \to \binom{V}{2} \cup V$ that assigns to each edge $e \in E$ one or two vertices of V, called the end vertices of e.*

Since we deal in this section with undirected graphs exclusively, we omit the word undirected and speak simply of graphs. Some authors write $G = (V, E, \phi)$ in order to specify the incidence function of the graph G. We prefer the abbreviated notation $G = (V, E)$ assuming the incidence function is known from the context. We write $e = \{u, v\}$ if u and v are the end vertices of the edge e. In this case, the edge e is said to be *incident* to the vertices u and v, the vertices u and v are called *adjacent* (or *neighbor vertices*) in G. An edge that has only one end vertex is called a *loop*. Two edges $e = \{u, v\}$ and $f = \{u, v\}$ sharing the same pair of end vertices are called *parallel*. A *simple graph* is a graph that contains neither loops nor parallel edges. A graph is called *finite* if its vertex and edge set are finite. All graphs considered in this book are finite.

The *degree* $\deg v$ of a vertex v is the number of edges that are incident to v. The *neighborhood* $N(v)$ is the set of all vertices adjacent to v. The *order* $v(G)$ of a graph $G = (V, E)$ is the number $|V|$ of vertices of G, the *size* $e(G)$ is the number of edges of G. A graph of size zero, i.e. a graph without any edges, is called an *empty graph*. A graph without any edges or vertices is called a *null graph*. A *walk* in a graph $G = (V, E)$ is an alternating sequence of vertices and edges of G,

$$v_1, e_1, v_2, e_2, v_3, ..., v_{k-1}, e_{k-1}, v_k,$$

such that v_i and v_{i+1} are the end vertices of e_i for $i = 1, ..., k-1$. The number of edges of a walk is called its *length*. A walk is *closed* if its first and last vertex coincide. A *path* is a walk that contains no vertex twice. A *cycle* is a closed walk for which all vertices, except the first and last one, are distinct. A graph G is *connected* if there exists a path from any vertex to any other vertex of G.

A graph $H = (W, F)$ is a *subgraph* of the graph $G = (V, E)$ if $W \subseteq V$ and $F \subseteq E$. The subgraph H is *spanning* (or an *edge subgraph*) if $W = V$, i.e. H contains all vertices of G. A maximal connected subgraph of a graph G is a *component* of G. A *tree* is a connected graph without any cycles. Figure 4.3 shows a tree with 16 vertices. One can easily verify that the order of a

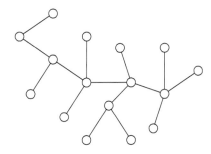

Figure 4.3: A tree

tree exceeds its size by one. A *spanning tree* of a graph G is a connected cycle-free spanning subgraph of G. A *forest* is a graph whose components are trees. A simple graph in which any two different vertices are adjacent is called a *complete graph*. Figure 4.4 shows a complete graph with six vertices. The complete graph with n vertices is denoted by K_n. A *clique* is a complete

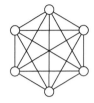

Figure 4.4: The complete graph K_6

subgraph of a graph.

Two graphs $G = (V, E)$ and $H = (W, F)$ are *isomorphic* if there exists a bijection $\phi : V \rightarrow W$ such that the number of edges between v and w in G is equal to the number of edges between $\phi(v)$ and $\phi(w)$ for all $v, w \in V$. Consequently, if G and H are isomorphic then we can obtain H from G by a renaming of vertices. Properties or functions of graphs that coincide for all isomorphic graphs are called *graph invariants*. A graph invariant that is important for network reliability is the number of spanning trees. Figure 4.5 shows two isomorphic graphs.

Exercise 4.1 *Draw all non-isomophic trees with seven vertices.*
Hint: You should obtain 11 different trees.

Exercise 4.2 *How many different paths are there in a complete graph of order 6 between any two vertices?*

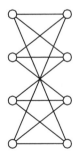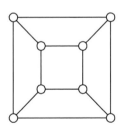

Figure 4.5: Isomorphic graphs

Exercise 4.3 *Show that for any graph G the following equation is satisfied:*

$$\sum_{v \in V(G)} \deg v = 2\,|E(G)|$$

Exercise 4.4 *Show that a simple graph of order $n \geq 2$ has two vertices with the same degree.*

Exercise 4.5 *A graph $G = (V, E)$ is called* bipartite *if the vertex set can be partitioned into two disjoint subsets V_1 and V_2 such that each edge of G links a vertex of V_1 with a vertex of V_2. Show that the size of simple bipartite graph of order n does not exceed $\frac{n^2}{4}$.*

4.2.2 Directed Graphs

This subsection gives a short introduction to directed graphs. The interested reader is referred to the monograph by *Bang–Jensen* and *Gutin* [14].

Definition 4.2 *A directed graph (or short digraph) $G = (V, E)$ consists of a vertex set V and an edge set E, together with an* incidence function $\psi : E \to V^2$ *that assigns an ordered pair of vertices to each edge of G.*

Remark 4.3 *There are alternative notions for an edge of digraph like* directed edge *or* arc.

If e is an edge of G such that $\psi(e) = (u, v)$ then we write $e = uv$ or $e = (u, v)$. In this case, the vertex u is called the *tail* and vertex v the *head* of e. The *indegree* $d^-(v)$ of a vertex v in a digraph G is the number incoming edges, i.e. the number of edges with head v. The *outdegree* $d^+(v)$ is the number of edges of G with tail v. Figure 4.6 shows a digraph with one vertex of outdegree zero.

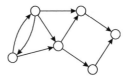

Figure 4.6: A digraph

Neglecting the orientation of the edges, we can assign to each digraph G an undirected graph G', which we call the *underlying graph* of G. Many concepts from undirected graphs apply to digraphs, too. A *walk* in a digraph $G = (V, E)$ is an alternating sequence of vertices and edges of G,

$$v_1, e_1, v_2, e_2, v_3, ..., v_{k-1}, e_{k-1}, v_k,$$

such that vertex v_{i+1} is the of head e_i and the tail of e_{i+1} for $i = 1, ..., k - 1$. The vertex v_1 is the tail of e_1, the head of e_{k-1} is v_k. A vertex v is *reachable* from a vertex u if there is a walk from u to v in G. A digraph $G = (V, E)$ is *strongly connected* if each vertex of V is reachable from any other vertex of V. A digraph is *connected* if the underlying graph is connected.

Let $G = (V, E)$ be an undirected graph. An *orientation* of G is a digraph obtained from G by assigning a direction (an orientation) to each edge of G.

Exercise 4.6 *A digraph without any directed cycles is called* acyclic. *Show that any (finite) acyclic digraph has a vertex of indegree zero.*

4.3 Deterministic Reliability Measures

In this section, we investigate *topological* (structural) properties of a network that are essential for its reliable operation. A network is *robust* if it withstands random failures of vertices and edges. The term *vulnerability* is used to describe the dependence of the network operation on malicious attacks. We can measure the vulnerability of a graph, for instance, by the minimum number of edges or vertices that have to be removed in order to destroy its connectedness. This idea leads us in a quite natural way to edge and vertex connectivity measures of graphs. *Bulteau* and *Rubino* [75] discuss some general properties a vulnerability measure should have. We summarize their thoughts:

1. A vulnerability measure ought to be able to compare networks, that is, its codomain should be a linearly ordered set or at least a partially ordered set.

2. Monotonicity: For each edge e, the vulnerability of $G-e$ should be equal or larger than the vulnerability of G. Here $G - e$ denotes the graph G with edge e removed.

3. A vulnerability measure should be sensitive to local differences, that is, it should not be restricted to the detection of weakest points such as bridges in a graph. *Bulteau* and *Rubino* call this property "globality."

4. Two networks in parallel connection should be less vulnerable than the same two networks in a series connection.

Let $G = (V, E)$ be a graph. For a vertex set $X \subseteq V$, we denote by $G - X$ the graph obtained from G by removing the vertices of X together with all edges that are incident to a vertex of X. A *separating vertex set* (or *separator*) is a vertex set $S \subseteq V$ such that $G-S$ has more components than G. If $S = \{v\}$ is a separating vertex set of G consisting of a single vertex then the vertex v is called an *articulation* (or *cut vertex*) of G. The *connectivity* $\kappa(G)$ is the minimum cardinality of a vertex set S such that $G - S$ is disconnected or consists of only one vertex. This definition implies $\kappa(K_n) = n - 1$ and

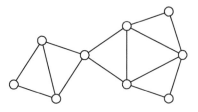

Figure 4.7: A graph with an articulation and edge connectivity 2

$\kappa(T) = 1$ for any tree T with at least two vertices. We call two st-paths in G, i.e. paths that connect the two vertices s and t, *internally disjoint* if they have no vertices except s and t in common. According to a famous theorem of *Karl Menger* [223], the connectivity of a graph G equals the minimum number of internally disjoint paths between any two vertices of G. There is also a *local* version of this theorem: If there exists no separator with less than k vertices in G that separates s from t (also called an st-separator) then there exist at least k internally disjoint st-paths in G. The *local (vertex) connectivity* $\kappa_{st}(G)$ equals k if there exist exactly k internally disjoint st-paths in G. The local connectivity can be efficiently computed by application of algorithms for the maximum-flow problem. Obviously, we obtain

$$\kappa(G) = \min_{s,t \in V} \{\kappa_{st}(G)\}.$$

We proceed by considering the effect of edge removal on the connectedness of a graph. If $G = (V, E)$ is a graph and $F \subseteq E$ is an edge subset then we denote by $G - F$ the graph obtained from G by the removal of all edges from F. The set F is a *cut* (*edge-cut* or *cut set*) if $G - F$ has more components than G. A *minimum cut* is a cut that contains no other cut as a proper subset. A *bridge* is an edge of G that taken by itself forms a cut. As an example, each edge of a tree is a bridge. The *edge connectivity* $\lambda(G)$ of graph G is the minimum cardinality of a cut of G. If G is disconnected, then we have $\lambda(G) = 0$. Figure 4.7 shows a graph with vertex connectivity 1 and edge connectivity 2. A connected graph G has edge connectivity 1 if and only if G has a bridge.

A minimum set of operating edges in a graph that ensures connectedness is the edge set of a spanning tree. Thus the number $t(G)$ of spanning trees can also be considered a measure of redundancy. Intuitively, we expect that the number of spanning trees of a graph increases with its edge density. Indeed the complete graph is, for all simple graphs of a given order, the graph with the maximum number of spanning trees. According to a theorem of *Arthur Cayley* [82], $t(K_n) = n^{n-2}$. The number $t(G)$ can be easily computed for any graph by applying the *Kirchhoff* determinant formula [181].

There is a class of measures for the vulnerability of graphs with respect to edge or vertex removal. The *integrity* $I(G)$ of a graph $G = (V, E)$ is the number

$$I(G) = \min_{X \subseteq V} \left\{ |X| + m(G - X) \right\},$$

where $m(G - X)$ is the maximum order of a component of the graph $G - X$, obtained from G by removal of all vertices of X, see [140], [101]. Let $k(G)$ be the number of components of a graph G. The *toughness* of a graph G is defined by

$$\text{tg}(G) = \min \left\{ \frac{|S|}{k(G - S)} : S \text{ is separator of } G \right\}.$$

A generalization of this concept, the *bounded fragmentation graphs* are introduced in [151]. A graph G has bounded fragmentation if after removal of arbitrary chosen k vertices the number of remaining components is bounded by function $f(k)$.

The *distance* $d(u, v)$ of two vertices u and v in a graph G is the length of a shortest path that connects the two vertices. If there is no uv-path in G then we define $d(u, v) = \infty$. We have $d(v, v) = 0$ for any vertex $v \in V(G)$. For any three vertices $u, v, w \in V$, the *triangle inequality* $d(u, w) \leq d(u, v) + d(v, w)$ is satisfied. For some communication services, the distance between start and destination vertex is bounded, since a great distance causes great signal delay. A global distance measure for a graph G is its *diameter*, which is defined by the greatest distance between any two vertices of G. The *mean distance* of a connected graph $G = (V, E)$ is

$$\mu(G) = \frac{2}{n(n - 1)} \sum_{\{u,v\} \in \binom{V}{2}} d(u, v).$$

We can now define the *edge vulnerability* as the ratio $\frac{\mu(G)}{\mu(G-e)}$. The definition of $\mu(G)$ yields no finite value for disconnected graphs. A modification of the definition of $\mu(G)$ avoids this problem:

$$\mu'(G) = \frac{2}{n(n - 1)} \sum_{\{u,v\} \in \binom{V}{2}} \frac{1}{d(u, v)}$$

The measure $\mu'(G)$ also has applications for the investigation of attack vulnerability of social networks [157].

During the last decades, mathematical methods for social network analysis have evolved which have many interesting interrelations to network reliability. We restrict our attention here to *centrality measures* in social networks. A *social network* is a mathematical model of a community, for instance as an undirected graph, where the vertices correspond to individuals (organizations, enterprises). The edges of a social network reflect social relations such as kinship (friendship, membership, trade relations). A centrality measure is a real number $c(v)$ assigned to a vertex v of the network that rates the importance,

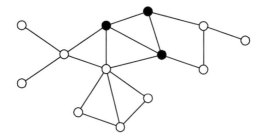

Figure 4.8: The center of a graph

the weight, the account, the significance, the power, or even the centrality of v. Vertices of high centrality are most important with respect to network vulnerability or robustness, since their removal or breakdown causes maximal damage for the whole network. A first simple centrality measure is the degree of a vertex. A vertex of high degree is considered more important in the network than one of low degree. The degree is easy to compute but of limited conclusiveness with respect to the importance of a vertex.

Another centrality measure, the *eccentricity* $\mathrm{ecc}(v)$ of a vertex $v \in V(G)$ is defined by

$$\mathrm{ecc}(v) = \max\{d(u,v) \mid u \in V(G)\}.$$

The reciprocal $c_e(v) = 1/\mathrm{ecc}(v)$ of the eccentricity is a centrality measure that rates vertices high when there are no other vertices "too far away." All vertices of a graph G that have minimum eccentricity form the *center* of G. Figure 4.8 depicts a graph with a center consisting of three vertices.

For two specified vertices $s, t \in V$, let $\sigma_{st}(G)$ be the number of shortest paths from s to t in G. We denote by $\sigma_{st}(G, v)$ the number of shortest paths from s to t in G that traverse the vertex $v \in V \setminus \{s, t\}$. Then the *betweenness centrality* [129], [65], [66], [244] of v is

$$c_b(v) = \sum_{s \in V \setminus \{v\}} \sum_{t \in V \setminus \{s,v\}} \frac{\sigma_{st}(G, v)}{\sigma_{st}(G)}.$$

A failure of a vertex with high betweenness centrality disrupts many shortest paths in the network, which leads to a considerable loss of communication capability.

Exercise 4.7 *Show that for any graph G, we have*

$$\kappa(G) \leq \lambda(G) \leq \min\{\deg v \mid v \in V(G)\}.$$

Exercise 4.8 *Compute the eccentricity for all vertices of the graph depictured in Figure 4.9. Which vertex has highest betweenness centrality? What is its edge and vertex connectivity?*

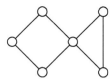

Figure 4.9: An undirected graph

4.4 Stochastic Reliability Measures

From now on we focus on random failures of edges and vertices of the network. We assume that edge or vertex failure probabilities are known and that all failures are stochastically independent. There is only one exception; if a vertex fails, then all edges incident to that vertex fail, too. We exclude all other kinds of dependencies for two reasons. In practical applications, it is often hard enough to obtain sufficiently precise data for each single component. It is hardly possible to describe all dependencies between component failures, since this could produce an amount of data that is exponentially increasing with the size of the network. On the other hand, the mathematical models needed to deal with stochastically dependent failures are much more complex than in the case of independent failures.

In principle, there is a simple way to introduce a reliability measure for a graph $G = (V, E)$. First we define a set $\mathcal{P}$ of path sets. Assume that only the edges are subject to random failure and that the reliability of edge e is given by p_e, for all $e \in E$. Then a path set F is a subset of the edge set E. For any path set $F \in \mathcal{P}$, let $P(F)$ be the probability that all edges of F but no other edges are operating, we obtain

$$P(F) = \prod_{e \in F} p_e \prod_{f \in E \setminus F} (1 - p_f).$$

The reliability $\mathrm{REL}(G)$ of G is the probability that at least one path set is operating, hence

$$\mathrm{REL}(G) = \sum_{F \in \mathcal{P}} P(F). \tag{4.1}$$

Usually, we have an additional property of path sets, namely the *monotonicity*. If F is a path set and $A \supseteq F$ then also A is a path set. However, we will learn in the next chapter that there are also reliability measures violating the monotonicity condition.

Remark 4.4 *We can introduce indicator variables for the states of the edges:*

$$z_e = \begin{cases} 1 \ \textit{if } e \textit{ is in operating state,} \\ 0 \ \textit{if } e \textit{ is in failed state.} \end{cases}$$

Then we find that $p_e = \Pr(\{z_e = 1\})$. *Analogously, the state of* G *can be defined by*

$$z_G = \begin{cases} 1 & \text{if at least one path set of } G \text{ is in operating state,} \\ 0 & \text{else.} \end{cases}$$

Consequently, we have $\mathrm{REL}(G) = \Pr(\{z_G = 1\})$, *where* z_G *is a function of the variables* z_e, $e \in E$. *For many interesting reliability measures of networks, we obtain a coherent binary system. In this case, all methods that are presented in Chapter 1 and in Chapter 2 can be applied to the reliability analysis of networks.*

A monotonic reliability measure defined by (4.1) can be calculated by means of the theory of monotone binary systems. The underlying network structure offers further possibilities for reliability analysis. For network reliability problems, the set of path sets of G is defined by graph properties or graph invariants. The most important reliability measure for a graph $G = (V, E)$ concerns its connectedness. A subset $F \subseteq E$ is a path set if and only F contains the edge set of any spanning tree of G. This definition of path sets leads to the all-terminal reliability, which is investigated in the next chapter. The two-terminal reliability is obtained in the following way. Let $s, t \in V$ be two specified terminal vertices of G and define a path set as any edge subset containing an st-path of G.

More general, any graph property can be used to define path sets. Let Θ be a *graph property* that a graph G may have. Here is a list of examples of graph properties that are of interest in network reliability:

- Is G is connected?

- Let $K \subseteq V$ be a set of terminal vertices. Do all vertices of K belong to one component of G?

- Let k be a given positive integer. Is the diameter of G at most k?

- Let s and t be two given vertices and $k \in \mathbb{Z}^+$. Are there k vertex- (edge-) disjoint st-paths in G?

- Let $G = (V, E)$ be a digraph and $s \in V$. Are all vertices of $V \setminus \{s\}$ reachable from s?

All these properties define a path set in a natural way. An edge set $F \subseteq E$ is a path set with respect to a given graph property Θ if the subgraph (V, F) satisfies Θ. In case the graph property Θ is *monotone*, which means that whenever (V, F) satisfies Θ and $F \subseteq F'$ then (V, F') also satisfies Θ, the corresponding system of path sets defines a monotone binary system.

The *resilience* $\mathrm{Res}(G)$ of an undirected graph $G = (V, E)$ with independently failing edges is the expected number of vertex pairs that can communicate via paths of operating edges. The probability that two specified vertices

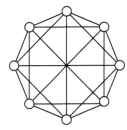

Figure 4.10: Optimally connected graph

$u, v \in V$ can communicate is the two-terminal reliability $R_{uv}(G)$. Hence, we obtain

$$\text{Res}(G) = \sum_{\{u,v\} \subseteq V} R_{uv}(G).$$

This reliability measure has been introduced by *Van Slyke* and *Frank* [301]. A trivial lower bound for the resilience of a graph with n vertices can be obtained by comparison with the all-terminal reliability $R(G)$ [94]:

$$\text{Res}(G) \geq \binom{n}{2} R(G)$$

Remark 4.5 *A more general definition of* network resilience *is given in [137], which is different from the notion of resilience introduced here.*

Now assume that in addition to edge reliabilities the edges of the undirected or directed graph $G = (V, E)$ are weighted with capacities. The *edge capacity* c_e of an edge e represents an upper bound for the amount of flow that an edge may carry. The *flow reliability problem*, see e.g. [201], is defined as follows. Let s and t be two given vertices and d_{st} a *demand* of flow from s to t. What is the probability that the demand can be satisfied? In case all capacities are equal to one and the demand is a positive integer k, we obtain the probability that G contains at least k edge-disjoint st-paths. As a generalization, we may consider a *demand matrix* that defines for any two vertices of the graph a demand value. Then we are interested in finding the probability that all demands can be realized simultaneously. In this case the sum value of all flows using an edge e is bounded by the edge capacity c_e. As an further requirement, we may ask for an *unsplittable flow* satisfying the demands. In this case, any flow between two vertices $u, v \in V$ has to use one single uv-path. The deterministic unsplittable flow problem was introduced in [104].

Let $G = (V, E)$ be a digraph with independently failing edges. The *strongly connected reliability* [71] of a digraph G is the probability that the operating edges of G induce a spanning strongly connected digraph.

We just mention here another huge area of research in network reliability theory, namely *reliability optimization*. The general problem in this field can be described as follows. Given a finite set of resources, for instance a number m of edges. What is the most reliable network (with respect to some given criterion) on n vertices? For readers interested in this field, we recommend [146] and many references cited within this nice survey. As an example, we can ask for the graph with n vertices and m edges that maximizes the edge connectivity, or reverse, for the least number of edges required to construct an n-vertex graph G with given edge connectivity $\lambda(G)$. Another problem is to find a minimal graph on n vertices with respect to the number of edges such that its diameter is at most k. These are typical problems of *extremal graph theory*. We refer to the excellent books by *Buckley* and *Harary* [74] and by *Bollobás* [63]. Figure 4.10 shows a graph with 8 vertices and edge connectivity 5 with a minimum number of edges.

Chapter 5

Reliability Analysis

In this chapter, reliability measures for graphs with randomly failing edges and/or vertices are defined. Sometimes we use the notion of the *stochastic graph* (*deterministic graph*) in order to indicate that the edges and vertices of the graph may fail (do not fail) randomly.

5.1 Connectedness

One of the most basic properties of a graph is connectedness. This section provides a first introduction to this topic. A deeper analysis of connectedness is presented in Chapter 6. In a connected graph $G = (V, E)$, there is a path between any two vertices of G. Hence we consider in this section a graph operational if it is connected. We assume that the edges of G fail randomly, whereas the vertices operate perfectly. The edge failures are presupposed to be stochastic independent. Let $q_e = 1 - p_e$ be the probability that edge e is in the failed state, $e \in E$. Let $\mathbf{p} = (p_e)_{e \in E}$ be the vector of the edge reliabilities of G for an arbitrary but fixed linear ordering of the edge set. Then the *all-terminal reliability* $R(G, \mathbf{p})$ is the probability that G is connected. If the edge reliabilities are known from the context, then we denote the all-terminal reliability simply by $R(G)$.

For a given graph $G = (V, E)$, we define the indicator function

$$\psi(G) = \begin{cases} 1 \text{ if } G \text{ is connected,} \\ 0 \text{ else.} \end{cases}$$

If we remove all edges that have failed from G then a subgraph of operating edges remains. The probability that a given edge subgraph $H = (V, F)$ of $G = (V, E)$ is realized is

$$\prod_{e \in F} p_e \prod_{f \in E \setminus F} q_f.$$

99

Consequently, we obtain for the all-terminal reliability

$$R(G) = \sum_{F \subseteq E} \psi(F) \prod_{e \in F} p_e \prod_{f \in E \setminus F} q_f, \tag{5.1}$$

where we use the abbreviated form $\psi(F)$ instead of $\psi((V, F))$ for the connectedness indicator of edge subgraph (V, F). The Equation (5.1) is from a computational point of view quite inefficient, because its application requires the enumeration of all $2^{|E|}$ subsets of the edge set of G. In later chapters we will discuss methods to overcome this problem. Figure 5.1 shows a graph with

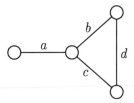

Figure 5.1: A graph with four edges

four vertices and four edges. The edge subsets of this graph that correspond to connected subgraphs are $\{a, b, c, d\}$, $\{a, b, c\}$, $\{a, b, d\}$, and $\{a, c, d\}$. The all-terminal reliability for the example graph is

$$R(G) = p_a p_b p_c p_d \left(1 + \frac{q_b}{p_b} + \frac{q_c}{p_c} + \frac{q_d}{p_c}\right)$$

$$= p_a p_b p_c + p_a p_b q_c p_d + p_a q_b p_c p_d.$$

The failure of a bridge in a graph destroys its connectedness. This observation gives immediately the all-terminal reliability of a tree $T = (V, E)$:

$$R(T) = \prod_{e \in E} p_e$$

A cycle C_n remains connected if and only if no edge or exactly one edge fails, resulting in

$$R(C_n) = \prod_{e \in E(C_n)} p_e \left(1 + \sum_{f \in E(C_n)} \frac{q_f}{p_f}\right).$$

In order to describe a more general method for the calculation of $R(G)$, we need some graph operations. Let $G = (V, E)$ be a given graph. The *deletion of an edge* $e \in E$ results in the new graph $G - e = (V, E \setminus \{e\})$. The *contraction of an edge* $e \in E$ is an operation that consists in removing edge $e = \{u, v\}$ and merging its end vertices u and v such that the resulting vertex is incident to all edges that were incident to u or v before the merge. The graph obtained from G as a result of the contraction of e is denoted by G/e. Figure 5.2 illustrates both graph operations.

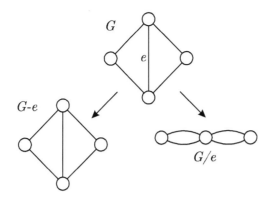

Figure 5.2: Deletion and contraction of an edge

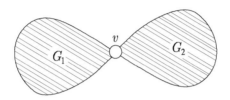

Figure 5.3: A graph with an articulation

Theorem 5.1 *Let e be an edge of the graph G. Then the all-terminal reliability of G satisfies*

$$R(G) = p_e R(G/e) + (1 - p_e) R(G - e).$$ (5.2)

Proof. Let C be the random event that G is connected and A be the event that edge e is operating. By the law of total probability, we have

$$\Pr(C) = \Pr(A)\Pr(C \mid A) + \Pr(\bar{A})\Pr(C \mid \bar{A}).$$

If $e = \{u, v\}$ is operating then for all problems of connectedness the two end vertices u and v of e can be considered as coinciding, which can be modeled by contraction of e. On the other hand, a failed edge can be considered as non-existent, thus we delete it.

∎

Theorem 5.2 *Let v be an articulation of $G = (V, E)$ such that G decomposes into two edge-disjoint subgraphs G_1 and G_2 that have exactly the vertex v in common (see Figure 5.3). Then the all-terminal reliability of G satisfies*

$$R(G) = R(G_1) R(G_2).$$

Proof. The graph G is connected if and only if both subgraphs G_1 and G_2 are connected. Since G_1 and G_2 are edge-disjoint, the connectedness of these two graphs forms stochastic independent events. ■

Let us return to Equation (5.1). We call an edge set $F \subseteq E$ that induces a connected spanning subgraph of G, i.e. an edge subset with the property $\psi(F) = 1$, a *path set* of G. Let $\mathcal{P}(G)$ be the set of all path sets of G. Then we can reformulate Equation (5.1) into

$$R(G) = \sum_{F \in \mathcal{P}(G)} \prod_{e \in F} p_e \prod_{f \in E \setminus F} q_f. \tag{5.3}$$

The set $\mathcal{P}(G - e)$ consists of all those path sets of G not containing e. Consequently, for any given edge e of G, we obtain $\mathcal{P}(G - e) \subseteq \mathcal{P}(G)$, which gives together with Equation (5.3) the inequality

$$R(G - e) \leq R(G). \tag{5.4}$$

A *cut set* of G is an edge subset whose removal disconnects G. The set of all cut sets of G is denoted by $\mathcal{C}(G)$. Hence, we obtain the cut set representation of the all-terminal reliability:

$$R(G) = 1 - \sum_{F \in \mathcal{C}(G)} \prod_{e \in F} q_e \prod_{f \in E \setminus F} p_f \tag{5.5}$$

For the example graph depicted in Figure 5.1, each subset of $\{a, b, c, d\}$ is a cut set, except $\emptyset$, $\{b\}$, $\{c\}$, and $\{d\}$. The set $\mathcal{C}(G/e)$ consists of all those cut sets of G not containing e, hence $\mathcal{C}(G/e) \subseteq \mathcal{C}(G)$. The last inclusion yields together with Equation (5.5) the inequality

$$R(G) \leq R(G/e) \tag{5.6}$$

for each edge e of G.

Let e be an edge of G. Then we obtain from Equation (5.2)

$$\frac{\partial R(G)}{\partial p_e} = R(G/e) - R(G - e).$$

The inequalities (5.4) and (5.6) yields $\frac{\partial R(G)}{\partial p_e} \geq 0$ for each edge $e \in E$, which shows that the all-terminal reliability is monotone increasing in p_e.

5.2 K-Terminal Reliability

In this section, we refine the definition of an operating state of a network, in order to meet practical requirements of communication networks. For many applications, we are not interested to know whether the whole network is connected. Let $G = (V, E)$ be an undirected graph whose edges are subject to

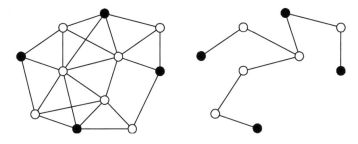

Figure 5.4: A graph with terminal vertices and a Steiner tree

random failure. Assume there is a special subset $K \subseteq V$ specified. We call the vertices of K *terminal vertices* of G. The *K-terminal reliability* $R(G, K)$ is the probability that all terminal vertices of G belong to one component of G. In other words, the K-terminal reliability is the probability that all vertices of K are connected by paths of operating edges. We call a deterministic graph $G = (V, E)$ *K-connected* if all vertices of K belong to one component of G. A *Steiner tree* of G with respect to K is a tree of G that contains all vertices of K such that each vertex of degree 1 (a so-called *leaf*) of the tree belongs to K. Consequently, a Steiner tree with respect to K is a minimal tree that connects all vertices of K. Figure 5.4 shows a graph with terminal vertices (in black) and an associated Steiner tree.

In case $K = \{s, t\}$, a Steiner tree of G is an st-path of G. The probability $R_{st}(G) = R(G, \{s, t\})$ is also called the *two-terminal reliability*. The two-terminal reliability $R_{st}(G)$ is just the probability that there exists at least one operating path between s and t in G. As a known special case of $R(G, K)$, we obtain for $K = V$ the all-terminal reliability of G.

Satyanarayana and *Tindell* [271] proposed a more general reliability measure that they call *(K, j)-reliability*. The *(K, j)-reliability* $R(G, K, j)$ is the probability that all terminal vertices of K are distributed across at most j components of G. Consequently, the K-terminal reliability corresponds to the $(K, 1)$-reliability. Another easy observation is $R(G, K, j) = 1$ whenever $j \geq |K|$. This measure was further investigated by *Rodriguez* and *Traldi* [256] in the context of domination theory. We restrict our attention here to the K-terminal reliability.

5.2.1 Some Basic Inequalities for the *K*-Terminal Reliability

The K-terminal reliability is monotone increasing with the edge reliabilities of G. The proof of this statement can be performed in the same vein as for the all-terminal reliability. Especially a *decomposition formula*, as given in Theorem 5.1, remains valid for the K-terminal reliability. Let $e = \{u, w\}$ be an edge of the graph $G = (V, E)$ and $K \subseteq V$ the set of terminal vertices of G. In order to define a terminal vertex set for the graph G/e, we denote the vertex that results from merging u and v while contracting e by v_e and set

$$K_e = \begin{cases} K \text{ if } K \cap \{u, w\} = \emptyset, \\ (K \setminus \{u, w\}) \cup \{v_e\} \text{ else.} \end{cases}$$

Then the decomposition formula reads

$$R(G, K) = p_e R(G/e, K_e) + (1 - p_e) R(G, K). \tag{5.7}$$

There are some interesting inequalities concerning the K-terminal reliability. Let K and L be vertex subsets of a given graph $G = (V, E)$ such that $K \subseteq L \subseteq V$. If G is L-connected then G is also K-connected, resulting in

$$K \subseteq L \implies R(G, K) \geq R(G, L). \tag{5.8}$$

The following result is presented in [68].

Theorem 5.3 *Let $G = (V, E)$ be a graph and $K, L \subseteq V$ with $K \cap L \neq \emptyset$. Then the following inequalities are satisfied:*

$$R(G, K \cup L) \geq R(G, K) + R(G, L) - 1 \tag{5.9}$$
$$R(G, K \cup L) \geq R(G, K)R(G, L) \tag{5.10}$$

Proof. Let $\psi(G, K)$ be an indicator function for the K-connectedness of G:

$$\psi(G, K) = \begin{cases} 1 \text{ if } G \text{ is } K\text{-connected}, \\ 0 \text{ else.} \end{cases}$$

Let K and L be vertex subsets of G with $K \cap L \neq \emptyset$. The condition $K \cap L \neq \emptyset$ implies that G is $(K \cup L)$-connected if and only if G is K-connected and L-connected. Elementary probability calculation shows

$$\begin{aligned} \Pr\left(\{\psi(G, K) = 1\} \cup \{\psi(G, L) = 1\}\right) &= \Pr\left(\{\psi(G, K) = 1\}\right) \\ &+ \Pr\left(\{\psi(G, L) = 1\}\right) \\ &- \Pr\left(\{\psi(G, K) = 1\} \cap \{\psi(G, L) = 1\}\right) \\ &= R(G, K) + R(G, L) - R(G, K \cup L). \end{aligned}$$

Since the probability of the left-hand side of this equation does not exceed 1, we obtain the first inequality of the theorem.

The monotonicity of the K-terminal reliability with respect to the edge reliabilities yields

$$\Pr\left(\{\psi(G, L) = 1\} \mid \{\psi(G, K) = 1\}\right) \geq \Pr\left(\{\psi(G, L) = 1\}\right). \tag{5.11}$$

The condition $\{\psi(G, K) = 1\}$ can be ensured by insertion of failure-free edges between all vertices of K or by increasing the edge reliabilities (from zero)

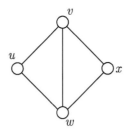

Figure 5.5: Example graph

to one. Hence inequality (5.11) is a direct consequence of the monotonicity of $R(G, K)$ in $\mathbf{p}$ (the vector of edge reliabilities). The inequality is valid independently of the choice of K and L, especially if we do not demand $K \cap L \neq \emptyset$. However, if we require $K \cap L \neq \emptyset$ then (5.11) can be applied in order to derive an inequality for $R(G, K \cup L)$:

$$
\begin{aligned}
R(G, K \cup L) &= \Pr\left(\{\psi(G, K) = 1\}\right) \ \Pr\left(\{\psi(G, L) = 1\} \mid \{\psi(G, K) = 1\}\right) \\
&\geq \Pr\left(\{\psi(G, K) = 1\}\right) \ \Pr\left(\{\psi(G, L) = 1\}\right) \\
&= R(G, K)\, R(G, L)
\end{aligned}
$$

This proves the second inequality of the theorem. ∎

Example 5.4 *Consider the graph depicted in Figure 5.5. Assuming all edge reliabilities to be 0.8, we obtain:*

$$
\begin{aligned}
R(G, V) &= 0.90112 \\
R(G, \{1, 2, 3\}) &= 0.93696 \\
R(G, \{1, 2, 4\}) &= 0.90624 \\
R(G, \{1, 2\}) &= 0.94848 \\
R(G, \{1, 3\}) &= 0.97408
\end{aligned}
$$

The results for all other terminal sets follow by symmetry. According to inequality (5.8), we observe

$$
\{1, 2\} \subseteq \{1, 2, 4\} \subseteq V \implies R(G, V) < R(G, \{1, 2, 3\}) < R(G, \{1, 2\}).
$$

As an example for (5.10), we find

$$
R(G, \{1, 2\})\, R(G, \{2, 3\}) = 0.89961 \leq R(G, \{1, 2, 3\}) = 0.93696.
$$

Exercise 5.1 *Let $r, s, t \in V(G)$. Show that*

$$
R(G, \{r, t\}) \geq R(G, \{r, s\}) R(G, \{s, t\}).
$$

Can you find a condition that makes this relation an equality?

Exercise 5.2 *What is the $(V, 2)$-reliability of a cycle C_n with vertex set V if all edges of the cycle fail independently with probability $1 - p$?*

Exercise 5.3 *In a cycle C_{2n} of even length each second vertex belongs to K and all edges have common reliability p. Compute $R(C_{2n}, K)$.*

Exercise 5.4 *Find an example graph with two disjoint vertex subsets K and L such that inequality (5.10) is violated.*

5.3   Vertex Failures

Now we generalize the model in order to include vertex failures. Assume the edges and vertices of a given graph $G = (V, E)$ fail independently with given probabilities. Even if we assume stochastic independent failures, one dependency remains. The failure of a vertex causes the failure of all edges incident to that vertex. Let p_e and p_v be the reliability of edge e and vertex v, respectively, for any $e \in E$ and $v \in V$. We redefine the K-terminal reliability in the following way. For a given set of terminal vertices $K \subseteq V$, let $R(G, K)$ be the probability that all terminal vertices are operating and contained in one component of G. Consequently, if $K = V$ then the operation of all vertices of G is required. Let $X \subseteq V$ and G_X be the graph obtained from G by setting all vertex reliabilities of vertices contained in X to 1, i.e. vertices of X are assumed to be failure-free. Then we obtain

$$R(G, V) = \prod_{v \in V} p_v \, R(G_V, V).$$

Thus we can consider vertex and edge failures separately. The vertex failures are taken into account by the first product. What remains is a "classic" K-terminal reliability problem without vertex failures. Unfortunately, this simple principle no longer works correctly when we have terminal vertex sets that are properly contained in V. Again, we have

$$R(G, K) = \prod_{v \in K} p_v \, R(G_K, V).$$

However, the graph G_K may have unreliable non-terminal vertices. The formula of total probability yields a vertex decomposition formula as an analogy to Equation (5.2). Let $v \in V \setminus K$ be a non-terminal vertex of $G = (V, E)$, then

$$R(G, K) = p_v R\left(G_{\{v\}}, K\right) + (1 - p_v) R(G - v, K). \tag{5.12}$$

This equation yields, together with the edge decomposition according to Equation (5.7), a recursive algorithm for the computation of the K-terminal reliability in graphs with unreliable edges and vertices.

Parallel and series reductions apply for unreliable vertices as introduced in Section 5.2, except that the new edge reliability in case of a series reduction is $p_c = p_a p_b p_v$ (compare Figure 6.12).

Figure 5.6: Transformation of unreliable edges into unreliable vertices

Bodlaender and *Wolle* [59] observed that many known reliability measures, including the K-terminal reliability, can be modeled by graphs with perfectly operating edges. Figure 5.6 shows the transformation of an unreliable edge $e = \{u, v\}$ into two perfectly operating edges $f = \{u, v\}$ and $g = \{v, w\}$. The additionally introduced vertex v is an unreliable non-terminal vertex with reliability $p_v = p_e$. Transforming all unreliable edges in this way results in a network in which only vertices are subject to random failure. Unreliable vertices are obviously also included in this model.

5.4 Residual Connectedness

In this section we investigate a rather special reliability measure of graphs whose vertices are subject to random failure whereas the edges are assumed to be perfectly reliable. Assume the vertices of a given graph $G = (V, E)$ fail independently with given probabilities $1 - p_v$, $v \in V$. The graph is considered operating if at least one vertex is operating and the subgraph induced by the set of operating vertices is connected. If all vertices fail with identical probability $1 - p$ then the residual connectedness reliability can be expressed by a polynomial in p. This polynomial can be easily derived when we know the sequence of numbers $\{s_k(G)\}$ of vertex induced connected subgraphs of G with exactly k vertices. We start this section with the investigation of the generating function of this sequence.

5.4.1 Vertex Connectivity Polynomial

Let $G = (V, E)$ be a simple undirected graph and let $X \subseteq V$ a vertex subset. The *vertex induced subgraph* $G[X]$ of G has vertex set X. Two vertices of X are adjacent in $G[X]$ if and only if these vertices are adjacent in G. The number of connected subgraphs of G that are induced by a vertex subset X of cardinality k is denoted by $s_k(G)$. Hence we obtain

$$s_1(G) = |V|,$$
$$s_2(G) = |E|.$$

In addition, we have $s_n(G) = 1$, $n = |V|$, if and only if G is connected, otherwise $s_n(G) = 0$ follows. We define $s_0(G) = 0$. The ordinary generating function of the sequence $(s_1(G), \ldots, s_n(G))$ is the polynomial

$$S(G, x) = \sum_{k=1}^{n} s_k(G)x^k,$$

which we call the *vertex connectivity polynomial* of G. The value of this polynomial at $x = 1$ equals the number of vertex induced connected subgraphs of G. The degree $\deg S$ yields the order of the largest component of G.

Theorem 5.5 *Let G be a graph with the components $G_1, \ldots, G_c$. Then*

$$S(G, x) = \sum_{i=1}^{c} S(G_i, x).$$

Proof. A vertex subset $X \subseteq V$ that induces a connected subgraph of G is contained completely in one component of G. Consequently, we obtain for each $k \in \mathbb{N}$ the sum representation $s_k(G) = s_k(G_1) + \ldots + s_k(G_c)$, which proves the theorem. ∎

Theorem 5.6 *Let $G = (V, E)$ be a simple undirected graph and $H = (V, F)$ a subgraph of G with the same vertex set. Then for each $k \in \mathbb{N}$ the following inequality is valid:*

$$s_k(G) \geq s_k(H)$$

Proof. Let $X \subseteq V$ such that $H[X]$ is connected. Then $G[X]$ is connected as G contains all edges of H. ∎

The vertex connectivity polynomial can be easily calculated for some special graphs. In a complete graph K_n, each nonempty vertex subset induces a connected subgraph. Consequently, we obtain

$$S(K_n, x) = \sum_{k=1}^{n} \binom{n}{k} x^k = (1 + x)^n - 1.$$

Any graph G with n vertices can have at most as many connected vertex induced subgraphs as the complete graph K_n, hence

$$s_k(G) \leq \binom{n}{k}.$$

We obtain for the edgeless graph $\bar{K}_n$ (the complement of the complete graph) the vertex connectivity polynomial

$$S(\bar{K}_n, x) = nx,$$

since in this case only singletons induce connected subgraphs.

In a path P_n with n vertices, a subset $X \subseteq V$ induces a connected subgraph if and only if the vertices of X are successively traversed in P_n. In this case,

$P_n[X]$ is a path itself. Hence we obtain $s_k(P_n) = n - k + 1$, since any subpath with k vertices within P_n can appear at $n - k + 1$ different places. Thus the vertex connectivity polynomial of a path is

$$S(P_n, x) = \sum_{k=1}^{n} (n - k + 1) x^k = \frac{x^{n+2} - (n+1) x^2 + nx}{(x-1)^2}.$$

For a given graph $G = (V, E)$ and a vertex $v \in V$, let $N[v] = N(v) \cup \{v\}$ be the *closed neighborhood* of v. For any graph $G = (V, E)$ and an arbitrary vertex $v \in V$, we define the graph G/v by removing v from G and inserting edges between all pairs of nonadjacent neighbor vertices of v. Consequently, the neighborhood $N(v)$ forms in G/v the vertex set of a clique. Let $G - X$ be the graph obtained from a graph $G = (V, E)$ by removal of a vertex subset $X \subseteq V$. We write briefly $G - v$ instead of $G - \{v\}$. The following theorem yields the foundation for the calculation of the vertex connectivity polynomial of arbitrary graphs. This result is based on an equivalent formula for the residual connectedness reliability derived by *Sutner, Satyanarayana*, and *Suffel* [289].

Theorem 5.7 *Let $G = (V, E)$ be a simple graph and $v \in V$. Then the vertex connectivity polynomial satisfies the equation:*

$$S(G, x) = x \left(S(G/v, x) - S(G - N[v], x) \right) + S(G - v, x) + x$$

Proof. Let v be an arbitrarily chosen vertex of G. We calculate the number $s_k(G)$ of connected vertex induced subgraphs of G. To this end, we consider the family $\mathcal{X}_k$ of all vertex subsets of cardinality k that induce subgraphs contributing to $s_k(G)$. The family $\mathcal{X}_k$ falls into two disjoint classes: $\mathcal{X}_k = \mathcal{Y}_k \cup \mathcal{Z}_k$. The class $\mathcal{Y}_k$ comprises all subsets of V excluding v. The subsets of this class are exactly those vertex sets with k vertices that induce connected subgraphs of $G - v$. Consequently, we obtain $|\mathcal{Y}_k| = s_k(G - v)$.

The second class, $\mathcal{Z}_k$, is the family of all vertex subsets of cardinality k that induce a connected subgraph of G and contain vertex v. In case $k = 1$, the class $\mathcal{Z}_k$ consists of the set $\{v\}$ only, which results in the term x of the generating function (the polynomial). Now assume $k > 1$, then each set $X \in \mathcal{Z}_k$ is the union of $\{v\}$ with a subset $X' \subseteq V$ of cardinality $k - 1$ that induces a connected subgraph of G/v. However, the subgraph induced by $X = X' \cup \{v\}$ is connected only if at least one vertex w of the neighborhood $N(v)$ of v belongs to X'. The generating function $S(G/v, x) - S(G - N[v], x)$ counts exactly those sets X'. The contribution of v itself is included by the last term, x, of the equation. ∎

A *separator* or a *separating vertex set* of a connected graph G is a set $X \subseteq V(G)$, such that $G - X$ is disconnected. Let $c_k(G)$ the number of separators of cardinality k of G. If $X \subseteq V$ is not a separator of G then $V \setminus X$ induces a connected subgraph of G. If, conversely, X induces a disconnected

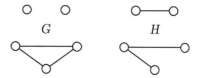

Figure 5.7: Non-isomorphic graphs with coinciding vertex connectivity polynomial

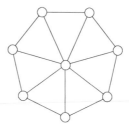

Figure 5.8: A wheel with eight vertices

subgraph of G then $V \setminus X$ is a separator of G. Hence we conclude

$$c_{n-k}(G) + s_k(G) = \binom{n}{k}$$

for all $k = 1, \ldots, n = |V|$. The number of all separators of G is, consequently, $2^n - S(G, 1) - 1$.

Two non-isomorphic graphs may have the same vertex connectivity polynomial. Moreover, there exist graphs with a different number of components sharing one vertex connectivity polynomial. An example is shown in Figure 5.7. For further results on residual connectedness reliability, the reader is refered to [288, 60, 96, 287, 87].

Exercise 5.5 *Find the vertex connectivity polynomial of a cycle with n vertices.*

Exercise 5.6 *Find the vertex connectivity polynomial of the star S_n with n vertices.*

Exercise 5.7 *Show that the complete bipartite graph $K_{m,n}$ satisfies*

$$S(K_{m,n}, x) = ((1+x)^m - 1)((1+x)^n - 1) + (m+n)x.$$

Exercise 5.8 *Let v be a vertex of degree 1 in G and let u be the only neighbor of v in G. Prove that the vertex connectivity polynomial satisfies the equation*

$$S(G, x) = (1+x) S(G - v, x) - x S(G - \{u, v\}, x) + x.$$

Exercise 5.9 *Determine the vertex connectivity polynomial of the graph that is depictured in Figure 5.8.*

Exercise 5.10 * *Are there two non-isomorphic trees sharing the same vertex connectivity polynomial?*

Exercise 5.11 *Let $G_{k,l} = P_k \times P_l$ be the $k \times l$ lattice graph. Determine the explicit representation of the terms of the sequence $(S(G_{2,l}, -1))_{l \in \mathbb{N}}$.*

Exercise 5.12 *Let $w(G)$ be the number paths of length 2 and $d(G)$ the number of triangles that are contained as subgraphs in G. Show that the relation $s_3(G) = w(G) - 2d(G)$ is valid for any graph G.*

Exercise 5.13 *Which coefficients of the vertex connectivity polynomial can we derive from the number of articulations of G?*

Exercise 5.14 *The (vertex) connectivity $\kappa(G)$ of G is the cardinality of a smallest separator of G. We set $\kappa(G) = 0$ if G is disconnected. For the complete graph, we define $\kappa(K_n) = n - 1$. Prove that each connected graph with n vertices satisfies $s_k(G) = \binom{n}{k}$ for all $k \in \{n - \kappa(G) + 1, \ ..., \ n\}$.*

5.4.2 Residual Network Reliability

Let $G = (V, E)$ be an undirected graph whose vertices are subject to random failure. The vertex failures are assumed to be stochastically independent. The failure probability of each vertex $v \in V$ is denoted by $q_v = 1 - p_v$. The edges of G are perfectly reliable. The *residual network reliability* $R_1(G)$ is the probability that a least one vertex of G is surviving and all surviving vertices lie in one common component of G. We call a (deterministic) graph satisfying this property *residual connected*.

Remark 5.8 *The concept of residual network reliability has been introduced by Boesch, Satyanarayana, and Suffel [60]. We use here the notation R_1 instead of R or R_n as employed by other authors in order to avoid confusion with different network reliability measures.*

The reliability measure $R_1(G)$ can be generalized as follows. The *k-residual network reliability* $R_k(G)$ is the probability that at least k vertices of G are in an operating state and that the subgraph induced by the set of all operating vertices is connected. Clearly, we assume $k \in \{1, 2, ..., n\}$, $n = |V|$, here.

We recall the definition of the indicator function for connectedness,

$$\psi(G) = \begin{cases} 1, \text{ if } G \text{ is connected,} \\ 0 \text{ else.} \end{cases}$$

From the definition of the k-residual network reliability, we obtain

$$R_k(G) = \sum_{\substack{X \subseteq V(G) \\ |X| \geq k}} \prod_{v \in X} p_v \prod_{w \in V \setminus X} q_w \, \psi(G\,[X]). \qquad (5.13)$$

This equation requires the enumeration of all 2^n vertex subsets of G. We observe, as a consequence of Equation (5.13), that the k-residual network reliability is monotone decreasing in k; if $k < l$ then $R_k(G) \geq R_l(G)$.

Recall that G/v is the graph obtained from $G - v$ by insertion of all edges between pairs of non-adjacent vertices of the neighborhood $N(v)$. Recall the definition of the Kronecker symbol δ_{ij} by

$$\delta_{ij} = \left\{ \begin{array}{l} 1, \text{ if } i = j \\ 0, \text{ if } i \neq j \end{array} \right. .$$

The following decomposition formula for the k-residual network reliability has been found by *Boesch, Satyanarayana,* and *Suffel* [60] for the special case R_1.

Theorem 5.9 *The k-residual network reliability $R_k(G)$ satisfies the following recurrence equation for each vertex $v \in V(G)$:*

$$R_k(G) = p_v \left[R_{k-1}(G/v) - \prod_{w \in N(v)} q_w \, R_{k-1}(G - N[v]) \right]$$
$$+ q_v R_k(G - v) + \delta_{1,k} p_v \prod_{u \in V \setminus \{v\}} q_u$$

Proof. Let $v \in V(G)$ be an arbitrary vertex of G. Consider first the case $k = 1$. In this case, G is residually k-connected if v is operating and all other vertices fail. The probability of this event is accounted for by the last term of the formula. If v fails then the remaining graph $G - v$ has to be k-residually connected in order to guarantee that G itself is k-residually connected. The probability of this event is $q_v R_k(G - v)$. The multiplication of probabilities is here and in all other cases justified by the presumed stochastic independence of vertex failures.

Now consider the case that vertex v is operating. Then the k-residual connectedness can be achieved if there are at least $k - 1$ vertices surviving in $G - v$. Since the operating vertex v ensures the connectedness of its neighbors, we use now G/v instead of $G - v$. The inserted edges of G/v emulate the connections of v. This case is realized with probability $q_v R_k(G - v)$. However, it may happen that G/v is $(k - 1)$-residual connected and v is operating but G is not k-residually connected. This event occurs if all neighbors of v fail but the remaining graph $G - N[v]$ is $(k - 1)$-residually connected. The corresponding probability is $\prod_{w \in N(v)} q_w \, R_{k-1}(G - N[v])$, which proves the theorem. ∎

We assume in the following that the failure probability $q = 1-p$ is identical for all vertices. In this case, $R_k(G)$ is a polynomial in p, which we call the *k-residual reliability polynomial* or, for $k = 1$, simply *residual reliability polynomial*. Theorem 5.9 leads to a recurrence equation for the residual reliability polynomial. For a graph $G = (V, E)$ with n vertices and $v \in V$, we obtain

$$R_1(G, p) = pR_1(G/v, p) - pq^{\deg v} R_1(G - N[v], p) + qR_1(G - v, p) + pq^{n-1}.$$

The coefficients of the vertex connectivity polynomial contain all the information we need in order to determine the residual reliability polynomial. The following theorem shows a simple transformation between these two polynomials.

Theorem 5.10 *The residual reliability polynomial $R_1(G, p)$ of a graph G with n vertices is given by the vertex connectivity polynomial $S(G, x)$ via the following transformation*

$$R_1(G, p) = (1 - p)^n S\left(G, \frac{p}{1 - p}\right).$$

Proof. We obtain by Equation (5.13)

$$R_1(G, p) = \sum_{\emptyset \subset X \subseteq V(G)} p^{|X|} (1 - p)^{n-|X|} \ \psi(G[X])$$

$$= (1 - p)^n \sum_{\emptyset \subset X \subseteq V(G)} \left(\frac{p}{1 - p}\right)^{|X|} \psi(G[X]). \qquad (5.14)$$

Using the indicator function $\psi(G)$, the vertex connectivity polynomial can be defined as

$$S(G, x) = \sum_{k=1}^{n} s_k(G)x^k$$

$$= \sum_{k=1}^{n} \sum_{\substack{X \subseteq V(G) \\ |X|=k}} \psi(G[X])x^k,$$

which provides by comparison with Equation (5.14) the statement of the theorem. ∎

The known results for vertex connectivity polynomials of special graphs may be applied to $R_1(G, p)$ by means of Theorem 5.10. We obtain the following

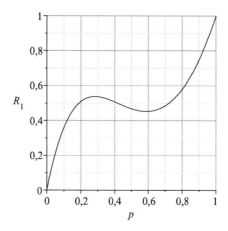

Figure 5.9: Residual reliability of a path

residual reliability polynomials:

$$R_1(P_n, p) = \frac{p^{n+2} + \left(np - p^2(2n+1)\right)(1-p)^n}{(1-2p)^2}$$

$$R_1(C_n, p) = p^n + n\left(\frac{1-p}{2p-1}(p^n - (1-p)^n) - (1-p)^n\right)$$

$$R_1(K_n, p) = 1 - (1-p)^n$$

$$R_1(K_{m,n}, p) = (1-p)^{m+n} \cdot$$
$$\cdot \left(\frac{p(m+n)}{1-p} + \left(\left(\frac{1}{1-p}\right)^m - 1\right)\left(\left(\frac{1}{1-p}\right)^n - 1\right)\right)$$

Example 5.11 *The residual reliability polynomial of a path with five vertices is*

$$R_1(P_5, p) = 3p^5 - 12p^4 + 21p^3 - 16p^2 + 5p.$$

The graph of this function in dependency on the vertex reliability p is shown in Figure 5.9. What is remarkable is that R_1 does not increase monotonously with p. However, the explanation of this phenomenon is easy: The failure of a vertex can cause the transition from a non-residually connected graph to a residually connected graph.

Theorem 5.12 *Let $G = (V, E)$ be a connected graph and C the set of all separators of G. Then*

$$R_1(G) = 1 - \sum_{X \in C \cup \{V\}} \prod_{v \in X} (1 - p_v) \prod_{v \in V \setminus X} p_v.$$

Proof. A vertex subset $X \subseteq V(G)$ induces a connected subgraph of G if and only if $V \setminus X$ is no separator of G. Consequently, G is residual connected if and only if not all vertices of one separator of G or all vertices of G fail. The product

$$\prod_{v \in X} (1 - p_v) \prod_{v \in V \setminus X} p_v$$

gives the probability that exactly the vertices of X and no others fail, hence the corresponding random events are disjoint. ■

The concept of residual connectedness reliability can be substantially generalized. What is the probability that a subgraph of G with exactly k components survives? The solution of this problem leads to the enumeration of vertex induced subgraphs of G with k components. Let $q_{ij}(G)$ be the number of vertex subsets $X \subseteq V$ with i vertices such that the vertex-induced subgraph $G[X]$ has exactly j components. The ordinary generating function for these numbers is the two-variable polynomial

$$Q(G; x, y) = \sum_{i=0}^{n} \sum_{j=0}^{n} q_{ij}(G) x^i y^j.$$

Having computed $Q(G; x, y)$, we obtain the probability $P_k(G)$ that a vertex induced subgraph of G has exactly k components easily:

$$P_k(G) = \frac{1}{k!} \frac{\partial^k}{\partial y^k} (1-p)^n Q\left(G; \frac{p}{1-p}, y\right)\Big|_{y=0}$$

For more results and applications concerning the polynomial $Q(G; x, y)$, the reader is referred to [294].

Exercise 5.15 *Show that the residual reliability polynomial of a graph G with n vertices and c components, $G_1, ..., G_c$, satifies the equation*

$$R_1(G, p) = \sum_{i=1}^{c} (1-p)^{n - |V(G_i)|} R_1(G_i, p).$$

5.5   Directed Graphs

In this section we consider digraphs whose edges fail randomly. Let $G = (V, E)$ be a digraph. The edges of G are assumed to fail stochastically independent, according to given probabilities. For each edge (arc) $e \in E$, let p_e denote the reliability (probability of operating) of edge e. Let $S, T \subseteq V$ be given nonempty vertex subsets. The *ST-reachability* $R(G, S, T)$ is the probability that there is a directed path (formed of operating edges) from any vertex of S to any vertex of T. The special case $S = \{s\}$ and $T = \{t\}$ is denoted by $R_{st}(G)$.

In case the vertices of G are subject to random failure too, an easy transformation of the network structure permits the replacement of unreliable vertices by unreliable edges without changing the reachability. Figure 5.10 illustrates the transformation.

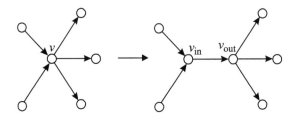

Figure 5.10: Transformation for unreliable vertices

Each unreliable vertex v is replaced by two new perfectly reliable vertices v_{in} and v_{out} such that all edges terminating in v are now redirected into v_{in}. All edges emanating from v are transformed into edges emanating from v_{out}. Finally the two new vertices are linked by an edge $e = v_{in}v_{out}$ that is weighted with reliability p_v. It is not too hard to see that this transformation preserves the reliabilities of all paths traversing v. Since digraphs with unreliable vertices can be transformed into networks with perfectly reliable vertices, we can restrict our attention to digraphs in which only the edges are subject to random failure.

Even more interesting, digraphs are able to mimic most common reliability measures of undirected graphs. *Nakazawa* [239] observed that sT-reachability in digraphs and $\{s\} \cup T$-reliability in undirected graphs coincide when all undirected edges are being replaced by two opposite directed edges with equal probabilities. Consequently, the investigation of reliability problems in digraphs would be sufficient to also solve the most important reliability problems in undirected graphs. However, the solution of reachability problems for digraphs is even more complex than the computation of undirected network reliability.

Clearly, the principal method of state or path set enumeration works for digraphs as already discussed for undirected graphs. The decomposition with respect to the state of a single edge is, however, not directly applicable to digraphs. The obstacle arising here is the contraction of a directed edge, which produces in general new directed paths that are not present in the original network. Thus the decomposition formula $R(G) = p_e R(G/e) + (1-p_e) R(G-e)$ is not valid for the calculation of reachability in digraphs.

Example 5.13 *Consider the digraph depicted in Figure 5.11. There does not exist an st-path in this digraph G. However, the digraph G/e obtained by the contraction of edge e contains an st-path.*

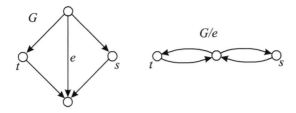

Figure 5.11: Contraction of an edge in a digraph

The last example shows that the decomposition formula fails in general for reachability problems in directed graphs. However, if we restrict our attention to the computation of the sT-reachability, then decomposition still applies to directed graphs. Let e be a directed edge emanating from s. Then we have

$$R(G, s, T) = p_e R(G/e, s, T) + (1 - p_e) R(G - e, s, T). \qquad (5.15)$$

The contraction of an edge e directed away from s does not create any improper directed paths. Now we are able to give an easy proof of Nakazawa's theorem.

Theorem 5.14 (Nakazawa, 1979) *Let $G = (V, E)$ be an undirected graph with given edge reliability p_e for each edge $e \in E$ and $K \subseteq V$. Construct a directed graph $\vec{G}$ from G through replacing each edge $e = \{u, v\}$ of G by two anti-parallel directed edges $e' = uv$ and $e'' = vu$ that operate with probability p_e each. Let $s \in K$, then*

$$R(G, K) = R(\vec{G}, s, K \setminus \{s\}).$$

Proof. The statement is obviously true for all graphs with one or two vertices. We proceed by induction on the number of edges, presupposing that the theorem is true for all graphs with at most m edges. Assume G is a graph with $m + 1$ edges and $\vec{G}$ is formed from G as described in the theorem. We choose a directed edge $e' = sv$ in $\vec{G}$ with tail s. Let $e'' = vs$ be the reversely directed edge of $\vec{G}$. If there does not exist any edge outgoing from s, then $R(G, K) = R(\vec{G}, s, K \setminus \{s\}) = 0$ and the statement follows. Otherwise, we apply the decomposition formula (5.15) to $\vec{G}$:

$$R(\vec{G}, s, K \setminus \{s\}) = p_{e'} R(\vec{G}/e', s, K \setminus \{s\}) + (1 - p_{e'}) R(\vec{G} - e', s, K \setminus \{s\})$$

Both of $\vec{G}/e'$ and $\vec{G} - e'$ are graphs with one edge less than $\vec{G}$. We can remove the edge e'' from $\vec{G}/e'$ without influencing the value $R(\vec{G}/e', s, K \setminus \{s\})$, since e'' does not appear in any directed path from s to another vertex of $\vec{G}$. The graph obtained from $\vec{G}/e'$ by removal of e'' is a directed graph that can also be obtained from an undirected graph via replacement of all edges by anti-parallel arc pairs, which could be symbolized by $\vec{G}/e' - e'' = \overrightarrow{G/e}$ with $e = \{s, v\} \in$

$E(G)$. Applying the induction hypothesis, we obtain $R(\vec{G}/e', s, K \setminus \{s\}) = R(G/e, K)$. An analog induction step for $\vec{G} - e'$ yields the statement of the theorem. ■

A digraph is *acyclic* if it does not contain any directed cycles. Acyclic graphs are of special interest, for instance in network routing. Let $G = (V, E)$ be an acyclic digraph and $s \in V$. We denote by $E^-(v)$ the set of all edges terminating in vertex v. *Ball* and *Provan* [12] observed that the problem of $s, V \setminus \{s\}$-reachability in acyclic digraphs offers a simple solution.

Theorem 5.15 *Let $G = (V, E)$ be an acyclic digraph and $s \in V$. Then*

$$R(G, s, V \setminus \{s\}) = \prod_{v \in V \setminus \{s\}} \left(1 - \prod_{e \in E^-(v)} (1 - p_e) \right).$$

Proof. There exists a path from s to any other vertex in G if and only if at least one incoming edge of each vertex except s is operating. Now the formula follows from the fact that the sets $E^-(v)$ and $E^-(w)$ are disjoint in case of $v \neq w$. ■

Exercise 5.16 *Let G be an acyclic orientation of the complete graph K_n. What is $R(G, s, V(G) \setminus s)$ if s is a vertex of indegree zero (a source) of G and all edges of G have reliability p?*

5.5.1 Algebraic Methods for Digraphs

The presentation of a powerful approach to the reliability of digraphs developed by *Shier* [278], *Gondran* and *Minoux* [142] requires some background from algebra. As an excellent introduction into path algebras, which are closely related to the algebraic structures introduced here, we recommend the book by *Carré* [81].

Definition 5.16 *Let A be a non-empty set, $\oplus : A \times A \to A$ and $\otimes : A \times A \to A$ two binary operations on A such that the following axioms are satisfied:*
(1) $\forall x, y, z \in A : (x \oplus y) \oplus z = x \oplus (y \oplus z)$,
(2) $\forall x, y, z \in A : (x \otimes y) \otimes z = x \otimes (y \otimes z)$,
(3) $\forall x, y \in A : x \oplus y = y \oplus x$,
(4) $\forall x \in A : x \oplus x = x$,
(5) $\exists \varepsilon \in A : \forall x \in A : x \oplus \varepsilon = \varepsilon \oplus x = x$,
(6) $\exists e \in A : \forall x \in A : x \otimes e = e \otimes x = x$,
(7) $\forall x \in A : x \otimes \varepsilon = \varepsilon \otimes x = \varepsilon$,
(8) $\forall x, y, z \in A : x \otimes (y \oplus z) = (x \otimes y) \oplus (x \otimes z)$,
(9) $\forall x, y, z \in A : (x \oplus y) \otimes z = (x \otimes z) \oplus (y \otimes z)$.
Then we call the structure $(A, \oplus, \otimes)$ a dioid.

Let $(A, \oplus, \otimes)$ be a dioid. We define a binary relation on A by

$$x \preceq y \iff x \oplus y = y.$$

We can easily verify that the relation "$\preceq$" is reflexive, antisymmetric, and transitive, hence it defines a partial order on A. We call this order relation the *canonical order* on A. From axiom (5) of the dioid definition, we conclude that ε is minimal with respect to "$\preceq$". We consider in the following exclusively *complete dioids*, i.e. dioids $(A, \oplus, \otimes)$ that form complete lattices with respect to the canonical order (which means that meet and join exists for any subset of elements from A) and if it satisfies for all $X \subseteq A$ and for any $y \in A$ the relations

$$\left(\bigoplus_{x \in X} x \right) \otimes y = \bigoplus_{x \in X} (x \otimes y) \quad \text{and}$$

$$y \otimes \left(\bigoplus_{x \in X} x \right) = \bigoplus_{x \in X} (y \otimes x) .$$

For any element $x \in A$ of a dioid, we define the power notation $x^0 = e$ and $x^{n+1} = x \otimes x^n$ for $n \geq 0$. In addition, for any integer $n \geq 0$, let

$$x^{(n)} = e \oplus x \oplus x^2 \oplus \dots \oplus x^n.$$

We call $x \in A$ an *n-stable element* $(n \geq 0)$ if $x^{(n)} = x^{(n+1)}$. An n-stable element is also an $(n + r)$-stable element for any $r > 0$, which can be shown by induction. An element $x \in A$ is called *stable* if x is n-stable for some n. For each n-stable $x \in A$, we define the *quasi-inverse* by

$$x^* = \lim_{k \to \infty} x^{(k)} = \bigoplus_{k \geq 0} x^k = \bigoplus_{k=0}^{n} x^k.$$

Consider the linear equation

$$x = a \otimes x \oplus b, \tag{5.16}$$

where a and b are given elements of A such that a is stable. Let x be a solution of (5.16) and $y \in A$ with $x \preceq y$. Then we find

$$a \otimes (x \oplus y) \oplus b = (a \otimes x) \oplus (a \otimes y) \oplus b$$
$$= a \otimes y \oplus b. \tag{5.17}$$

The second equation follows from $a \otimes x \preceq a \otimes y$. Equation (5.17) together with $x \oplus y = y$ shows that y is a solution for (5.16), too. Hence, if x is a solution of (5.16) and there are infinitely many $y \in A$ with $x \preceq y$, then the linear equation (5.16) has infinitely many solutions. The canonical order in a complete dioid ensures that there exists a unique minimal solution to (5.16), which is given by

$$x_0 = a^* \otimes b.$$

Assume a is n-stable. Then we obtain by substitution of $x = a^{(n)} \otimes b$ in (5.16)

$$a \otimes (a^{(n)} \otimes b) \oplus b = (a \otimes a^{(n)} \otimes b) \oplus (e \otimes b)$$
$$= (a \otimes a^{(n)} \oplus e) \otimes b$$
$$= a^{(n+1)} \otimes b$$
$$= a^{(n)} \otimes b$$

There are a lot of interesting dioid structures especially in combinatorial optimization. We consider here only two dioids that are appropriate for network reliability applications. We denote the set of all polynomials with m idempotent commuting variables $p_1, ..., p_m$ and coefficients in $\mathbb{Z}$ by $\mathbb{Z}[p_1, ..., p_m]$. The set $\mathbb{Z}[\mathbf{p}] = \mathbb{Z}[p_1, ..., p_m]$ forms together with the usual sum and product operations for polynomials a ring. The null polynomial (with all coefficients equal to zero) is the additive identity in $\mathbb{Z}[p_1, ..., p_m]$ and 1 is the multiplicative identity. We define two new operations $\otimes : \mathbb{Z}[\mathbf{p}] \times \mathbb{Z}[\mathbf{p}] \to \mathbb{Z}[\mathbf{p}]$ and $\oplus : \mathbb{Z}[\mathbf{p}] \times \mathbb{Z}[\mathbf{p}] \to \mathbb{Z}[\mathbf{p}]$, where $f \otimes g$ is the usual multiplication respecting idempotence of all variables and

$$f \oplus g = f + g - f \otimes g.$$

Let S_m be a subset of $\mathbb{Z}[\mathbf{p}]$ recursively defined by

(1) $0, 1 \in S_m$,

(2) $\forall i \in \{1, ..., m\} : p_i \in S_m$,

(3) if $f, g \in S_m$ then $f \oplus g \in S_m$ and $f \otimes g \in S_m$.

We can easily verify that for each polynomial $f \in S_m$ the inequality

$$0 \preceq f \preceq 1 \tag{5.18}$$

is valid with respect to the canonical order.

Now we return to network reliability. Let $G = (V, E)$ be a digraph with m edges $e_1, ..., e_m$. The reliability of edge e_i is denoted by p_i for $i = 1, .., m$. Let H, K be spanning subgraphs of G. We denote the random event that all edges of H and K are operating by A_H and A_K, respectively. The probability $P(H) = \Pr(A_H)$ that all edges of the subgraph $H = (V, E(H))$ are operating is

$$P(H) = \prod_{e_i \in E(H)} p_i = \bigotimes_{e_i \in E(H)} p_i.$$

The event $A_H \cap A_K$ occurs if and only if all edges of $H \cup K$ are operating, which yields

$$\Pr(A_H \cap A_K) = P(H \cup K) = P((V, E(H) \cup E(K)))$$
$$= \bigotimes_{e_i \in E(H) \cup E(K)} p_i$$
$$= P(H) \otimes P(K).$$

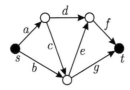

Figure 5.12: A digraph

Observe that in general $P(H \cup K) \neq P(H) \cdot P(K)$, due to the idempotence of the variables. The probability of the union of A_H and A_K follows by inclusion-exclusion:

$$\Pr(A_H \cup A_K) = \Pr(A_H) + \Pr(A_K) - \Pr(A_H \cap A_K)$$

$$= \bigotimes_{e_i \in E(H)} p_i + \bigotimes_{e_i \in E(K)} p_i - \bigotimes_{e_i \in E(H) \cup E(K)} p_i$$

$$= P(H) + P(K) - P(H) \otimes P(K)$$

$$= P(H) \oplus P(K)$$

Consequently, we may consider the operator "$\oplus$" as a kind of *inclusion-exclusion operator*. Let $s, t \in V(G)$ be two given vertices of G and let $\mathcal{W}_{st}(G)$ be the set of all directed st-paths of G represented as edge subsets. For any path $W \in \mathcal{W}_{st}(G)$ let A_W be the random event that all edges of W are in operating state. Then the st-reachability of G is given by

$$R_{st}(G) = \Pr\left(\bigcup_{W \in \mathcal{W}_{st}(G)} A_W \right)$$

$$= \bigoplus_{W \in \mathcal{W}_{st}(G)} \Pr(A_W)$$

$$= \bigoplus_{W \in \mathcal{W}_{st}(G)} \bigotimes_{e \in W} p_e. \tag{5.19}$$

Remark 5.17 *The idempotence-respecting product $\otimes$ in (5.19) could be replaced by normal product $\prod$. However, the given version is more general, as it allows paths containing the same edge several times, which is important for some applications.*

Example 5.18 *The digraph depictured in Figure 5.12 has the following directed st-paths:*

$$\mathcal{W}_{st}(G) = \{\{a, c, e, f\}, \{a, c, g\}, \{a, d, f\}, \{b, e, f\}, \{b, g\}\}$$

In order to save space, we use the same letters denoting the edges of G as variables for the corresponding edge reliabilities. The application of Equation (5.19) yields

$$\begin{aligned}
R_{st}(G) &= acef \oplus acg \oplus adf \oplus bef \oplus bg \\
&= acef + acg + adf + bef + bg \\
&\quad - abcef - abdef - abdfg - acdef - acdfg - acefg - abcg - befg \\
&\quad + abcdef + abcdfg + abcefg + abdefg + acdefg \\
&\quad - abcdefg
\end{aligned}$$

Next we try to express Equation (5.19) in matrix form. Let $G = (V, E)$ be a digraph without loops or parallel edges. However, we allow antiparallel edges in G, i.e. edges $e = (u, v)$ and $f = (v, u)$ such that the head of e is the tail of f and vice versa. Without loss of generality, we assume the vertex set of G given as $V = \{1, 2, ..., n\}$. Let $A = (a_{ij})_{n,n}$, $n = |V|$, be a matrix with the entries

$$a_{ij} = \begin{cases} p_e & \text{if } e = (i, j) \in E, \\ 0 & \text{else.} \end{cases}$$

Let $A^k = (a_{ij}^{(k)})_{n,n}$ be the k-th matrix power with respect to "$\otimes$," i.e.

$$A^k = \underbrace{A \otimes A \otimes ... \otimes A}_{k \text{ times}}.$$

Let A and B be $n \times n$ matrices. The entries of the product matrix $C = A \otimes B$ are computed by

$$c_{ij} = \bigoplus_{k=1}^{n} a_{ik} \otimes b_{kj} \text{ for } i, j \in \{1, ..., n\}.$$

For the k-th power of A, we obtain

$$a_{ij}^{(k)} = \bigoplus_{1 \leq r_1, ..., r_{k-1} \leq n} a_{ir_1} \otimes a_{r_1 r_2} \otimes a_{r_2 r_3} \otimes ... \otimes a_{r_{k-1} j}.$$

Let $\mathcal{W}_{st}^{(k)}(G)$ the set of all directed walks of length k from vertex s to vertex t. Then we obtain

$$a_{st}^{(k)} = \bigoplus_{W \in \mathcal{W}_{st}^{(k)}(G)} \bigotimes_{(i,j) \in W} a_{ij}. \tag{5.20}$$

Due to the indempotence the value of $a_{st}^{(k)}$ remains unchanged when we restrict $\mathcal{W}_{st}^{(k)}(G)$ to the set of all directed paths from s to t. Finally, the comparison of (5.19) and (5.20) yields

$$R_{st}(G) = \bigoplus_{k \geq 0} a_{st}^{(k)} = \left(\bigoplus_{k \geq 0} A^k \right)_{st}, \tag{5.21}$$

where the notation $(A)_{st}$ is the element of the matrix A located in row s and column t. Let d be the length of a longest directed path from s to t in G. Then we have

$$\bigoplus_{k \geq 0} a_{st}^{(k)} = \bigoplus_{k=0}^{d} a_{st}^{(k)},$$

which reflects the property that A is d-stable.

Remark 5.19 Shier [278] showed that $R_{st}(G)$ can be expressed as a solution of a linear equation of the form (5.16), which can be solved iteratively by a Jacobi method. The iteration finds within finitely many steps the exact solution (due to stability).

We can easily redesign the dioid $(A, \oplus, \otimes)$ in order to solve a different reliability problem. Instead of polynomials, we work now with real numbers from the interval $A = [0, 1]$. The multiplication is the usual multiplication of real numbers. The sum is defined as the maximum: $x \oplus y = \max\{x, y\}$. The neutral element with respect to addition is 0; the neutral element with respect to multiplication is 1. The (i, j)-entry of $A^* = \bigoplus_{k \geq 0} A^k$ gives now the reliability of the most reliable directed path between vertex i and vertex j.

Exercise 5.17 Let $(A, \oplus, \otimes)$ be a dioid. Show that the canonical order relation is compatible with the addition and multiplication in A, i.e. that $x \preceq y$ implies $x \oplus z \preceq y \oplus z$ and $x \otimes z \preceq y \otimes z$.

Exercise 5.18 Prove that the above defined set S_m of polynomials forms together with the two polynomial operations ($\oplus$ and $\otimes$) a dioid.

Exercise 5.19 Show that for all $q \geq 0$ the relation

$$\bigoplus_{k=0}^{q} a_{st}^{(k)} \preceq \bigoplus_{k=0}^{q+1} a_{st}^{(k)}$$

is satisfied.

Exercise 5.20 Is there a way to simplify the computation of $R_{st}(G)$ according to (5.21) if we assume that all edge reliabilities of G are equal to p?

Exercise 5.21 Verify that the matrix A defined for the most-reliable-path problem is $(n - 1)$-stable for a graph of order n.

5.6 Domination and Covering

A vertex subset $X \subseteq V$ of a graph $G = (V, E)$ is called a *dominating set* of G if each vertex of $V \setminus X$ is adjacent to a vertex in X. If we assume that we can monitor from a given vertex $v \in V$ all the neighbors of v in G and v then the occupation of a dominating set offers the possibility to monitor all vertices of the graph. As a contrast, assume we want to control all edges of G, assuming that we can monitor an edge $e = \{u, v\}$ from one of its end vertices u or v. In this case, we look for a vertex subset $Y \subseteq V$ such that Y contains at least one vertex of any edge of G. Such a subset Y is called a *vertex cover* of G. If the vertices of G fail randomly, then the existence of a dominating set and a vertex cover become random events, too, which gives rise to the definition of corresponding reliability measures that we investigate in this section.

5.6.1   Domination Reliability

We review some elementary properties of dominating sets in graphs. The first properties follow directly from the definition.

1. If X is a dominating set of $G = (V, E)$ and Y is a vertex set satisfying $X \subseteq Y \subseteq V$ then Y is also a dominating set of G.

2. Let $H = (V, F)$ be a spanning subgraph of $G = (V, E)$, which means $F \subseteq E$. Then each dominating set of H is also a dominating set of G.

3. If G consists of k components $G_1, ..., G_k$ then each dominating set of G is the union of k dominating sets of $G_1, ..., G_k$.

Since neither parallel edges nor loops have any effect on dominating sets, we assume in the following that all considered graphs are simple. As a prerequisite for the computation of a domination reliability measure, we count dominating sets of a graph with respect to its cardinality. Let $d_k(G)$ be the number of dominating sets of the graph G with exactly k vertices. The following properties of the numbers $d_k(G)$ are easily verified:

- For each graph with n vertices, we have $d_n(G) = 1$. This statement remains true for the null graph, which is the only graph with $d_0(G) = 1$.

- If G is a graph with l components then $d_k(G) = 0$ for all $k < l$.

- Let G be a graph of order n. Then $d_1(G) > 0$ if and only if G has a vertex v with $\deg v = n - 1$.

Definition 5.20 *Let $G = (V, E)$ be a simple graph. The domination polynomial of G is*

$$D(G, x) = \sum_{k=0}^{n} d_k(G)x^k.$$

Here and in following statements, we denote by n the order of G.

Remark 5.21 *A similar polynomial has been defined in [8]. The roots of the domination polynomial are considered in [3].*

For the empty graph $E_n = (V, \emptyset)$ of order n, there is but one dominating set, namely V itself, hence:

$$D(E_n, x) = x^n$$

In a complete graph K_n, any non-empty vertex subset is dominating, which gives

$$D(K_n, x) = (1 + x)^n - 1.$$

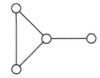

Figure 5.13: Example graph

Example 5.22 *Figure 5.13 shows a graph G with four vertices and $d_1(G) = 1$, $d_2(G) = 5$, $d_3(G) = 4$, $d_4(G) = 1$, which gives*

$$D(G, x) = x + 5x^2 + 4x^3 + x^4.$$

Lemma 5.23 *If G is a graph with l components $G_1, ..., G_l$ then*

$$D(G, x) = \prod_{i=1}^{l} D(G_i, x).$$

Proof. Assume $l = 2$ such that G_1 and G_2 are the components of G. Then each dominating set of G is the union of dominating sets of G_1 and G_2. Consequently, we obtain

$$d_k(G) = \sum_{j=0}^{k} d_j(G_1)d_{k-j}(G_2),$$

which gives also $D(G, x) = D(G_1, x)D(G_2, x)$. The lemma now easily follows by induction with respect to the number of components.

■

The *join* $G + H$ of two graphs $G = (V, E)$ and $H = (W, F)$ is obtained from the disjoint union of G and H by introducing edges from each vertex of V to each vertex of W. In case G and H are not vertex-disjoint, we produce a disjoint copy of H. In this way also $G + G$ is well-defined. Consider as an example the wheel presented in Figure 5.14, which can be obtained as a join of a cycle C_8 and a single vertex. The following lemma is presented in [8].

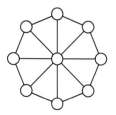

Figure 5.14: A wheel

Lemma 5.24 *The domination polynomial of a join of two graphs G and H is given by*

$$D(G + H, x) = \left[(1 + x)^{v(H)} - 1\right]\left[(1 + x)^{v(G)} - 1\right] + D(G, x) + D(H, x).$$

Proof. A dominating set of $G + H$ is obtained if we choose either at least one vertex from each of the graphs G and H, or a dominating set of one of the two graphs, which gives the three terms of the sum. ∎

In order to establish a more general method for the computation of the domination polynomial, we refine the counting of dominating sets. Let $G = (V, E)$ be a given graph and $U, W \subseteq V$. The *closed neighborhood* $N[A]$ of a set $A \subseteq V$ is defined as the union of A and the set of vertices that are adjacent to a vertex of A in G. We denote by $d_k(G; U, W)$ the number of vertex subsets of size k of U that contain in their closed neighborhood all vertices of the set W:

$$d_k(G; U, W) = |\{A \mid A \subseteq U \wedge |A| = k \wedge W \subseteq N[A]\}|.$$

The interpretation of these numbers is as following. We want to monitor all vertices of a subset $W \subseteq V(G)$, but we are allowed to use as a *partial dominating set* only vertices from the given set U. As a consequence, if $W \not\subseteq N[U]$ then we obtain

$$d_k(G; U, W) = 0 \text{ for all } k \geq 0. \tag{5.22}$$

We say that a vertex u *is dominated* by w (or w *dominates* u) if and only if u is contained in the closed neighborhood of w. The *extended domination polynomial* of $G = (V, E)$ is defined for all $U, W \subseteq V$ as

$$D(G; U, W, x) = \sum_{k=0}^{n} d_k(G; U, W)x^k.$$

This definition implies

$$D(G, x) = D(G; V, V, x). \tag{5.23}$$

Theorem 5.25 *Let $G = (V, E)$ be a graph, $U, W \subseteq V$, and $v \in U$. Then the extended domination polynomial satisfies*

$$D(G; U, W, x) = \begin{cases} 0 \text{ if } W \setminus N[U] \neq \emptyset, \\ (1+x)^{|U|} \text{ if } W = \emptyset, \\ xD(G - v; U \setminus \{v\}, W - N[v], x) \\ \quad + D(G, U - \{v\}, W, x) \text{ else.} \end{cases}$$

Proof. The first statement, namely $D(G; U, W, x) = 0$ when $W \setminus N[U] \neq \emptyset$, follows immediately from (5.22). In case of an empty set W, there are no vertices to dominate. Hence, any subset of U is dominating, which gives the generating function $(1+x)^{|U|}$.

Now let $v \in U$. Then the set of all partial dominating sets of G contained in U and dominating W can be divided into two disjoint classes. The first class $\mathcal{D}_{+v}$ contains all partial dominating sets that include v, the second class $\mathcal{D}_{-v}$ all those partial dominating sets not containing v. For each vertex $v \in U$, we have

$$D(G; U, W, x) = \sum_{k=0}^{n} (|\{X \in \mathcal{D}_{+v} : |X| = k\}| + |\{X \in \mathcal{D}_{-v} : |X| = k\}|) \, x^k.$$

(5.24)

A set X belongs to $\mathcal{D}_v$ if and only if $X \setminus \{v\}$ is a partial dominating set of G with respect to $W - N[v]$. Consequently, for $k = 1, ..., n$, we have

$$|\{X \in \mathcal{D}_{+v} \mid |X| = k\}| = d_{k-1}(G; U \setminus \{v\}, W - N[v]).$$

If $v \in V$ is any vertex with $v \notin U \cup W$ then $D(G; U, W, x) = D(G - v; U, W, x)$. Hence we obtain

$$\sum_{k=1}^{n} |\{X \in \mathcal{D}_{+v} \mid |X| = k\}| x^k = \sum_{k=1}^{n} d_{k-1}(G; U \setminus \{v\}, W - N[v]) x^k$$

$$= \sum_{k=0}^{n-1} x d_k(G; U \setminus \{v\}, W - N[v]) x^k$$

$$= x D(G; U \setminus \{v\}, W - N[v], x)$$

$$= x D(G - v; U \setminus \{v\}, W - N[v], x).$$

The cardinality of the second class can be easily determined. We just exclude v from U and obtain

$$|\{X \in \mathcal{D}_{-v} \mid |X| = k\}| = d_k(G; U \setminus \{v\}, W)$$

or in polynomial form

$$\sum_{k=1}^{n} |\{X \in \mathcal{D}_{-v} \mid |X| = k\}| x^k = D(G; U \setminus \{v\}, W, x),$$

which yields, substituted in (5.24), the theorem. ∎

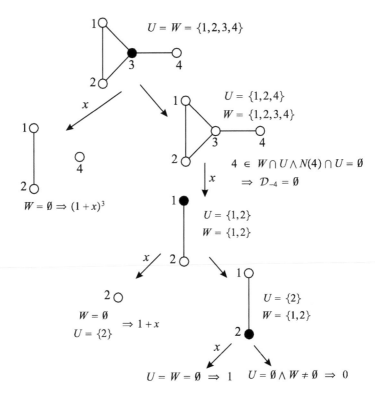

Figure 5.15: Computation of an extended domination polynomial

Example 5.26 *We demonstrate the application of Theorem 5.25 to the graph presented in Figure 5.13. The resulting computation tree is depicted in Figure 5.15. In order to obtain the domination polynomial of this graph, we start the computation of the extended domination polynomial with $U = W = \{1,2,3,4\}$, according to (5.23). The vertex used for the decomposition in each step is blackened in the figure. In the first step we use vertex 3. The final domination polynomial results from multiplying the weights along the branches with the polynomials in the leaves and adding up all terms:*

$$D(G,x) = x \cdot (1+x)^3 + x^2 \cdot (1+x) + x^2 \cdot 1 + x \cdot 0$$
$$= x + 5x^2 + 4x^3 + x^4$$

The computation of the extended domination polynomial can be simplified by applying *reductions*. A first kind of reduction arises when there is a vertex v that is contained in $U \cap W$ but no neighbor of v is contained in U. In this case, there is only one way to dominate v, namely by v itself. Hence we include v in the dominating set, remove v and its neighborhood from G, and multiply the polynomial of the reduced graph with x so as to consider the contribution of the vertex. This reduction is already applied in the last example for vertex 4 in the right branch of the computation tree.

We can cut a branch of the computation tree as soon as we find a vertex $v \in W \setminus U$ with $N(v) \cap U = \emptyset$. In this case, the vertex v cannot be dominated by any other vertex. Consequently, the resulting extended domination polynomial of this branch is identically zero.

Assume now that all vertices of a given subset $U \subseteq V$ of a graph $G = (V, E)$ operate independently with probability $p = 1 - q$. What is the probability that a given subset $W \subseteq V$ is dominated by vertices from U? We call this probability, denoted by $R_d(G; U, W, p)$, the *domination reliability* of G with respect to U and W. In case of $U = W = V$, we speak simply of the domination reliability of G and denote this measure by $R_d(G, p)$, see [108].

Remark 5.27 *If we say here that a vertex is* operating, *then we mean that this vertex is available for a dominating set of the graph G. A failed vertex remains still in the graph (it cannot be removed) but it must not be included in any dominating set of G. This distinction from other kinds of vertex failures considered in previous sections is important, since vertices that are not available for dominating sets have nevertheless to be dominated by other vertices of G.*

Theorem 5.28 *Let $D(G; U, W, x)$ be the extended domination polynomial of the graph $G = (V, E)$. Then the domination reliability satisfies*

$$R_d(G; U, W, p) = (1 - p)^{|U|} D(G; U, W, \frac{p}{1 - p}).$$

Proof. Let $X \subseteq U$ be a given set of cardinality k that dominates all vertices of W. The probability that exactly the vertices of X operate and all vertices of $U \setminus X$ fail is

$$p^k (1 - p)^{|U| - k} = (1 - p)^{|U|} \left(\frac{p}{1 - p} \right)^k.$$

The number of partial dominating sets of size k in G is $d_k(G; U, W)$, which is also the coefficient of x^k in $D(G; U, W, x)$. ∎

The domination reliability has the following monotonicity properties that are easily verified.

1. For all $U_1, U_2, W \subseteq V$ with $U_1 \subseteq U_2$ and for all $p \in [0, 1]$:

$$R_d(G; U_1, W, p) \leq R_d(G; U_2, W, p)$$

2. For all $U, W_1, W_2 \subseteq V$ with $W_1 \subseteq W_2$ and for all $p \in [0, 1]$:

$$R_d(G; U, W_1, p) \geq R_d(G; U, W_2, p)$$

3. If $H = (V, F)$ is a spanning subgraph of $G = (V, E)$ then for all $U, W \subseteq V$ and for all $p \in [0, 1]$:

$$R_d(H; U, W, p) \leq R_d(G; U, W, p)$$

4. For all $U, W \subseteq V$ and for all $p_1, p_2 \in [0, 1]$ with $p_1 < p_2$:

$$R_d(G; U, W, p_1) \leq R_d(G; U, W, p_2)$$

Let $G = (V, E)$ be a graph with vertex set $V = \{1, 2, ..., n\}$. We assign the following polynomial to G:

$$f_G(p; y_1, y_2, ..., y_n) = (1 - p)^n \prod_{v \in V} \left(1 + \frac{p}{1-p} \prod_{i \in N[v]} y_i \right) \qquad (5.25)$$

Theorem 5.29 *Let $G = (V, E)$ be a graph with vertex set $V = \{1, 2, ..., n\}$. For each subset $A \subseteq V$ and for $x \in V$, let*

$$I_A(x) = \begin{cases} 1 \ \ if \ x \in A, \\ 0 \ \ if \ x \notin A \end{cases}$$

be the indicator function for the set A. Then the domination reliability of G is given by

$$R_d(G, p) = \sum_{A \subseteq V} (-1)^{|V| - |A|} f_G(p; 1_A(1), \dots, 1_A(n)), .$$

Proof. Let us write the polynomial (5.25) differently:

$$f_G(p; y_1, y_2, ..., y_n) = \prod_{v \in V} \left(1 - p + p \prod_{i \in N[v]} y_i \right) .$$

Each monomial $(1 - p)^{n-k} p^k y_1^{\alpha_1} y_2^{\alpha_2} \cdots y_n^{\alpha_n}$ of the expanded form of f_G corresponds to a choice of a k-element subset X of V that dominates the vertex $i \in V$ exactly α_i times, $i = 1, ..., n$. Consequently, X is a dominating set of G if and only if $\alpha_i > 0$ for $i = 1, ..., n$. The term $f_G(p; 1_A(1), \dots, 1_A(n))$ is the probability of the random event $B_{V \setminus A} = \{$"no vertex of $V \setminus A$ is dominated"$\}$.

Consequently, we obtain

$$R_d(G, p) = P\left(\bigcap_{v \in V} \overline{B_{\{v\}}}\right) = P\left(\overline{\bigcup_{v \in V} B_{\{v\}}}\right) = 1 - P\left(\bigcup_{v \in V} B_{\{v\}}\right)$$

$$= 1 - \sum_{A \subseteq V, A \neq \emptyset} (-1)^{|A|+1} P\left(\bigcap_{v \in A} B_{\{v\}}\right)$$

$$= \sum_{A \subseteq V} (-1)^{|A|} P\left(\bigcap_{v \in A} B_{\{v\}}\right)$$

$$= \sum_{A \subseteq V} (-1)^{|A|} P\left(B_A\right)$$

$$= \sum_{A \subseteq V} (-1)^{|V|-|A|} P\left(B_{V \setminus A}\right)$$

$$= \sum_{A \subseteq V} (-1)^{|V|-|A|} f_G(p; 1_A(1), \ldots, 1_A(n)).$$

∎

Remark 5.30 *The domination reliability can easily be generalized to set systems. Let E be a finite set and $\mathcal{F} = \{F_1, ..., F_k\}$ be a family of subsets of E. Now assume that each subset F_i emerges independently with probability p_i. What is the probability that the ground set E is covered by randomly appearing subsets of $\mathcal{F}$? This problem is known as the* reliability covering problem *[278].*

5.6.2   Vertex Coverings

In this section we again consider unreliable vertices of a graph that are assumed to operate stochastically independent, with a given probability p_v for each vertex v. Let $R_C(G)$ be the probability that the set of all operating vertices forms a vertex cover of G. We call $R_C(G)$ the *vertex cover reliability* of G. Let us consider the complete graph K_n as a first example. We observe that each vertex cover of K_n must contain at least $n - 1$ vertices. Otherwise, assume that the two vertices $u, v \in V(K_n)$ are not included in a vertex cover. Since any two vertices are adjacent within a complete graph, we conclude that the edge connecting u and v is not covered, which is a contradiction. Hence we obtain

$$R_C(K_n) = \prod_{v \in V(K_n)} p_v \left(1 + \sum_{w \in V(K_n)} \frac{1 - p_w}{p_w}\right).$$

If we assume further that all vertex probabilities are equal, $p_v = p$ for all $v \in V$, then the above formula simplifies to

$$R_C(K_n) = np^{n-1} - (n-1)p^n.$$

In order to find a more general approach to the calculation of $R_C(G)$, we list some elementary properties of vertex covers:

1. A subset $C \subseteq V$ is a vertex cover of G if and only if $V \setminus C$ is an independent set of G. A vertex subset $X \subseteq V$ is called *independent* in $G = (V, E)$ if any two vertices of X are non-adjacent.

2. If C is a vertex cover of G then each vertex set B with $C \subseteq B$ is also a vertex cover of G.

3. If C is a vertex cover of $G = (V, E)$ then C is also a vertex cover for each spanning subgraph $H = (V, F)$ of G.

Let $A \subseteq V$ be a vertex subset of $G = (V, E)$. We denote by $R_C(G, A)$ the probability that G contains a vertex cover that is a subset of A. Consequently, we obtain the vertex cover reliability as $R_C(G) = R_C(G, V)$. The following lemma is a simple application of the formula of total probability.

Lemma 5.31 (Decomposition) *Let v be a vertex of $G = (V, E)$ and $A \subseteq V$. Then*
$$R_C(G, A) = p_v R_C(G - v, A - v) + (1 - p_v) R_C(G, A - v).$$

Here $A - v$ stands as an abbreviation for $A \setminus \{v\}$. Lemma 5.31 forms the basis for a recursive algorithm for the computation of the vertex cover reliability.

The next lemma follows immediately from independence of vertex operation.

Lemma 5.32 *Let G be a graph consisting of k components $G_1, ..., G_k$. Then*
$$R_C(G) = \prod_{i=1}^{k} R_C(G_i).$$

The following theorem provides an alternative to Lemma 5.31 for the computation of the vertex cover reliability.

Theorem 5.33 *Let $G = (V, E)$ be a graph and $v \in V$. Then*
$$R_C(G) = (1 - p_v) \prod_{w \in N(v)} p_w R_C(G - N[v]) + p_v R_C(G - v).$$

Proof. Each vertex of G can be in exactly one of two states: "failed" or "operating." Assume the vertex $v \in V$ fails. This event occurs with probability $1 - p_v$. In this case, in order to cover all edges incident to v, all neighboring vertices of v must be in operating state. This happens with probability $\prod_{w \in N(v)} p_w$. In case all vertices of the neighborhood $N(v)$ operate perfectly, only edges of $G - N[v]$ have to be covered. The corresponding probability is $R_C(G - N[v])$.

The vertex v operates with probability p_v and covers in this case all edges incident to v, which gives the last term of the equation. ∎

Exercise 5.22 *Let u and v be two adjacent vertices of G such that $N(u) \setminus \{v\} \subseteq N(v)$. Show that if X is a dominating set of G that contains both vertices u and v then $X \setminus \{u\}$ is a dominating set of G, too.*

Exercise 5.23 *Show that the domination polynomial of the complete bipartite graph $K_{r,s}$ is*

$$D(K_{r,s}, x) = [(1+x)^r - 1] [(1+x)^s - 1] + x^r + x^s.$$

Exercise 5.24 *Let $p_n = D(P_n, x)$ be the domination polynomial of the path P_n. Show that the polynomials p_n satisfy for $n \geq 3$ the recurrence equation*

$$p_n = x (p_{n-1} + p_{n-2} + p_{n-3})$$

with initial values

$$p_0 = 1$$
$$p_1 = x$$
$$p_2 = 2x + x^2.$$

Exercise 5.25 *For any subset $X \subseteq V$, let its closed neighborhood in G be defined by*

$$N[X] = \bigcup_{v \in X} N[v]$$

Show that the domination reliability $R_d(G, p)$ satisfies

$$R_d(G, p) = \sum_{X \subseteq V} (-1)^{|X|} q^{N[X]}.$$

Hint: *Observe that $q^{N[X]}$ is the probability that no vertex of X is dominated by any vertices of G and apply the principle of inclusion-exclusion.*

Exercise 5.26 *Let v be a vertex of degree 1 in a graph G and u its only neighbor vertex. Show that the vertex cover reliability satisfies*

$$R_C(G) = p_v R_C(G - v) + (1 - p_v) p_u R_C(G - u - v).$$

Derive a recurrence relation for the vertex cover reliability of the path P_n.

Chapter 6

Connectedness in Undirected Graphs

6.1 Reliability Polynomials

The *reliability polynomial* $R(G, p)$ of a graph $G = (V, E)$ is the the all-terminal reliability of G under the assumption that all edges fail independently with identical probability p. Hence all properties of the all-terminal reliability introduced in Section 5.1 apply directly to the reliability polynomial. From Equation (5.1), we obtain

$$R(G, p) = \sum_{F \subseteq E} \psi(G[F]) p^{|F|} (1-p)^{|E \setminus F|}. \qquad (6.1)$$

Let $G = (V, E)$ be a graph with n vertices and m edges. Since any path set of G must contain at least $n - 1$ edges and at most m edges, we obtain a polynomial of the following form

$$R(G, p) = \sum_{i=n-1}^{m} a_i p^i. \qquad (6.2)$$

If the given graph G is not connected, then there exists no path set and the reliability polynomial of G is identically zero. If G is connected then G contains at least one spanning tree, i.e. a path set of cardinality $n - 1$. The complete edge set $E(G)$ forms a path set for each connected graph G. Consequently, $R(G, p)$ is a polynomial of degree m and of low degree $n - 1$ (smallest power of p). The coefficient a_{n-1} is the number of spanning trees of G, which can be computed efficiently (in polynomial time) by the matrix tree theorem of Kirchhoff, see e.g. [64].

6.1.1   Representations of the Reliability Polynomial

A second representation of the reliability polynomial gives a direct combinatorial interpretation of all coefficients of the polynomial. Let n_i be the number of path sets (of connected spanning subgraphs) with exactly i edges. Then from (6.1), we conclude

$$R(G,p) = \sum_{i=n-1}^{m} n_i p^i (1-p)^{m-i}. \tag{6.3}$$

A similar formulation is possible in terms of the cut sets of G. Let c_i be the number of cut sets of size i of G, i.e. the number of edge cuts with exactly i edges. Then we obtain

$$R(G,p) = 1 - \sum_{i=\lambda(G)}^{m} c_i (1-p)^i p^{m-i}. \tag{6.4}$$

Remember that $\lambda(G)$ denotes the edge connectivity of G, i.e. the cardinality of a minimum edge cut of G. This representation of the reliability polynomial is an immediate conclusion from (5.5).

Let f_i be the number of edge subsets of size i such that the complement of the set induces a connected subgraph. The sequence $\{f_i\}$ yields the F-form of the reliability polynomial,

$$R(G,p) = \sum_{i=0}^{m-n+1} f_i (1-p)^i p^{m-i}. \tag{6.5}$$

Finally, if we factor out p^{n-1} and write the residual polynomial in powers of $(1-p)$, then we obtain the H-form,

$$R(G,p) = p^{n-1} \sum_{i=0}^{m-n+1} h_i (1-p)^i. \tag{6.6}$$

If we know one representation of the reliability polynomial then we obtain by an easy calculation all the other forms. Table 6.1 provides an overview of transformations between different forms of the reliability polynomial. Why do there exist so many different representations of one and the same object? The answer is that each of the above given representations has its own advantage. The C-form and N-form permit an easy combinatorial interpretation of the coefficients as number of cut sets and path sets, respectively. We will see later that the F-form and the H-form are the starting point for the construction of reliability bounds.

Some entries of Table 6.1 are easily obtained. If F is an edge set whose complement induces a connected subgraph of $G = (V, E)$ then clearly $E \setminus F$ is a path set of G, hence $f_k = n_{m-k}$. If an edge set F is not a path set of G

$$a_k = \sum_{i=n-1}^{m} \binom{m-i}{k-i} (-1)^{k-i} \left(\binom{m}{i} - c_{m-i} \right)$$

$$= \sum_{i=n-1}^{m} \binom{m-i}{k-i} (-1)^{k-i} f_{m-i}$$

$$= (-1)^{k-n+1} \sum_{i=0}^{m-n+1} \binom{i}{k-n+1} h_i = \sum_{i=n-1}^{m} \binom{m-i}{k-i} (-1)^{k-i} n_i$$

$$c_k = \binom{m}{k} - \sum_{i=n-1}^{m} \binom{m-i}{m-k-i} a_i = \binom{m}{k} - \sum_{i=0}^{k} \binom{m-n+1-i}{k-i} h_i$$

$$= \binom{m}{k} - f_k = \binom{m}{k} - n_{m-k}$$

$$f_k = \sum_{i=n-1}^{m} \binom{m-i}{m-k-i} a_i = \binom{m}{k} - c_k = \sum_{i=0}^{k} \binom{m-n-i+1}{k-i} h_i = n_{m-k}$$

$$h_k = (-1)^{k} \sum_{i=0}^{m-n+1} \binom{i}{k} a_i = \sum_{i=0}^{k} \binom{m-n+1-i}{k-i} (-1)^{k-i} \left(\binom{m}{i} - c_i \right)$$

$$= \sum_{i=0}^{k} \binom{m-n+1-i}{k-i} (-1)^{k-i} f_i = \sum_{i=0}^{k} \binom{m-n+1-i}{k-i} (-1)^{k-i} n_{m-i}$$

$$n_k = \sum_{i=n-1}^{m} \binom{m-i}{k-i} a_i = \binom{m}{k} - c_{m-k} = \sum_{i=0}^{m-k} \binom{m-n-i+1}{m-k-i} h_i = f_{m-k}$$

Table 6.1: Transformations between different representations of the reliability polynomial

then its complement is a cut set of G, giving $c_k + n_{m-k} = \binom{m}{k}$. By binomial expansion, we obtain

$$\sum_{i=n-1}^{m} n_i p^i (1-p)^{m-i} = \sum_{i=n-1}^{m} \sum_{j=0}^{m-i} n_i (-1)^j \binom{m-i}{j} p^{i+j}$$

$$= \sum_{i=n-1}^{m} \sum_{k=i}^{m} n_i (-1)^{k-i} \binom{m-i}{k-i} p^k$$

$$= \sum_{k=n-1}^{m} \sum_{i=n-1}^{k} n_i (-1)^{k-i} \binom{m-i}{k-i} p^k$$

$$= \sum_{k=n-1}^{m} a_k p^k,$$

which yields via comparison of coefficients

$$a_k = \sum_{i=n-1}^{k} (-1)^{k-i} \binom{m-i}{k-i} n_i.$$

All the other entries of Table 6.1 are obtained in a similar fashion.

6.1.2   Reliability Functions

In the following, we assume that $G = (V, E)$ is a connected undirected graph with at least one edge. Then we deduce the following properties or the reliability polynomial, which we now consider as a function of the real variable p:

- $R(G, 0) = 0$,

- $R(G, 1) = 1$,

- $p_1 < p_2 \implies R(G, p_1) < R(G, p_2)$.

If we compare all graphs with the same number of edges then we can identify special graphs with a minimum and a maximum number of path sets. A tree $T = (V, E)$ with m edges has exactly one path set, namely the edge set E itself. On the other hand, in a graph H consisting of two vertices and m parallel edges connecting these two vertices, each nonempty subset of E is a path set. Consequently, for each connected graph with m edges and for any $p \in [0, 1]$, we have

$$p^m \le R(G, p) \le 1 - (1 - p)^m. \tag{6.7}$$

The first and last term in (6.7) are the reliability polynomial of a tree and of a graph consisting of m parallel edges, respectively. If G has at least three vertices then the reliability polynomial $R(G, p)$ has degree 2 or more, which gives

$$\left. \frac{dR(G, p)}{dp} \right|_{p=0} = 0.$$

For any graph G with $\lambda(G) \ge 2$, we obtain from (6.4)

$$\left. \frac{dR(G, p)}{dp} \right|_{p=1} = 0.$$

Thus the slope of the reliability function at point 0 and point 1 vanishes for any sufficiently complex graph (meaning $n \ge 3$ and $\lambda(G) \ge 2$). Moreover, *Birnbaum*, *Esary*, and *Saunders* [52] showed that the reliability function is always an s-shaped monotonic function connecting the points $(0, 0)$ and $(1, 1)$.

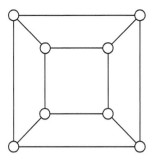

Figure 6.1: Cube graph

Example 6.1 *Figure 6.1 shows the cube graph (the skeleton of a three-dimensional cube). The reliability polynomial,*

$$R(G, p) = 384p^7 - 1512p^8 + 2420p^9 - 1962p^{10} + 804p^{11} - 133p^{12},$$

of the cube graph as a function of the edge reliability p is depicted in Figure 6.2 as a solid curve. The dotted curve shows the reliability of a graph consisting of twelve parallel edges between two vertices, whereas the dashed curve represents the reliability polynomial of a tree with 12 edges. Corresponding to Inequality (6.7), the function graph of a reliability polynomial of an arbitrarily chosen connected graph with 12 edges is always located in the region between the dashed and the dotted curve of Figure 6.2.

Is there, for a given number of vertices and edges, a graph that is uniformly most reliable, which means for all values of the edge reliability p? It has been shown [178, 95] that reliability functions can cross, hence we cannot find a uniformly most reliable graph in general.

The basic tool for the computation of the reliability polynomial is the decomposition equation,

$$R(G, p) = p\, R(G/e, p) + (1 - p)\, R(G - e, p), \qquad (6.8)$$

which follows directly from Equation (5.2). This decomposition formula is valid for each edge e of G. We can even use (6.8) in order to *define* the reliability polynomial

$$R(G, p) = \begin{cases} 0 \text{ if } G \text{ is disconnected,} \\ 1 \text{ if } G \text{ consists of a single vertex,} \\ p\, R(G/e, p) + (1 - p)\, R(G - e, p), \ e \in E \text{ else.} \end{cases}$$

It is important to note that this recursive definition of the reliability polynomial $R(G, p)$ does not require any interpretation of the variable p as being a probability. There is indeed a purely combinatorial interpretation of the reliability polynomial.

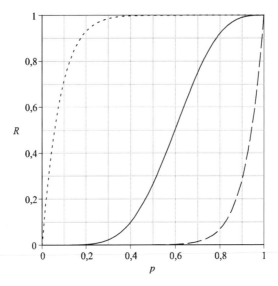

Figure 6.2: Reliability function of the cube graph

6.1.3 Inclusion–Exclusion and Domination

Let $\mathcal{T}$ be the collection of all spanning trees of $G = (V, E)$. We denote by A_T the random event that all edges of a spanning tree $T \in \mathcal{T}$ are operating. Then the reliability polynomial of G gives the probability that at least one of the events $\{A_T : T \in \mathcal{T}\}$ occurs:

$$R(G, p) = \Pr\left(\bigcup_{T \in \mathcal{T}} A_T\right)$$

Applying the principle of inclusion–exclusion, we obtain

$$R(G, p) = \sum_{k=1}^{|\mathcal{T}|} (-1)^{k+1} \sum_{\substack{\mathcal{U} \subseteq \mathcal{T} \\ |\mathcal{U}|=k}} \Pr\left(\bigcap_{T \in \mathcal{U}} A_T\right). \tag{6.9}$$

The practical application of this formula is, even for rather small networks, out of the question. The number $t(G)$ of spanning trees increases exponentially with the size of the graph G; in addition, the principle of inclusion–exclusion causes a number of terms that grows exponentially with $t(G)$.

Fortunately, the reliability polynomial does not contain so many terms as expected by the application of the principle of inclusion–exclusion, even in case we generalize the reliability polynomial to a multivariate polynomial by allowing individual reliabilities for the edges. The reason is the cancellation of many terms of different sign. *Satyanarayana* and *Prabhakar* [270] (see also [61]) found an elegant way to describe the surviving terms of the sum using the notion of domination. Each intersection of the form $\bigcap_{T \in \mathcal{U}} A_T$ corresponds to the edge set of a connected subgraph of G. If $|\mathcal{U}|$ is odd (even) then the probability of this intersection contributes with positive (negative) sign to the sum. Let H be a connected subgraph of G. We call a set $\mathcal{U}$ of spanning trees of G with $|\mathcal{U}| = k$ a *k-formation* of H if the union of all edge sets of spanning trees from $\mathcal{U}$ is $E(H)$. A k-formation is *odd* and *even* when k is odd and even, respectively. The *signed domination* $d(H)$ of H is the number of odd formations of H minus the number of even formations of H. The reliability polynomial can now be expressed as

$$R(G, p) = \sum_{H \subseteq G} d(H) p^{|E(H)|}. \tag{6.10}$$

Here $H \subseteq G$ means that H is a spanning subgraph of G. We can restrict the summation to connected spanning subgraphs of G, since the domination of any disconnected subgraph is zero.

Figure 6.3: A graph with five edges

Example 6.2 *Consider the graph pictured in Figure 6.3. This graph has 14 connected spanning subgraphs. Eight of these subgraphs are spanning trees having the domination $+1$. The spanning subgraph H with edge set $\{a, b, c, e\}$ can be presented by the following formations*

$$\{\{a, b, c\}, \{a, c, e\}\}, \{\{a, b, c\}, \{b, c, e\}\}, \{\{a, c, e\}, \{b, c, e\}\},$$
$$\{\{a, b, c\}, \{a, b, e\}, \{b, c, e\}\}.$$

We count three even and one odd formation, which results in $d(H) = -2$. By symmetry, we argue that the spanning connected subgraphs with the edge sets $\{a, b, d, e\}$, $\{a, c, d, e\}$ and $\{b, c, d, e\}$ have also domination -2. The 4-cycle $\{a, b, c, d\}$ provides 6 formations consisting of two spanning trees, four 3-formations and one 4-formation, resulting in $d(H) = -3$. Finally, we can verify (with some patience) that the source graph itself yields a domination of 4. From (6.10), we obtain

$$R(G, p) = 8p^3 - 11p^4 + 4p^5.$$

6.1.4　Combinatorial Properties of the Reliability Polynomial

Consider the ordinary generating function for the numbers n_i of connected spanning subgraphs of G with exactly i edges,

$$f(G, x) = \sum_{i=n-1}^{m} n_i x^i. \tag{6.11}$$

Example 6.3 *The cycle C_n contains one spanning connected subgraph with n edges (the cycle itself) and n connected spanning subgraphs with $n-1$ edges, namely the spanning trees. Thus we have $n_n = 1$, $n_{n-1} = n$ and $n_k = 0$ for all $k \notin \{n, n-1\}$. The generating function for the number of connected spanning subgraphs is, consequently,*

$$f(C_n, x) = x^n + n\, x^{n-1}.$$

The polynomial $f(G, x)$ satisfies a recurrence equation similar to (6.8), namely

$$f(G, x) = f(G - e, x) + x\, f(G/e, x).$$

We have also $f(G, x) = 0$ in case G is disconnected and $f(K_1, x) = 1$, $K_1 = (\{v\}, \emptyset)$. Comparing (6.3) and (6.11), we find

$$f(G, x) = (1 + x)^m\, R\left(G, \frac{x}{1 + x}\right).$$

Let $k(G)$ be the number of components and let $v(G)$ be the order of G. The *Tutte polynomial* is defined by

$$T(G; x, y) = \sum_{F \subseteq E} (x - 1)^{k(G[F]) - k(G)} (y - 1)^{|F| - v(G) + k(G[F])}. \tag{6.12}$$

The Tutte polynomial has been introduced by *William Tutte* [297]. Since then it gained a still ongoing interest and found applications in matroid theory, knot theory, and statistical physics. There exists a huge literature on the Tutte polynomial, which is of special interest for us, since we will show that the reliability polynomial is an evaluation of the Tutte polynomial. As a consequence, all results known for the Tutte polynomial can be easily applied to the reliability polynomial. If $G[F]$ is connected then $k(G[F]) = k(G) = 1$, presupposing G itself is a connected graph. Consequently, we obtain from (6.12)

$$T(G; 1, y + 1) = y^{-v(G) + 1} \sum_{\substack{F \subseteq E \\ G[F]\ \text{conn.}}} y^{|F|},$$

where the sum is taken over all connected edge-induced subgraphs of G. Hence we conclude

$$x^{v(G) - 1} T(G; 1, x + 1) = \sum_{i=n-1}^{m} n_i x^i = f(G, x)$$

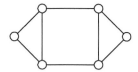

Figure 6.4: An example graph

and, for a connected graph with n vertices and m edges,

$$R\left(G,p\right) = (1-p)^{m-n+1}\,p^{n-1}T\left(G;1,\frac{1}{1-p}\right). \qquad (6.13)$$

Example 6.4 *The Tutte polynomial of the graph depicted in Figure 6.4 is*

$$T(G;x,y) = x+3x^2+4x^3+3x^4+x^5+4xy+3xy^2+5x^2y+2x^3y+y+2y^2+y^3.$$

The substitution according to (6.13) yields

$$R(G,p) = 30p^5 - 65p^6 + 48p^7 - 12p^8.$$

Assume the edge set of the graph $G = (V, E)$ is linearly ordered. Let $T = (V, F)$ be a spanning tree of G. Adding of an edge $e \in E \setminus F$ to T results in a graph T' with exactly one cycle C. We call the edge e *externally active* with respect to T if e is the smallest edge of C (with respect to the given linear order). We say T has external activity i if there are exactly i externally active edges with respect to T. Let t_i be the number of spanning trees of G with external activity i. Then as a consequence of a known theorem on Tutte polynomials, see [298], we obtain

$$R\left(G,p\right) = (1-p)^{m-n+1}\,p^{n-1}\sum_{i=0}^{n}\frac{t_i}{(1-p)^i}.$$

Exercise 6.1 *Determine the reliability polynomial of the graph depicted in Figure 6.5. This graph consists of two cycles of length k and l, such that the intersection of this cycles is a path P_j of length $j-1$ ($k \geq j$, $l \geq j$).*

Exercise 6.2 *Compute the reliability polynomial of the graph shown in Figure 6.6. How many connected spanning subgraphs does this graph contain?*

Exercise 6.3 *Show that the representations (6.2) and (6.4) of the reliability polynomial can be transformed into each other by*

$$c_k = \binom{m}{k} - \sum_{i=n-1}^{m}\binom{m-i}{m-k-i}a_i$$

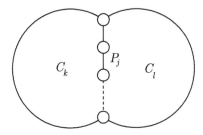

Figure 6.5: A graph composed of two cycles

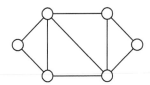

Figure 6.6: A graph with 6 vertices

and

$$a_k = \sum_{i=n-1}^{m} \binom{m-i}{k-i} (-1)^{k-i} \left(\binom{m}{i} - c_{m-i} \right).$$

Exercise 6.4 *Let $G = (V, E)$ be a graph and $e, f \in E$. Show that the two graphs $(G - e)/f$ and $(G/f) - e$ are isomorphic. Deduce that the reliability polynomial of a graph, when calculated by (6.8), is independent of the order of the edges.*

Exercise 6.5 *Show that the domination of a cycle is $d(C_n) = 1 - n$ and that the domination of a complete graph is given by $d(K_n) = (-1)^{n+1}(n-1)!$*

Exercise 6.6 *Let $G = (V, E)$ be a graph with n vertices and m edges. Show that the number of connected spanning subgraphs of G is given by $2^m R\left(G, \frac{1}{2}\right)$.*

Exercise 6.7 *Prove Equation (6.13).*

6.2 Special Graphs

6.2.1 Complete Graphs

The high symmetry of the complete graph can be used to derive a recurrence relation for its reliability polynomial. The first result in this direction was established by *Gilbert* [138].

Let $R(K_n, q)$ be the reliability polynomial of the complete graph K_n with n vertices as a function of the edge failure probability $q = 1 - p$. In order to simplify the following presentation, we use the abbreviation $r_n := R(K_n, q)$. Obviously, we have $r_1 = 1$. Let $v \in V(K_n)$ be an arbitrarily chosen vertex of the complete graph K_n. We denote by A_k $(k = 1, ..., n)$ the random event that v forms together with $k - 1$ vertices out of $V(K_n) \setminus \{v\}$ a connected component. Since these events A_k are disjoint for different k, we obtain

$$\sum_{k=1}^{n} \Pr(A_k) = 1. \tag{6.14}$$

Now we calculate the probabilities $\Pr(A_k)$. There are $\binom{n-1}{k-1}$ possibilities to select a subset $X \subseteq V(K_n) \setminus \{v\}$ of size $k-1$ that forms with v one component. The component forms a local complete graph with the vertex set $X \cup \{v\}$. Hence the probability that this component is connected is r_k. To ensure that this component has exactly size k, all edges of the K_n connecting the component on $X \cup \{v\}$ with vertices outside of this set must fail. There are $k(n-k)$ edges of this kind. Consequently, the probability of A_k is

$$\Pr(A_k) = \binom{n-1}{k-1} q^{k(n-k)} r_k.$$

Substituting the probability $\Pr(A_k)$ in (6.14) results in

$$\sum_{k=1}^{n} \binom{n-1}{k-1} q^{k(n-k)} r_k = 1, \tag{6.15}$$

which gives the recurrence equation

$$r_n = 1 - \sum_{i=k}^{n-1} \binom{n-1}{k-1} q^{k(n-k)} r_k \tag{6.16}$$

with initial value $r_1 = 1$. We obtain form $n = 1, ..., 5$:

$$r_1 = 1$$
$$r_2 = 1 - q$$
$$r_3 = 1 - 3q^2 + 2q^3$$
$$r_4 = 1 - 4q^3 - 3q^4 + 12q^5 - 6q^6$$
$$r_5 = 1 - 5q^4 - 10q^6 + 20q^7 + 30q^8 - 60q^9 + 24q^{10}$$

There is also an explicit representation for the reliability polynomial of the complete graph. In order to present this result, we need some definitions. A *partition* of a number $n \in \mathbb{Z}^+$ is a representation of n as a sum of positive integers without regarding the order of the summands. We denote partitions of n by small Greek letters. The terms (summands) of a partition λ are called the *parts* of λ. If λ is partition of n then we write $\lambda \vdash n$. A partition λ is written as a non-increasing sequence $(\lambda_1, ..., \lambda_r)$ of its parts. Let $|\lambda|$ be the number of parts of a partition λ.

Example 6.5 *The partitions of 5 are*

$$(5), \ (4,1), \ (3,2), \ (3,1,1), \ (2,2,1), \ (2,1,1,1), \ (1,1,1,1,1).$$

To each partition $\lambda \vdash n$, we assign a sequence $k_\lambda = (k_1, k_2, ..., k_n)$, where k_i denotes the number of parts of size i of λ. Consequently, we have $\sum_{i=1}^{n} i \, k_i = n$. We define a function that assigns to each partition $\lambda \vdash n$ an integer $a_2(\lambda)$ by

$$a_2(\lambda) = \frac{1}{2}\left(n^2 - \sum_{i=1}^{|\lambda|} \lambda_i^2 \right).$$

Multinomial coefficients concerning integer partitions are abbreviated in the following way:

$$\binom{n}{\lambda} = \binom{n}{\lambda_1, \, ..., \, \lambda_r} \quad \text{and} \quad \binom{|\lambda|}{k_\lambda} = \binom{|\lambda|}{k_1, \, k_2, \, ..., \, k_n}$$

Using this notation, we can give the following formula for the reliability polynomial of the complete graph:

$$r_n = \sum_{\lambda \vdash n} (-1)^{|\lambda|+1} \binom{n}{\lambda} \binom{|\lambda|}{k_\lambda} q^{a_2(\lambda)}$$

The proof of this equation can be performed by a careful expansion of the recurrence equation (6.16). Since this procedure is rather troublesome and technical, we omit the proof here.

Equation (6.16) can be employed in order to derive an exponential generating function for the sequence $\{r_n\}_{n=0,1,2,...}$ of reliability polynomials of complete graphs. We start with substituting $Q = \sqrt{q}$, resulting in

$$q^{k(n-k)} = q^{\frac{1}{2}\left(n^2 - k^2 - (n-k)^2\right)} = Q^{n^2 - k^2 - (n-k)^2}.$$

From (6.15), we obtain now

$$\sum_{k=1}^{n} \binom{n-1}{k-1} \frac{r_k}{Q^{k^2}} \frac{1}{Q^{(n-k)^2}} = \frac{1}{Q^{n^2}}$$

or

$$\sum_{k=1}^{n+1} \binom{n}{k-1} \frac{r_k}{Q^{k^2}} \frac{1}{Q^{(n+1-k)^2}} = \frac{1}{Q^{(n+1)^2}}.$$

The index transformation $j = k - 1$ yields

$$\sum_{j=0}^{n} \binom{n}{j} \frac{r_{j+1}}{Q^{(j+1)^2}} \frac{1}{Q^{(n-j)^2}} = \frac{1}{Q^{(n+1)^2}}. \tag{6.17}$$

We define $r_0 := 0$ and introduce the exponential generating functions

$$R(z) = \sum_{n \geq 0} \frac{r_n}{Q^{n^2}} \frac{z^n}{n!} \quad \text{and} \quad T(z) = \sum_{n \geq 0} \frac{1}{Q^{n^2}} \frac{z^n}{n!}.$$

We denote by $\mathbf{D}F(z)$ the formal derivative of a formal power series $F(z)$ with respect to z. Notice that the derivative of $F(z) = \sum_{n \geq 0} f_n \frac{z^n}{n!}$ is $\mathbf{D}F(z) = \sum_{n \geq 0} f_{n+1} \frac{z^n}{n!}$. The product of two exponential generating functions $F(z) = \sum_{n \geq 0} f_n \frac{z^n}{n!}$ and $G(z) = \sum_{n \geq 0} g_n \frac{z^n}{n!}$ is $FG(z) = \sum_{n \geq 0} \sum_{k=0}^{n} \binom{n}{k} f_k g_{n-k} \frac{z^n}{n!}$. These properties together with (6.17) yield

$$\mathbf{D}R(z) \cdot T(z) = \mathbf{D}T(z).$$

and hence

$$\frac{\mathbf{D}T(z)}{T(z)} = \mathbf{D}\ln T(z) = \mathbf{D}R(z).$$

We obtain by integration

$$R(z) = \ln T(z).$$

The integration constant vanishes as $R(0) = 0$ and $T(0) = 1$.

6.2.2 Recurrent Structures

Recurrent structures are graphs that have "regular shape" consisting of a "repeating pattern." To make this vague definition of a recurrent structure more precise requires an algebraic and more difficult description of graphs. We will restrict our attention to examples that can perhaps explain this notion best. Consider the open wheel, presented in Figure 6.7. This graph can be constructed recursively, beginning with the vertices 0 and 1 and the edge linking these two vertices; call this graph W_1. Assume the graph W_k is already generated. We obtain W_{k+1} from W_k by introducing a new vertex $k + 1$ and two edges, namely $\{0, k + 1\}$ and $\{k, k + 1\}$. The ladder graph depicted in Figure 6.8 can be obtained recursively, starting from a single edge, by repeated attachment of a u-shaped subgraph.

In order to determine the reliability polynomial of the ladder graph, we use the decomposition formula (6.8). In the upper part of Figure 6.9 the ladder graph is presented with the last three edges named e, f, g. In order to ensure connectedness of the graph, at least two of these three edges must operate. First assume there are exactly two of the edges e, f, g operating. According to the decomposition equation, we contract the two operating edges and delete the failed edge. In each case we obtain a ladder L_{n-1} with one stave less.

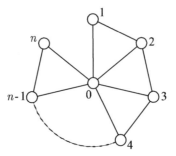

Figure 6.7: The open wheel

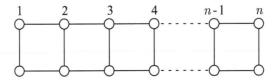

Figure 6.8: The ladder graph

The probability of this event is $3p^2 - 3p^3$. In case all three edges operate, which happens with probability p^3, we obtain a degenerated ladder that has instead of the last stave only a single vertex. We call this graph H_n. The decomposition of H_n with respect to the last two edges f and g, shown in Figure 6.9 (lower part), results again in one of our two graphs L_n and H_n, however, with different probabilities $2p - 2p^2$ and p^2, respectively. Combining these results, we obtain the system of recurrence equations:

$$R\left(L_n, p\right) = \left(3p^2 - 3p^3\right) R\left(L_{n-1}, p\right) + p^3 R\left(H_{n-1}, p\right)$$
$$R\left(H_n, p\right) = \left(2p - 2p^2\right) R\left(L_{n-1}, p\right) + p^2 R\left(H_{n-1}, p\right)$$
$$R\left(L_1, p\right) = p$$
$$R\left(H_1, p\right) = 1$$

With the friendly help of a computer algebra system, we obtain the explicit solution

$$R\left(L_n, p\right) = \frac{p^{2n-1}}{2^n \alpha} \left[(4 - 3p + \alpha)^n - (4 - 3p - \alpha)^n\right]$$
$$\text{with } \alpha = \sqrt{12 - 20p + 9p^2}.$$

Exercise 6.8 *Determine the reliability polynomial of the open wheel presented in Figure 6.7.*

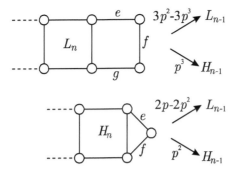

Figure 6.9: Decomposition of the ladder graph

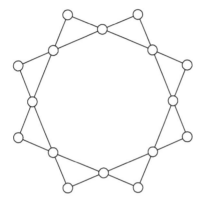

Figure 6.10: A symmetric graph

Exercise 6.9 *Show that the complete bipartite graph $K_{2,t}$ satisfies*

$$R(K_{2,t}, p) = p^t \left((2 - p)^t - (2 - 2p)^t \right).$$

Exercise 6.10 *Show that the reliability polynomial of the graph depicted in Figure 6.10 is*

$$p^{15} (3 - 2p)^7 (16 - 29p + 14p^2).$$

Exercise 6.11 *Let $R(G, p)$ be the reliability polynomial of a given graph G. We construct a new graph H by replacing each edge of G by a path of length k $(k > 0)$. Prove that the reliability polynomial of H is given by*

$$R(H, p) = p^{k-1} (p + k(1 - p)) R\left(G, \frac{p}{p + k(1 - p)} \right).$$

Exercise 6.12 *Determine the reliability polynomial of the (closed) wheel that arises from the open wheel according to Figure 6.7 by insertion of the edge $\{1, n\}$.*

6.3   Reductions for the K-Terminal Reliability

In this section, we consider methods for the computation of the K-terminal reliability $R(G, K)$ of an undirected graph $G = (V, E)$. The K-terminal reliability $R(G, K)$ was introduced in the last chapter. It is the probability that a specified vertex subset (the terminal vertices of G) belongs to one component of G, i.e. that all vertices of K are connected by paths of operating edges. In case all edges of the graph $G = (V, E)$ operate with identical probability p, the K-terminal reliability $R(G, K) = R(G, K, p)$ is a polynomial that coincides with the reliability polynomial when $K = V$. Let $n_i(G, K)$ be the number of subgraphs of G with exactly i edges that contain a Steiner tree with respect to K. We use the representation

$$R(G, K, p) = \sum_{i=0}^{m} n_i(G, K) p^i (1 - p)^{m-i}$$

of the K-terminal reliability polynomial. We denote the number of terminal vertices by k. Since any Steiner tree of G with respect to K contains at least $k - 1$ edges, we conclude that $n_i(G, K) = 0$ for $i \leq k - 2$. However, in general the computation of the numbers n_i is a hard task.

Some special graphs offer combinatorial methods for the computation of the K-terminal reliability. Let us consider a cycle C_n with k terminal vertices. We start from an arbitrarily chosen terminal vertex $w_1 \in K$ and traverse the cycle. We denote the terminal vertices of the cycle by $w_1, w_2, ..., w_k$ where the indices correspond to the order of traversal. Let l_i and l_k denote the number of edges traversed between the terminal vertex w_i and w_{i+1} for $i = 1, ..., k-1$ and between w_k and w_1, respectively. Thus these numbers l_i give the "gap lengths" between successively along the cycle traversed terminal vertices. Then any edge set containing a Steiner tree can have missing edges from at most one gap. Consequently, we obtain

$$R(C_n, K, p) = \sum_{i=1}^{k} (1 - p^{l_i}) p^{n - l_i} + p^n.$$

The term $(1 - p^{l_i}) p^{n - l_i}$ gives the probability that at least one edge in gap i fails whereas all other edges operate. The last term (p^n) corresponds to the event that all edges of the cycle are in operating state.

A *reduction* is a simplification of the graph G resulting in a new graph G'. Usually a reduction of G is performed by replacing a (small) subgraph H of G by a simpler graph H'. The reduction involves also the assignment of probabilities to the edges and/or vertices of H' in such a way that the K-terminal reliability of G can be obtained by computing the K-terminal reliability of G'. In the following, we assume that the graphs we are considering do not contain any *irrelevant edges*, which means edges that do not belong to any Steiner tree with respect to the terminal vertex set K. Note that irrelevant

Figure 6.11: Parallel reduction

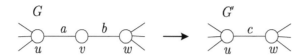

Figure 6.12: Series reduction

edges may arise during the application of the decomposition formula (5.7) even in case the input graph is free of irrelevant edges. However, irrelevant edges are easy to detect (and to remove). Biconnected graphs with at least two terminal vertices do not contain any irrelevant edges.

6.3.1 Simple Reductions

In order to reduce the computational effort when applying the decomposition formula, we can perform *simple reductions*. We call a reduction *reliability-preserving* if the *K*-terminal reliability (or other reliability measures of interest) coincides for the original graph G and the reduced graph G'. The first reduction of this kind is the *parallel reduction*, which replaces two edges $a = \{u, v\}$ and $b = \{u, v\}$ linking the same pair of vertices by one single edge $c = \{u, v\}$ such that $p_c = p_a + p_b - p_a p_b$. Figure 6.11 illustrates the parallel reduction. The *series reduction* replaces two edges $a = \{u, v\}$, $b = \{v, w\}$ and a non-terminal vertex v of degree 2 by one edge $c = \{u, w\}$ with $p_c = p_a p_b$. Figure 6.12 shows the principle of the series reduction.

Example 6.6 *Figure 6.13 presents a graph G with two (blackened) terminal vertices. If we decompose the graph according to Equation (5.7) with respect*

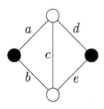

Figure 6.13: Graph with terminal vertices (blackened)

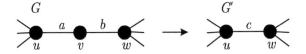

Figure 6.14: Degree-2-reduction

*to edge c, and apply series and parallel reductions to the graphs G/c and $G - c$
then we obtain*

$$R(G, K) = p_c (p_a + p_b - p_a p_b) (p_d + p_e - p_d p_e)$$
$$+ (1 - p_c) (p_a p_d + p_b p_e - p_a p_b p_d p_e).$$

The expansion of the last expression yields

$$R(G, K) = p_a p_d + p_b p_e + p_a p_c p_e + p_b p_c p_d$$
$$- p_a p_b p_c p_d - p_a p_b p_c p_e - p_a p_b p_d p_e - p_a p_c p_d p_e - p_b p_c p_d p_e$$
$$+ 2 p_e p_a p_b p_c p_d.$$

*The first four terms of the resulting polynomial correspond to the four paths
connecting the two terminal vertices.*

Observe that the series reduction can be performed only in case the vertex
of degree 2 that is to be replaced is a non-terminal vertex. There is another
reduction, called *degree-2-reduction*, that can be applied for terminal vertices
of degree two. This reduction, however, is no longer reliability-preserving. In
general we are satisfied with a reduction that produces a new graph G' with
a new terminal vertex set K' out of a given graph G such that $R(G, K) = f(R(G', K'))$, presupposing we know (or can compute) the function $f : [0, 1] \to [0, 1]$. In case of the degree-2-reduction, the function f is linear. We set

$$R(G, K) = h R(G', K') \qquad (6.18)$$

as a *reduction approach*, where G' is obtained from G by replacing a vertex v
of degree 2 and its incident edges by a single edge as shown in Figure 6.14. We
require for this reduction that also the two neighbor vertices of the degree-2-
vertex v are terminal vertices, i.e. $u, v, w \in K$. The new terminal vertex set in
G' is $K' = K \setminus \{v\}$. Let $G_{u/w}$ be the graph obtained from G by the removal
of v, i.e. $G_{u/v} = G - v$. We denote by G_{uw} the graph $(G/a)/b$ obtained
from G by contraction of a and b. Observe that the order of contraction does
not change the result. The two graphs $G_{u/w}$ and G_{uw} can be obtained from
G', too; we have $G_{u/w} = G' - c$ and $G_{uw} = G'/c$. The application of the
decomposition formula (5.7) to G with respect to the edges a and b results in

$$R(G, K) = (p_a(1 - p_b) + (1 - p_a)p_b) R(G_{u/w}, K') + p_a p_b R(G_{uw}, K'). \quad (6.19)$$

The decomposition of G' with respect to edge c gives

$$R(G', K') = (1 - p_c)R(G_{u/w}, K') + p_c R(G_{uw}, K'). \qquad (6.20)$$

The substitution of the Equations (6.19) and (6.20) in (6.18) gives

$$(p_a + p_b - 2p_a p_b) R(G_{u/w}, K') + p_a p_b R(G_{uw}, K')$$
$$= h(1 - p_c)R(G_{u/w}, K') + hp_c R(G_{uw}, K').$$

This equation has to be valid for an arbitrary graph G and hence for any $G_{u/w}$ and G_{uw}. Consequently, the coefficients of $R(G_{u/w}, K')$ and $R(G_{uw}, K')$ on both sides of the equation must agree, which yields

$$p_a + p_b - 2p_a p_b = h(1 - p_c),$$
$$p_a p_b = hp_c.$$

Solving for h and p_c, we obtain

$$h = p_a + p_b - p_a p_b,$$
$$p_c = \frac{p_a p_b}{p_a + p_b - p_a p_b}.$$

Using the same reduction approach (6.18), we can establish the *bridge reduction*. Let $e = \{u, v\}$ be a bridge of G, i.e. an edge whose removal disconnects G. We denote the vertex of G/e that results from merging u and v by uv. Then we obtain by decomposition $R(G, K) = p_e R(G/e, K')$ with

$$K' = \begin{cases} K \text{ if } \{u, v\} \cap K = \emptyset, \\ (K \smallsetminus \{u, v\}) \cup \{uv\} \text{ if } u \in K \text{ or } v \in K. \end{cases}$$

Consequently, the contraction of a bridge e in G corresponds to a reduction according to (6.18) with $h = p_e$ and $G' = G/e$.

6.3.2 Polygon-to-Chain Reductions

Let us consider again the graph depicted in Figure 6.13. This graph does not permit any simple reductions introduced so far. The reason is not its structure, but the distribution of its terminal vertices. The exchange of terminal and non-terminal vertices would result in a reducible graph. Let $G = (V, E)$ be a graph that contains a series structure as depicted in Figure 6.12. The vertex v of degree 2 has the two neighbor vertices u and w. There are eight possibilities for the three vertices u, v, w with respect to membership to the terminal vertices. Only five of them yield reducible structures, namely $\{u, v, w\} \subseteq K$ or $v \notin K$. If the degree-2-vertex is a terminal vertex, but at least one of its neighbors is not, then no simple reduction can be applied. To overcome this obstacle, *Wood* [309, 310] introduced so-called *polygon-to-chain reductions*. A *polygon* in a graph G is a subgraph of G consisting of two paths with common end vertices such that all inner vertices of the paths have degree two in G.

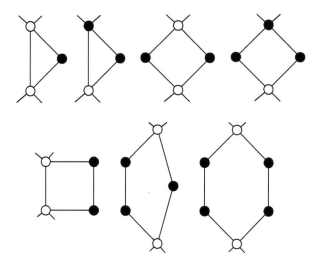

Figure 6.15: Seven different typs of a polygon

Definition 6.7 *Let $G = (V, E)$ be a connected (deterministic) graph. A* series replacement *in G is the replacement of a vertex of degree 2 and its two incident edges by a single edge as shown in Figure 6.12. A* parallel replacement *is the substitution of two parallel edges by a single one as shown in Figure 6.11. The graph G is called a* series-parallel graph *if G can be transformed into a single vertex by a sequence of series replacements, parallel replacements, and bridge contractions.*

Remark 6.8 *There are alternative definitions of a series-parallel graph, not all of these coinciding with the definition given here.*

In the following, we subsume series reductions, parallel reductions, degree-2-reductions and bridge reductions by the term *simple reductions*.

Theorem 6.9 (Wood, 1982) *Let G be a series-parallel graph with at least four vertices that does not permit any simple reductions with respect to a given terminal vertex set K. Then G contains a polygon that corresponds to one of the seven types illustrated in Figure 6.15.*

Proof. The graph G does not permit any simple reductions. Hence G has no parallel edges and no bridges. Consequently, G possesses a vertex v of degree 2 for G being a series-parallel graph. The vertex v belongs to the terminal vertex set K. Otherwise G would be reducible by a series reduction. Let u and w be the two neighbor vertices of v in G. At least one of these two vertices is a non-terminal vertex, since no degree-2-reduction is possible in G. If u and w are adjacent in G then we obtain one of the first two polygons (upper left corner of Figure 6.15) and the theorem is true.

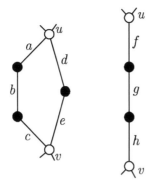

Figure 6.16: An example of a polygon-to-chain reduction

Now assume that u and w are non-adjacent in G. At most one of the vertices u and w is a vertex of degree two, since otherwise a series reduction or a degree-2-reduction would result. Hence we obtain a path with one or two inner terminal vertices of degree 2. If the path has two inner vertices and the end vertices u and x are adjacent, then we obtain the fifth type of polygon (second row left in Figure 6.15). Now consider the case that u and x are not adjacent in G. The end vertices of the path, say u and x, are vertices of degree at least 3. Observe that in case the vertices u and x coincide, we obtain again one of the first two polygon types. Hence we assume that u and x are different vertices. The series replacement of the inner vertices introduces an edge between u and x. However, in order to reduce the degrees of u and x there is a parallel reduction necessary, which can emerge only in case there is a second path connecting u and x. For this second path the same arguments apply. This path can have at least two inner vertices of degree 2 that are necessarily terminal vertices. Both paths together form one polygon of our list.

■

Theorem 6.9 is of great value because all seven polygons shown in Figure 6.15 can be reduced to a path, which is called a *chain* in [272] and [309], hence the name polygon-to-chain reduction. Figure 6.16 shows a polygon (left) and a chain (right). We use this example in order to explain the principle of the polygon-to-chain reduction. We start again with the reduction approach $R(G, K) = hR(G', K')$, where the graph G' is obtained from G by substitution of the polygon containing the five edges a, b, c, d, e by the chain with three edges f, g, h. The polygon as well as the chain have only the vertices u and v in common with the rest of the graph. First we compute the K-terminal reliability of the graph G using decomposition with respect to the states of the five edges of the polygon. We neglect all states that isolate at least one of the three terminal vertices of the polygon, since these states do not contribute to the K-connectedness. Assume only the two edges a and d of the polygon fail. Then all terminal vertices of the polygon are connected to vertex v. We denote the graph obtained from G by removal of the edges a and b and contraction of c, d, and e by $G_{u/\underline{v}}$. This notation symbolizes the fact that u and v are separated within the polygon and that only the vertex v is connected to terminal vertices, which means we add v to the set of terminal vertices of this graph. We define the graphs $G_{\underline{u}/v}$, $G_{\underline{u}/\underline{v}}$, and G_{uv} in an analogous way, where the last one denotes the graph obtained from G by contraction of all edges of the polygon. In this case, the vertices u and v are merged into a single vertex denoted by uv. We define the following probabilities

$$\alpha = (1 - p_a)p_b p_c (1 - p_d)p_e$$
$$\beta = p_a p_b (1 - p_c)p_d (1 - p_e)$$
$$\gamma = (1 - p_a)p_b p_c p_d (1 - p_e) + p_a p_b (1 - p_c)(1 - p_d)p_e$$
$$\quad + p_a (1 - p_b)p_c (p_d + p_e - 2p_d p_e)$$
$$\delta = p_a p_b p_c (p_d + p_e - p_d p_e) + (p_a p_b + p_a p_c + p_b p_c - 2p_a p_b p_c)p_d p_e.$$

The graph $G_{u/\underline{v}}$ is induced with probability α. More general, we obtain

$$R(G, K) = \alpha R(G_{u/\underline{v}}, K \cup \{v\}) + \beta R(G_{\underline{u}/v}, K \cup \{u\})$$
$$\quad + \gamma R(G_{\underline{u}/\underline{v}}, K \cup \{u, v\}) + \delta R(G_{uv}, K \cup \{uv\}). \qquad (6.21)$$

In a similar vein, we obtain for the reduced graph

$$(G', K') = (1 - p_f)p_g p_h R(G_{u/\underline{v}}, K \cup \{v\}) + p_f p_g (1 - p_h)R(G_{\underline{u}/v}, K \cup \{u\})$$
$$\quad + p_f (1 - p_g)p_h R(G_{\underline{u}/\underline{v}}, K \cup \{u, v\}) + p_f p_g p_h R(G_{uv}, K \cup \{uv\}). \qquad (6.22)$$

Substituting $R(G, K)$ and $R(G\prime, K\prime)$ in the reduction approach (6.18) by (6.21) and (6.22), respectively, yields via comparison of coefficients the following system of equations:

$$\alpha = h(1 - p_f)p_g p_h$$
$$\beta = hp_f p_g (1 - p_h)$$
$$\gamma = hp_f (1 - p_g)p_h$$
$$\delta = hp_f p_g p_h$$

Solving this equations yields finally

$$p_f = \frac{\delta}{\alpha + \delta}$$
$$p_g = \frac{\delta}{\gamma + \delta}$$
$$p_h = \frac{\delta}{\beta + \delta}$$
$$h = \frac{(\alpha + \delta)(\beta + \delta)(\gamma + \delta)}{\delta^2}.$$

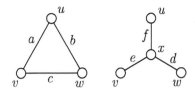

Figure 6.17: Delta-star transformation

Wood presents in [310] reduction formulae for all types of polygons. A closer look at the polygon-to-chain reductions shows that the polygon structure itself is not essential for this kind of reduction approach. A similar reduction is possible as long as we can define the four probabilities $\alpha, \beta, \gamma, \delta$ for the realization of the corresponding graphs $G_{u/\underline{v}}, G_{\underline{u}/v}, G_{\underline{u}/\underline{v}}, G_{\underline{uv}}$. Consequently, a sufficient condition is the presence of a subgraph H in G that has exactly two vertices, say u and v, in common with the rest of G. Then α can be characterized as the probability that all terminal vertices of H are connected to v but none to u. For the details of this more general reduction technique, the reader is referred to [311].

6.3.3 Delta–Star and Star–Delta Transformations

A *delta* in a graph $G = (V, E)$ is nothing but a cycle of length three. A *star* means in the context of this reduction a vertex of degree three with three different neighbor vertices. The transformation between delta and star substructures in graphs has long been known in electric network analysis. Even in network reliability analysis, the delta–star and star–delta transformations have a long history in which a lot of papers, for instance [198, 260, 131, 254, 295, 122, 86] emerged.

Figure 6.17 shows a delta (a triangle) with three edges a, b, c and a star with the edges d, e, f. The vertex x of the star is assumed to be a nonterminal vertex. We consider first the *delta–star transformation* (or short ΔY-transformation), which replaces a delta by a star. We speak of *transformation* rather than *reduction* because the graph is not reduced in size by this replacement. In fact, we have the same number of edges and one vertex more after this transformation. We assume that in the given graph G the delta has exactly the three vertices u, v, w in common with the rest of the graph. There naturally arise some questions concerning the delta–star transformation:

- Why should we perform the transformation if it does not simplify the graph?

- Can the transformation be performed in such a way that the K-terminal reliability remains unchanged?

The delta–star transformation itself does not simplify the graph. However, the structure simplifies when we use a triangle where two or three of the corner vertices have degree three. Then those vertices become degree-2-vertices in the transformed graph and may be subsequently removed by simple reductions. In order to answer the second question, we establish a system of equations that ensures a reliability preserving reduction (transformation). Let us first consider the delta. Its three edges can take altogether eight different states. If two or three edges are operating, which happens with probability

$$p_{uvw}^{\triangle} = p_a p_b + p_a p_c + p_b p_c - 2 p_a p_b p_c,$$

then the contraction of operating edges leads to the merging of the three vertices u,v and w. We denote the arising graph by G_{uvw}. If only edge a is operating then u and v are merged, resulting in the graph $G_{uv/w}$ and so on. The lower index here denotes a partition of the vertex set $\{u, v, w\}$. The states of the edges of the delta or star can induce one of the partitions of the set

$$S = \{uvw, uv/w, uw/v, u/vw, u/v/w\}.$$

The corresponding probabilities for these events are

$$
\begin{aligned}
p_{uvw}^{\triangle} &= p_a p_b + p_a p_c + p_b p_c - 2 p_a p_b p_c \\
p_{uv/w}^{\triangle} &= p_a (1 - p_b)(1 - p_c) \\
p_{uw/v}^{\triangle} &= (1 - p_a) p_b (1 - p_c) \\
p_{u/vw}^{\triangle} &= (1 - p_a)(1 - p_b) p_c \\
p_{u/v/w}^{\triangle} &= (1 - p_a)(1 - p_b)(1 - p_c).
\end{aligned}
\tag{6.23}
$$

However, these five events are not independent, so we obtain

$$\sum_{s \in S} p_s^{\triangle} = 1. \tag{6.24}$$

In order to find a reliability preserving transformation to the star depicted in Figure 6.17, four of these probabilities have to be equated with the corresponding probabilities resulting from the states of the edges d, e and f. The validity of the fifth equation follows then from (6.24). In order to satisfy the four equations, we need a fourth parameter in addition to the edge reliabilities p_d, p_e and p_f of the star. This parameter is a vertex weight p_x for the central vertex x of the star. Using the usual decomposition, we obtain

$$
\begin{aligned}
p_{uvw}^{Y} &= p_d p_e p_f p_x \\
p_{uv/w}^{Y} &= (1 - p_d) p_e p_f p_x \\
p_{uw/v}^{Y} &= p_d (1 - p_e) p_f p_x \\
p_{u/vw}^{Y} &= p_d p_e (1 - p_f) p_x \\
p_{u/v/w}^{Y} &= 1 - p_x (p_d p_e + p_d p_f + p_e p_f - 2 p_d p_e p_f).
\end{aligned}
\tag{6.25}
$$

Substituting the corresponding probabilities in $p_s^\Delta = p_s^Y$ for $s \in S$ and solving for the unknown parameters of the star, we obtain

$$p_d = \frac{p_{uvw}^\Delta}{1 - p_{uv/w}^\Delta},$$

$$p_e = \frac{p_{uvw}^\Delta}{1 - p_{uw/v}^\Delta},$$

$$p_f = \frac{p_{uvw}^\Delta}{1 - p_{u/vw}^\Delta},$$

$$p_x = \frac{\left(1 - p_{uv/w}^\Delta\right)\left(1 - p_{uw/v}^\Delta\right)\left(1 - p_{u/vw}^\Delta\right)}{\left(p_{uvw}^\Delta\right)^2}.$$

The inverse transformation from star to delta leads to a system of equations that has in general no solution. The reason is that the delta offers only three parameters, which is insufficient in order to satisfy four independent equations.

The application of the delta–star transformation introduces vertices with an additional parameter, which complicates the further process of reliability calculation. The inverse transformation, from star into delta, can be performed as an approximation only. The reason why we are nevertheless interested in these two transformations is the following theorem by *Epifanov* [114].

Theorem 6.10 (Epifanov, 1966) *A connected planar graph with exactly two terminals can be reduced to a single edge by simple reductions, delta–star and star–delta transformations.*

Translated into the language of network reliability, this theorem states that the two-terminal reliability of any planar graph can be calculated by application of the above introduced reductions and transformations. However, since the star–delta transformation is only approximative, the result will be an approximation, too. In order to improve the quality of approximation, *Chari, Feo* and *Provan* [86] developed methods to ensure minimal error bounds for delta–star and star–delta transformations. The application of these transformations requires sophisticated algorithms in order to find a suitable sequence of transformations. In some cases, a generalization to reliability problems with more than two terminal vertices or even to non-planar graphs is possible [7].

Exercise 6.13 *Compute the K-terminal reliability of the graph in Example 6.6 again, this time by reduction techniques only.*

Exercise 6.14 *Derive the formula for the polygon-to-chain reduction presented in Figure 6.18.*

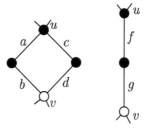

Figure 6.18: Another polygon-to-chain reduction

Figure 6.19: A special polygon

Exercise 6.15 *Show that the polygon shown in Figure 6.19 can be reduced to a chain with three edges. Performing this reduction we have to distinguish whether G has exactly two terminal vertices or more than two. Why?*

6.4   Inequalities and Reliability Bounds

In this section we consider lower and upper bounds for the all-terminal reliability of networks. The methods presented here can be easily generalized to the K-terminal reliability. Let $G = (V, E)$ be a connected graph with n vertices and m edges. The simplest bounds are obtained by comparing the all-terminal reliability of the given graph G with a series system (a tree with m edges and $n+1$ vertices) and a parallel system (denoted by $P_{2,m}$) consisting of two vertices that are linked by m parallel edges:

$$\prod_{e \in E} p_e \le R(G) \le 1 - \prod_{e \in E}(1 - p_e)$$

In a tree T, we have exactly one path set, namely the edge set $E(T)$ itself. On the other hand, any non-empty subset of $E(P_{2,m})$ is a path set. We can easily improve the lower bound by observing that the all-terminal reliability of G is at least as large as the all-terminal reliability of any spanning tree of G. Let $T(G)$ be the set of all spanning trees of G. Then we obtain

$$\max\{R(T) \mid T \in T(G)\} \le R(G). \tag{6.26}$$

Since

$$R(T) = \prod_{e \in E(T)} p_e = \exp\left(-\sum \ln\frac{1}{p_e}\right),$$

a minimum spanning tree of G with respect to the edge weights $\ln\frac{1}{p_e}$ maximizes $R(T)$ in (6.26). Such a minimum spanning tree can be found efficiently by a greedy algorithm [247]. If we consider the set of all cut sets of G, denoted by $\mathcal{C}(G)$, then we obtain, in the same vein, an upper bound:

$$R(G) \leq 1 - \max\left\{\prod_{e \in C}(1 - p_e) \mid C \in \mathcal{C}(G)\right\} \tag{6.27}$$

Consequently, a cut C for which $\prod_{e \in C}(1 - p_e)$ is maximal yields the best upper bound of this form. Such a cut may be found by max-flow algorithms.

Both bounds, (6.26) and (6.27), can be improved by involving more than one path or cut set. Let $H_1 = (V, E_1), ..., H_k = (V, E_k)$ be a collection of connected subgraphs of $G = (V, E)$ such that for $i \neq j$ always $E_i \cap E_j = \emptyset$ follows. Let $F \subseteq E$ be an edge set such that (V, F) is disconnected. Then clearly all the graphs $(V, E_i \cap F)$, $i = 1, ..., k$, are disconnected, too. This gives

$$1 - R(G) \leq \prod_{i=1}^{k}(1 - R(H_i)),$$

which results in the lower bound

$$1 - \prod_{i=1}^{k}(1 - R(H_i)) \leq R(G). \tag{6.28}$$

Bounds of this kind are called *edge-packing bounds* [93]; they were introduced by *Polesskii* in [250]. The lower bound (6.28) will be improved when we use many graphs H_i of high all-terminal reliability. In order to make the computation of the bounds easy, the chosen subgraphs H_i should be of simple structure, for instance, trees.

Again, an upper bound can be obtained using edge-disjoint cuts in a similar way. Let $C_1, ..., C_k$ be a set of edge-disjoint cuts of G. Let $F \subseteq E$ be an edge subset such that (V, F) is connected. Then F contains at least one edge from each cut C_i, $i = 1, ..., k$. Consequently,

$$R(G) \leq \prod_{i=1}^{k}\left(1 - \prod_{e \in C_i}(1 - p_i)\right). \tag{6.29}$$

Lomonosov and *Polesskii* [209] proved that the upper bound (6.29) applies also to certain systems of non-disjoint cut sets — so-called noncrossing cuts. A cut of a graph can be defined by a vertex partition (X, Y) with $Y = V \setminus X$. Two

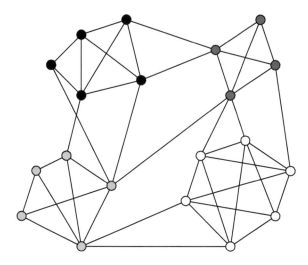

Figure 6.20: A graph with dense subgraphs

cuts (X_1, Y_1) and (X_2, Y_2) are called *noncrossing* if at least one of the four sets $X_1 \cap X_2$, $X_1 \cap Y_2$, $Y_1 \cap X_2$, and $Y_1 \cap Y_2$ is empty. A system of $n-1$ noncrossing cuts (more precisely a cut basis of the graph) can be found efficiently by an algorithm of *Gomory* and *Hu* [141] (see also [64]). The interested reader may find further results concerning edge-packing bounds in the book by *Colbourn* [92].

Another class of bounds is obtained by means of elementary edge and vertex operations. Applying the monotonicity of the all-terminal reliability, $R(G)$ increases (decreases) if we raise (lower) the reliability p_e of an edge e of G. In case p_e reaches zero, we obtain $R(G) \geq R(G-e)$. When we set $p_e = 1$, we obtain $R(G) \leq R(G/e)$. Thus edge deletion gives lower bounds, edge contraction leads to upper bounds. We consider edge contraction first. More generally, we may contract an arbitrary selected subgraph H of G, which means that we merge (identify) all vertices of H in G. The resulting graph is denoted by G/H. For any subgraph H of G, we conclude $R(G) \leq R(G/H)$. Clearly, the accuracy of this upper bound depends highly on the chosen subgraph. In practical application we deal often with high reliable edges (say $p_e > 0.99$). In this case a dense subgraph H (for instance a clique) has an all-terminal reliability very close to 1, i.e. the difference between $R(G)$ and $R(G/H)$ becomes small. Consider, as an example, the graph depicted in Figure 6.20. We discover without much effort four dense subgraphs within this graph. The vertices of each subgraph are presented in the same gray level.

Figure 6.21 shows the graph $G' = G/H_1/H_2/H_3/H_4$ obtained from the graph G presented in Figure 6.20 by contraction of the four dense subgraphs $H_1,...,H_4$. The all-terminal reliability satisfies $R(G) \leq R(G')$. In a similar

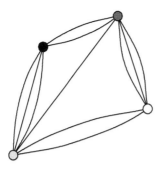

Figure 6.21: Graph with contracted subgraphs

way, we obtain the lower bound

$$R(G) \geq \prod_{i=1}^{4} R(H_i) \cdot R(G').$$

In case we are able to list all path sets or all cut sets of the network, we can apply (in connection with the principle of inclusion–exclusion) classical Bonferroni inequalities or modern improvements [105, 106]. Unfortunately, the number of paths and cuts increases exponentially with the size of the network, which restricts the applicability of inclusion–exclusion techniques to rather small graphs.

Much work has been done to establish bounds for the reliability polynomial, i.e. the case where all edge reliabilities are identical. A lot of beautiful theory from order theory, matroids, extremal sets, and discrete topology has found application in this field. Consider the form (6.3) of the reliability polynomial, i.e.

$$R(G, p) = \sum_{i=n-1}^{m} n_i p^i (1 - p)^{m-i}.$$

The coefficient n_i counts the path sets of size i of G. The first coefficient different from zero is the number of spanning trees, $n_{n-1} = t(G)$, which can be easily obtained via Kirchhoff's determinant formula [181]. The set of all path sets forms an upper semi-lattice of the Boolean lattice, i.e. a set system $\mathcal{F}$ with the property that $X \in \mathcal{F}$ and $X \subseteq Y$ implies $Y \in \mathcal{F}$. We know already that there are $t(G)$ subsets at level (size) $n-1$ within this set system and that all sets are subsets of a set E of cardinality m. Can we estimate the number of sets at level n, $n+1$, and so on from this data? Let $\mathcal{X}$ be a collection of subsets of E of size k. We define the following families of subsets of E with respect to $\mathcal{X}$:

$$\Delta \mathcal{X} = \{A \mid |A| = k - 1 \text{ and } \exists B \in \mathcal{X} : A \subseteq B\},$$
$$\nabla \mathcal{X} = \{A \mid |A| = k + 1 \text{ and } \exists B \in \mathcal{X} : B \subseteq A\}.$$

We call $\Delta \mathcal{X}$ the *shadow* and $\nabla \mathcal{X}$ the *shade* of $\mathcal{X}$. The following result due to *Sperner* [282] gives bounds for the size of shade and shadow.

Theorem 6.11 ([282]) *Let $\mathcal{X}$ be a family of k-subsets of an m-set with $k \in \{1, ..., m-1\}$. Then*

$$|\nabla \mathcal{X}| \geq \frac{m-k}{k+1}|\mathcal{X}| \;\; and \;\; |\Delta \mathcal{X}| \geq \frac{k}{m-k+1}|\mathcal{X}|.$$

For a proof of this statement, see [4].

This theorem yields for the coefficients of the reliability polynomial

$$n_{i+1} \geq \frac{m-i}{i+1}n_i. \tag{6.30}$$

We can state equivalent inequalities for the coefficients of the F-form (6.5) and the C-form (6.4) of the reliability polynomial. There are substantial improvements of this bound possible [301] by using a result of *Kruskal* [188]. However, all these bounds show an error that increases with the difference of the levels.

The set systems defining the F-form and the H-form of the reliability polynomial offer even more combinatorial structure, namely a *matroid* structure (an independence structure with exchange property) and a *shellable complex* (an interval-decomposable set system), respectively. For details and applications to reliability bounds, the reader is referred to the excellent book by *Colbourn* [92] on this subject.

Further reliability bounds are investigated in later chapters of this book in connection with the splitting of graphs along vertex separators.

Exercise 6.16 *Find lower bounds corresponding to (6.28) for $R(K_5)$ by decomposing the complete graph K_5 into two edge-disjoint cycles. What is the maximum error with respect to a common edge availability $p \in [0,1]$?*

Exercise 6.17 *Show that we can derive the following result from the upper bound (6.29):*
The all-terminal reliability of a graph G is not larger than the product of the reliabilities of all its bridges.

Chapter 7

Partitions of the Vertex Set and Vertex Separators

A subset F of the edge set E of a graph $G = (V, E)$ defines in a natural way a partition of the vertex set V. The blocks (subsets) of the partition are formed by vertices that belong to the same component of (V, F). Partitions can be employed in order to define more general reliability measures. We will pursue this idea in the last section of this chapter.

7.1 Combinatorics of Set Partitions

This section gives a brief introduction to the theory of set partitions, especially to their combinatorial and order-theoretic properties. For more detailed treatment of this subject, we refer the reader to the textbooks [2, 5, 285].

Let V be a finite set with $|V| = n$ (in the following usually the vertex set of a graph). A *partition* of V is a set of nonempty disjoint subsets of V whose union is V. We denote the set of all partitions of V by $\mathbb{P}(V)$. The subsets $A \in \pi$ of a partition $\pi \in \mathbb{P}(V)$ are called the *blocks* of π. The partitions of $\mathbb{P}(V)$ are naturally ordered by refinement. A partition σ is a *refinement* of a partition π if each block of σ is a subset of a block of π. An order relation in $\mathbb{P}(V)$ is given in the following way. We set $\sigma \leq \pi$ if and only if σ refines π. The set $\mathbb{P}(V)$ together with the refinement order forms the *partition lattice* of V. Figure 7.1 shows the Hasse diagram of the partition lattice $\mathbb{P}(\{a, b, c, d\})$. Here and in following examples, we use the abbreviated form $ab/c/d$ instead of $\{\{a, b\}, \{c\}, \{d\}\}$ to specify a partition. The unique *least element* $\hat{0} = \{\{1\}, ..., \{n\}\}$ in $\mathbb{P}(V)$ is the finest partition consisting entirely of singletons. The *greatest element* in $\mathbb{P}(V)$ is the partition $\hat{1} = \{V\}$ with the only block V. For two partitions $\sigma, \pi \in \mathbb{P}(V)$ let $\sigma \vee \pi$ denote the *join* of σ and π, i.e. the smallest partition $\tau \in \mathbb{P}(V)$ with $\tau \geq \sigma$ and $\tau \geq \pi$. In an analogous way we define $\sigma \wedge \pi = \max\{\tau \in \mathbb{P}(V) : \tau \leq \sigma\}$ as the *meet* of

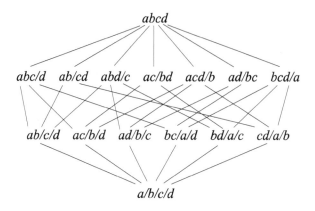

Figure 7.1: The partition lattice $\mathbb{P}\left(\{a,b,c,d\}\right)$

σ and π. We denote by $|\pi|$ the number of blocks of the partition π. The *rank* of a partition $\pi \in \mathbb{P}(V)$ is defined by $r(\pi) = n - |\pi|$. The number of partitions of an n-element set with exactly k blocks is called the *Stirling number of the second kind* and denoted by $\left\{{n \atop k}\right\}$. Using the rank function, we obtain

$$\left\{{n \atop k}\right\} = |\{\pi \in \mathbb{P}(V) : r(\pi) = n - k\}|.$$

The Stirling numbers of the second kind satisfy the recurrence equation

$$\left\{{n \atop k}\right\} = \left\{{n-1 \atop k-1}\right\} + k\left\{{n-1 \atop k}\right\}, n > 0, k > 0.$$

The number of all partitions of an n-element set is the *Bell number* defined by

$$B(n) = \sum_{k=1}^{n} \left\{{n \atop k}\right\} = |\mathbb{P}(\{1,...,n\})|.$$

The first Bell numbers are:

n	1	2	3	4	5	6	7	8	9	10
$B(n)$	1	2	5	15	52	203	877	4140	21147	115975

7.1.1   Connected Partitions

Let $G = (V, E)$ be an undirected graph with n vertices. Each vertex subset $X \subseteq V$ determines a subgraph $G[X]$ consisting of the vertex set X and all edges of G that have both end vertices in X. We call $G[X]$ the subgraph *induced by* X. A partition $\pi \in \mathbb{P}(V)$ is called *connected* if each block of π induces a connected subgraph of G. The set of all connected partitions of G is denoted by $\mathbb{P}_c(G)$. The least partition $\hat{0}$ is a connected partition of any graph. Let $e = \{u, v\}$ be an edge of G. The edge e *induces* a partition $\pi_e \in \mathbb{P}_c(G)$ consisting of singletons exclusively, except the two-element block $\{u, v\}$. We write in the following simply e instead of π_e and refer to e as an *edge partition*. Indeed, we may consider each subset $X \subseteq V$ also as a partition of V by augmenting X with all singletons formed from elements of $V \setminus X$. Consequently, we can interpret expressions such as $e \vee \pi$ or $X \leq \pi$, the latter meaning that there is a block $Y \in \pi$ with $X \subseteq Y$.

Let us first consider some properties of the join of two partitions $\sigma, \pi \in \mathbb{P}(V)$. Each block of σ and π is a subset of a block of $\sigma \vee \pi$. Consequently, if $X \in \sigma$ and $Y \in \pi$ with $X \cap Y \neq \emptyset$ then $\sigma \vee \pi$ contains a block Z with $X \cup Y \subseteq Z$. We conclude analogously that the conditions $W \in \sigma$, $X, Y \in \pi$, $X \cap W \neq \emptyset$, and $Y \cap W \neq \emptyset$ imply the existence of a block $Z \in \sigma \vee \pi$ with $W \cup X \cup Y \subseteq Z$. This observation can be generalized in the following way.

Let $X_1, ..., X_k$ be blocks of σ or π such that $X_j \cap \bigcup_{i=1}^{j-1} X_i \neq \emptyset$ for $j = 2, ..., k$.

Then there exists a block $Z \in \sigma \vee \pi$ containing $\bigcup_{i=1}^{j} X_i$. Hence each block of $\sigma \vee \pi$ is built by a process of "mutual chaining" blocks of σ and π. The application of this construction to the edge partitions of a graph yields the following statement.

Lemma 7.1 *A graph $G = (V, E)$ with at least two vertices is connected if and only if $\bigvee_{e \in E} e = \hat{1}$.*

Let $H = (V, F)$ be a subgraph of $G = (V, E)$. The components of H *induce* a partition $\pi(H) \in \mathbb{P}_c(G)$. Two vertices $u, v \in V$ are elements of the same block of $\pi(H)$ if and only if these vertices belong to one component of H. We observe that a partition $\pi \in \mathbb{P}(V)$ belongs to $\mathbb{P}_c(G)$ if and only if there is an edge subset $F \subseteq E$ such that $\pi = \pi(V, F)$. Each vertex partition of a complete graph is a connected partition. Consequently, we obtain $|\mathbb{P}_c(K_n)| = B(n)$.

Lemma 7.2 *Let $G = (V, E)$ be a forest with m edges. Then $|\mathbb{P}_c(G)| = 2^m$.*

Proof. Each edge subset $F \subseteq E$ of the forest $G = (V, E)$ generates a partition $\pi(V, F)$. Conversely, if $\pi \in \mathbb{P}_c(G)$ then there exists a subset $F \subseteq E$ with $\pi(V, F) = \pi$. Assume there is a different edge set $F' \subseteq E$ with $\pi(V, F') = \pi$. In this case there are components of (X, J) of (V, F) and (X, J') of (V, F') with the same vertex set but different edge sets. Consequently, $(X, J \cup J')$ contains a cycle which contradicts the premise that G is a forest. Hence each edge subset induces a unique partition of $\mathbb{P}_c(G)$. ∎

The next statement follows in the same vein as Lemma 7.1 from the construction of the meet of partitions.

Lemma 7.3 *Let $G = (V, E)$ and $H = (V, F)$ be two graphs with the same vertex set V. Then the partition induced by $(V, E \cup F)$ is $\pi(V, E \cup F) = \pi(G) \vee \pi(H)$.*

7.1.2   Incidence Functions

Let P be a finite *poset* (a partially ordered set) and let K be any field (for instance the real numbers). A function $f : P \times P \to K$ is called an *incidence function* on P if for $x, y \in P$ the relation $x \not\leq y$ implies $f(x, y) = 0$. We consider here only the special case $P = \mathbb{P}(V)$ and $K = \mathbb{R}$. Hence, we have real valued functions $f(\pi, \sigma)$ that vanish as soon as π is not a refinement of σ. With the usual scalar multiplication and the pointwise addition of functions, the set $I(P)$ of all incidence functions on P becomes a vector space. Moreover, we can multiply incidence functions as follows:

$$(f * g)(x, y) = \sum_{z \in P} f(x, z) g(z, y)$$

With this additional operation, $I(P)$ is an algebra — the *incidence algebra* on P. The identity with respect to multiplication is the *Kronecker function*,

$$\delta(x, y) = \begin{cases} 1 \text{ if } x = y \\ 0 \text{ else.} \end{cases}$$

The *Riemann zeta function* is defined by

$$\zeta(x, y) = \begin{cases} 1 \text{ if } x \leq y \\ 0 \text{ else.} \end{cases}$$

Consequently, the zeta function determines the order relation on P completely. The inverse of the zeta function is the *Möbius function* μ, defined by

$$\mu * \zeta = \delta.$$

The Möbius function in the partition lattice $\mathbb{P}(V)$ is given by (see [263])

$$\mu(\pi, \sigma) = (-1)^{|\pi|-|\sigma|} \prod_{i=1}^{|\sigma|} (p_i - 1)! , \quad \pi, \sigma \in \mathbb{P}(V), \ \pi \leq \sigma.$$

Here p_i denotes the number of blocks of π that are contained in the i-th block of σ. As a consequence of this representation, we obtain

$$\mu(\hat{0}, \sigma) = (-1)^{n-|\sigma|} \prod_{X \in \sigma} (|X| - 1)!$$

and

$$\mu(\pi, \hat{1}) = (-1)^{|\pi|-1} (|\pi| - 1)! \tag{7.1}$$

The most important application of the Möbius function is the *Möbius inversion* [263]. Let $f, g : P \to \mathbb{C}$ be functions from a poset into the field of complex numbers. Then for all $x \in P$,

$$f(x) = \sum_{y \leq x} g(y)$$

if and only if

$$g(x) = \sum_{y \leq x} \mu(y, x) f(y).$$

There is a second (dual) form of the Möbius inversion, namely

$$f(x) = \sum_{y \geq x} g(y) \Leftrightarrow g(x) = \sum_{y \geq x} \mu(x, y) f(y).$$

Another function introduced by *Doubilet* [109] and *Lindström* [202] that has applications in network reliability is related to the meet of partitions, see [49]. For any two partitions $\pi, \sigma \in \mathbb{P}(V)$, we define

$$a(\pi, \sigma) = \begin{cases} 1 \text{ if } \pi \vee \sigma = \hat{1} \\ 0 \text{ else.} \end{cases} \tag{7.2}$$

It can be shown that a has a multiplicative inverse that is given by

$$a^{-1}(\pi, \sigma) = \sum_{\tau \in \mathbb{P}(V)} \frac{\mu(\tau, \sigma) \mu(\tau, \pi)}{\mu(\tau, \hat{1})}. \tag{7.3}$$

7.2 Separating Vertex Sets — Splitting Formulae

Let $G = (V, E)$ be a connected undirected graph, further let $G^1 = (V^1, E^1)$ and $G^2 = (V^2, E^2)$ be two subgraphs of G with nonempty edge sets such that $G^1 \cup G^2 = G$ and $G^1 \cap G^2 = (U, \emptyset)$, which means that G^1 and G^2 are edge-disjoint and have only vertices contained in U in common. We call U a *separating vertex set* of G. The subgraphs G^1 and G^2 are called the *split components* of G. Figure 7.2 illustrates the idea of a separating vertex set.

Can we determine the reliability of G by separate reliability calculations of G^1 and G^2 and combining the partial results? *Rosenthal* presented in [262] the first ideas for applications of separating vertex sets in network reliability calculations. More than ten year later, *Bienstock* [49] developed an order-theoretic framework for splitting methods.

Remark 7.4 *The notation G^i for subgraphs written with superscripts instead of subscripts is used in order to keep the subscript place free for graph operations that are later introduced. Other interpretations of superscripts, like graph powers, are not used in this book. Hence there should not arise any confusion.*

7.2.1 All-Terminal Reliability

We know already from Theorem 5.2 that the all-terminal reliability of G is simply the product of $R(G^1)$ and $R(G^2)$, in case U is a one-element set —

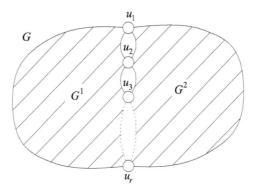

Figure 7.2: A separating vertex set

an articulation of G. In order to generalize this result, we have to investigate partitions of U. Let $P\left(G^{i}, \pi\right)$ be the probability that the components of G^{i} $(i = 1, 2)$ induce the partition $\pi \in \mathbb{P}\left(U\right)$ of the separating vertex set *and* that each vertex of G^{i} is connected by a path to a vertex of U. The second condition is satisfied if the number of components of G^{i} equals the number of blocks of π. Now assume that G^{1} induces a given partition $\pi \in \mathbb{P}\left(U\right)$. Then we can presuppose in G^{2} that all vertices belonging to one block of π are connected by external paths. This presupposition can be realized by merging the blocks of π in G^{2}. Let G_{π}^{i} be the graph obtained from G^{i} $(i = 1, 2)$ by merging all vertices of the respective blocks of π. The induction of a given partition $\pi \in \mathbb{P}\left(U\right)$ in G^{1} is a random event A_{π} that corresponds to a collection of states of the edge set E^{1}. Two such events A_{π} and A_{σ} are disjoint for different partitions $\pi, \sigma \in \mathbb{P}\left(U\right)$, which leads to the following statement.

Theorem 7.5 *Let $G = (V, E)$ be a connected undirected graph and U a separating vertex set of G such that G^{1} and G^{2} are the split components of G with respect to U. Then the all-terminal reliability of G satisfies*

$$R\left(G\right) = \sum_{\pi \in \mathbb{P}(U)} P\left(G^{1}, \pi\right) R\left(G_{\pi}^{2}\right). \tag{7.4}$$

Clearly, this formula remains valid if we exchange the roles of G^{1} and G^{2}. However, there are some problems concerning the application of (7.4). The first problem is caused by the vastly growing number of terms of the sum. The number of terms increases exponentially with the size of the separating vertex set. We have no other way to surmount this obstacle than to choose a separating vertex set of minimum cardinality. There are efficient algorithms that find minimum size separating vertex sets in graphs. Unfortunately, not all graphs possess separating vertex sets of small size. If we encounter a graph without any suitably small vertex separators then the splitting approach fails. The second problem is easier to resolve. It is caused by the asymmetry of our Splitting Formula (7.4). We already have a lot of different methods at our disposal in order to compute for G^2 and its derived graphs G^2_π the all-terminal reliability values. The situation is less convenient with respect to the computation of the probabilities $P\left(G^1, \pi\right)$. The next theorem shows that the direct computation of the probabilities $P\left(G^1, \pi\right)$ can be avoided. The statement and its proof uses the a-function and its inverse defined in (7.2) and (7.3), see [49].

Theorem 7.6 *Let $G = (V, E)$ be a connected undirected graph and U a separating vertex set of G such that G^1 and G^2 are the split components of G with respect to U. Then the all-terminal reliability of G satisfies*

$$R\left(G\right) = \sum_{\pi \in \mathbb{P}(U)} \sum_{\sigma \in \mathbb{P}(U)} a^{-1}\left(\pi, \sigma\right) R\left(G^1_\sigma\right) R\left(G^2_\pi\right). \qquad (7.5)$$

Proof. We define for each $\sigma \in \mathbb{P}\left(U\right)$ a simple graph H_σ with vertex set U, where two vertices $u, v \in U$ are adjacent in H_σ if and only if u and v belong to the same block of σ. According to Lemma 7.3, if G^1 induces the partition $\pi \in \mathbb{P}\left(U\right)$ then the graph $G^1 \cup H_\sigma$ induces the partition $\pi \vee \sigma$. Consequently, $G^1 \cup H_\sigma$ is connected if and only if $\pi \vee \sigma = \hat{1}$. If we contract all edges of H_σ then we also obtain the following statement. The graph G^1_σ is connected if and only if G^1 induces a partition $\pi \in \mathbb{P}\left(U\right)$ satisfying $\pi \vee \sigma = \hat{1}$. Using (7.2), we obtain

$$R\left(G^1_\pi\right) = \sum_{\sigma : \pi \vee \sigma = \hat{1}} P\left(G^1, \sigma\right) = \sum_{\sigma \in \mathbb{P}(U)} a\left(\pi, \sigma\right) P\left(G^1, \sigma\right)$$

and hence

$$P\left(G^1, \pi\right) = \sum_{\sigma \in \mathbb{P}(U)} a^{-1}\left(\pi, \sigma\right) R\left(G^1_\sigma\right).$$

The substitution of $P\left(G^1, \pi\right)$ in (7.4) yields the desired result. ∎

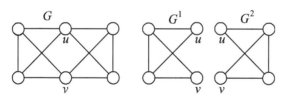

Figure 7.3: An example graph and its splitting

Example 7.7 *Consider the graph G that is shown in Figure 7.3. The graph contains a separating vertex pair $\{u, v\}$ allowing a symmetric splitting into two isomorphic split components G^1 and G^2 as presented on the right-hand side of the figure. Equation (7.5) specializes for $U = \{u, v\}$ to*

$$R\left(G\right) = R\left(G^1\right) R\left(G^2_{uv}\right) + R\left(G^2\right) R\left(G^1_{uv}\right) - R\left(G^1\right) R\left(G^2\right).$$

If all edge reliabilities are identical to p then we obtain $R\left(G^1\right) = R\left(G^2\right) = 8p^3 - 11p^4 + 4p^5$ and $R\left(G^1_{uv}\right) = R\left(G^2_{uv}\right) = 8p^2 - 14p^3 + 9p^4 - 2p^5$, which gives, via application of the splitting formula, the reliability polynomial of the example graph as

$$R\left(G, p\right) = 128p^5 - 464p^6 + 692p^7 - 527p^8 + 204p^9 - 32p^{10}.$$

7.2.2 K-Terminal Reliability

The computation of the K-terminal reliability by the splitting approach requires an extension of the notion of set partition. Assume again that we have a graph G that can be split along a separating vertex set U into two split components G^1 and G^2. The set of terminal vertices K of G splits also into subsets $K^1 = V^1 \cap K$ and $K^2 = V^2 \cap K$. We call a component of a graph G that contains at least one terminal vertex a K-*component* of G. A *labeled set partition* of U is a set partition of U whose blocks may carry a *label*. We denote a labeled block by an underline. For instance, the set of labeled partitions of $\{a, b\}$ is

$$\{a/b, \ a/\underline{b}, \ \underline{a}/b, \ \underline{a}/\underline{b}, \ ab, \ \underline{ab}\}.$$

The set of all labeled partition of a set U is denoted by $\mathbb{P}\left(U\right)$. If $\pi \in \mathbb{P}\left(U\right)$ is a labeled partition then π denotes the (unlabeled) partition of U obtained from $\underline{\pi}$ by removing all labels. Let $\underline{\pi}, \underline{\sigma} \in \mathbb{P}\left(U\right)$; we define $\underline{\pi} \vee \underline{\sigma}$ as the labeled partition obtained from $\pi \vee \sigma$ by labeling all blocks that contain a labeled block of $\underline{\pi}$ or $\underline{\sigma}$. As an example, $\underline{ac}/b/dfg/\underline{eh} \vee ae/bc/d/fg/\underline{h} = \underline{abceh}/dfg$. For $i = 1, 2$ and $\underline{\pi} \in \mathbb{P}\left(U\right)$, let $P\left(G^i, \underline{\pi}\right)$ be the probability that G^i induces the partition $\pi \in \mathbb{P}\left(U\right)$ and that component C of G^i contains all vertices of a labeled block of $\underline{\pi}$ if and only if C is a K-component of G^i. We say that G^i *induces* the labeled partition $\underline{\pi}\left(G^i\right)$ if two vertices of G^i are in one component if and only if they belong to the same block of π and if exactly the K-components of G^i correspond to labeled blocks of $\underline{\pi}$. Consequently, if G^i induces $\underline{\pi} \in \mathbb{P}\left(U\right)$ then G^i does not contain any K-component that is vertex-disjoint to U.

Lemma 7.8 *Let $G = (V, E)$ be an undirected graph with terminal vertex set $K \subseteq V$; let $G^1 = \left(V^1, E^1\right)$ and $G^2 = \left(V^2, E^2\right)$ be two subgraphs of G with nonempty edge sets such that $G^1 \cup G^2 = G$ and $G^1 \cap G^2 = (U, \emptyset)$. Suppose the two subgraphs G^1 and G^2 contain each at least one terminal vertex. Then the graph G is K-connected if and only if $\underline{\pi}\left(G^1\right) \vee \underline{\pi}\left(G^2\right)$ has exactly one labeled block.*

Proof. We consider here G as a deterministic graph, i.e. a graph without any edge failures, and which is not necessarily connected. According to Lemma 7.3, G induces the partition $\pi\left(G^1\right) \vee \pi\left(G^2\right) \in \mathbb{P}\left(U\right)$. A graph is K-connected if it contains exactly one K-component. The graph G contains exactly one K-component if exactly one block of $\underline{\pi}\left(G^1\right) \vee \underline{\pi}\left(G^2\right)$ is labeled, since all K-components of G^1 and G^2 contain a vertex of U. ■

We generalize the a-function (7.2) for labeled partitions as arguments in the following way. For all $\underline{\pi}, \underline{\sigma} \in \mathbb{P}(U)$, we define

$$a\left(\underline{\pi}, \underline{\sigma}\right) = \begin{cases} 1 \text{ if } \underline{\pi} \vee \underline{\sigma} \text{ has exactly one labeled block} \\ 0 \text{ else.} \end{cases} \tag{7.6}$$

Then we obtain as an immediate consequence of Lemma 7.8

$$R\left(G, K\right) = \sum_{\underline{\pi} \in \mathbb{P}(U)} \sum_{\underline{\sigma} \in \mathbb{P}(U)} P\left(G^1, \underline{\pi}\right) a\left(\underline{\pi}, \underline{\sigma}\right) P\left(G^2, \underline{\sigma}\right), \ K^1 \neq \emptyset, K^2 \neq \emptyset. \tag{7.7}$$

Example 7.9 *Let us consider the special case of a separating vertex pair $U = \{u, v\}$. If both subgraphs G^1 and G^2 contain at least one terminal vertex then the set of relevant labeled partitions is*

$$\mathbb{P}\left(U\right) = \{\underline{u}/v, u/\underline{v}, \underline{u}/\underline{v}, \underline{uv}\}.$$

We can neglect vertex partitions without any labeled block as they yield no contribution to the K-connectedness of G. The a-function (7.6) can be presented as a matrix $A = (a\left(\underline{\pi}, \underline{\sigma}\right))$, where we use the order $[\underline{u}/v, u/\underline{v}, \underline{u}/\underline{v}, \underline{uv}]$ of the labeled partitions:

$$A = \begin{pmatrix} 1 & 0 & 0 & 1 \\ 0 & 1 & 0 & 1 \\ 0 & 0 & 1 & 1 \\ 1 & 1 & 1 & 1 \end{pmatrix}$$

The application of the splitting formula (7.7) requires the computation of four probabilities for each subgraph.

Let $u = |U|$ be the cardinality of the separating vertex set. If we neglect vertex partitions without any labeled block, then the number n_u of terms of each sum in (7.7) is obtained by

$$n_u = \sum_{k=1}^{u} \begin{Bmatrix} u \\ k \end{Bmatrix} \left(2^k - 1\right).$$

The first numbers of this sequence are

u	1	2	3	4	5	6	7	8	9
n_u	1	4	17	79	402	2227	13337	85778	589035

Now we consider the graph G^i_π obtained by merging the blocks of $\pi \in \mathbb{P}(U)$ in G^i, $i = 1, 2$. The merge operation transforms the separating vertex set U into a new vertex set U_π of G^i_π, where the vertices of U_π correspond to blocks of π. If $\underline{\pi} \in \mathbb{P}(U)$ is a labeled set partition then we form the graph G^i_π analogously; in addition we define K_π as the subset of vertices of U_π that

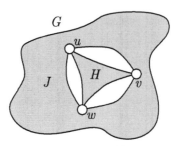

Figure 7.4: A three-attached subgraph

arises from labeled blocks of $\underline{\pi}$. Using this notation, we can reformulate the splitting formula for the K-terminal reliability as

$$R\left(G, K\right) = \sum_{\underline{\pi} \in \underline{\mathbb{P}}(U)} P\left(G^1, \underline{\pi}\right) R\left(G^2_{\underline{\pi}}, \left(K^2 \setminus U\right) \cup K_{\underline{\pi}}\right). \qquad (7.8)$$

The two-terminal reliability is a special case of the K-terminal reliability for $K = \{s, t\}$. We assume here that $s \in V^1$ and $t \in V^2$. Then a labeled partition induced by G^1 or G^2 can have only one labeled block. Since labeled partitions without any labeled blocks do not contribute to $R_{st}(G)$, we exclude them here. Let $\underline{\mathbb{P}}_1\left(U\right)$ be the set of all labeled partitions of U containing exactly one labeled block. Hence we obtain from Equation (7.7)

$$R_{st}\left(G\right) = \sum_{\underline{\pi} \in \underline{\mathbb{P}}_1(U)} \sum_{\underline{\sigma} \in \underline{\mathbb{P}}_1(U)} P\left(G^1, \underline{\pi}\right) a\left(\underline{\pi}, \underline{\sigma}\right) P\left(G^2, \underline{\sigma}\right), \; s \in V^1, t \in V^2. \quad (7.9)$$

Remark 7.10 *Numerical computations suggested that the a-function restricted to $\underline{\mathbb{P}}_1\left(U\right)$ has always an inverse with respect to convolution, whereas in the general case (for $\underline{\mathbb{P}}\left(U\right)$) for $|U| > 2$ no inverse exists. Finally, Frank Simon [281] showed that with a suitable chosen subset of $\underline{\mathbb{P}}\left(U\right)$ a symmetric splitting formulae for the K-terminal reliability equivalent to (7.5) can always be achieved.*

7.2.3 Splitting and Reduction

We met vertex partitions already in the last chapter in the context of delta-star and star-delta transformations. Now we will use the splitting formulae in order to derive more general reductions. Consider the graph G depicted in Figure 7.4. The graph G is the union of two graphs J and H such that $J \cap H = (\{u, v, w\}, \emptyset)$. We call the graph H *three-attached* to J. Our aim is to replace H by a simpler subgraph H' such that $R(G) = R(J \cup H')$. According to the splitting formula (7.4), this equality holds if $R(H'_{\pi}) = R(H_{\pi})$ for all $\pi \in \mathbb{P}\left(\{u, v, w\}\right)$. Consequently, we look for a replacement graph with at least

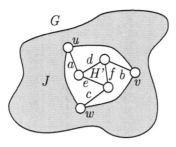

Figure 7.5: Replacement graph

Figure 7.6: Graph K_3

five edges, in order to have sufficiently many variables in the resulting system of equations. The replacement graph H' is depicted in Figure 7.5. We assign six parameters p_a, p_b, p_c, p_d, p_e, p_f to the edges of the graph H'. Observe that we speak of *edge parameters* rather than edge reliabilities or edge probabilities. These edge parameters are no longer restricted to the real interval $[0,1]$. We now use the edge decomposition formula for the all-terminal reliability,

$$R(G) = p_e R(G/e) + (1 - p_e) R(G - e), \qquad (7.10)$$

as the *definition* of $R(G)$. Indeed, the value of $R(G)$ is uniquely defined by (7.10) when we impose the usual initial conditions, i.e. $R(K_1) = 1$ for the one-vertex graph K_1 and $R(G) = 0$ for any disconnected graph G.

Example 7.11 *Consider the complete graph presented in Figure 7.6. The all-terminal reliability of this graph is*

$$R(G) = p_a p_b + p_a p_c + p_b p_c - 2 p_a p_b p_c.$$

If all three edges operate with probability 0.326352 then the all-terminal reliability equals 0.25. Applying the decomposition (7.10) together with the initial conditions, we obtain the same result using the edge parameters

$$p_a = 1.5, \ p_b = -0.5, \ p_c = 0.4.$$

Hence we consider these three edge parameters as a valid representation for our reliability problem. Even the partially complex parameters,

$$p_a = 1 + i, \ p_b = 1 - i, \ p_c = \frac{7}{8},$$

lead to the same result $R(G) = \frac{1}{4}$.

The following consideration shows that all results for the all-terminal reliability that we obtained by edge decomposition remain valid for graphs with more general edge parameters. Let $\mathbb{K}$ be a given field (for instance real or complex numbers) and let $\mathcal{G}_m$ be the set of all finite undirected graphs with m edges that are weighted with edge weights from $\mathbb{K}$. For each edge e of a graph G, let $x_e \in \mathbb{K}$ be the weight assigned to e and denote by $\mathbf{x} = (x_e)_{e \in E(G)}$ the vector of edge weights. We define a function $f : \mathcal{G}_m \times \mathbb{K}^m \to \mathbb{K}$ that assigns, to each graph with edge parameters from $\mathbb{K}$, the value

$$f(G, \mathbf{x}) = \sum_{F \subseteq E(G)} \psi\left(G\left[F\right]\right) \prod_{f \in F} x_f \prod_{g \in E \setminus F} (1 - x_g). \qquad (7.11)$$

We can easily verify that this definition implies the decomposition formula

$$f(G, \mathbf{x}) = x_e f(G/e, \mathbf{x}) + (1 - x_e) f(G - e, \mathbf{x}), \qquad (7.12)$$

which is valid for each edge $e \in E(G)$. To show Equation (7.12), we split the sum of the right-hand side of (7.11):

$$f(G, \mathbf{x}) = \sum_{F \subseteq E(G) \setminus \{e\}} \psi\left(G\left[F \cup \{e\}\right]\right) \prod_{f \in F \cup \{e\}} x_f \prod_{g \in E(G) \setminus (F \cup \{e\})} (1 - x_g)$$

$$+ \sum_{F \subseteq E(G) \setminus \{e\}} \psi\left(G\left[F\right]\right) \prod_{f \in F} x_f \prod_{g \in E \setminus F} (1 - x_g)$$

$$= x_e \sum_{F \subseteq E(G/e)} \psi\left((G/e)\left[F\right]\right) \prod_{f \in F} x_f \prod_{g \in E(G/e) \setminus F} (1 - x_g)$$

$$+ (1 - x_e) \sum_{F \subseteq E(G-e)} \psi\left((G - e)\left[F\right]\right) \prod_{f \in F} x_f \prod_{g \in E(G-e) \setminus F} (1 - x_g)$$

$$= x_e f(G/e, \mathbf{x}) + (1 - x_e) f(G - e, \mathbf{x})$$

Now let us return to the reduction from H to H' depicted in Figure 7.4 and Figure 7.5. Our objective is to find a solution of the system

$$R(H'_{uvw}) = R(H_{uvw})$$
$$R(H'_{u/vw}) = R(H_{u/vw})$$
$$R(H'_{uv/w}) = R(H_{uv/w})$$
$$R(H'_{uw/v}) = R(H_{uw/v})$$
$$R(H'_{u/v/w}) = R(H_{u/v/w}),$$

where the right-hand side is given by the corresponding probabilities of the original graph H. The left-hand side of this system consists of polynomials in six parameters $p_a, p_b, p_c, p_d, p_e, p_f$. Simulations show that there seem to exist numerical solutions for all possible right-hand sides. However, a rigorous proof

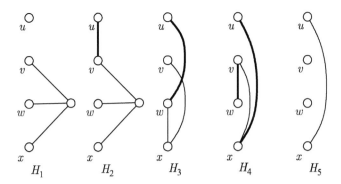

Figure 7.7: Reduction graphs for a separating vertex set of cardinality 4

for solvability is not (yet) known. This problem becomes even harder when we consider subgraphs to be reduced with more than three attachment vertices. An alternative approach presented in [46] uses more than one replacement graph, which means that H is replaced by a collection $\{H_1, ..., H_k\}$ of graphs such that

$$\sum_{i=1}^{k} \omega_i P(H_i, \pi) = P(H, \pi) \qquad (7.13)$$

is satisfied for any partition π of the separating vertex set (set of attachment vertices). The introduction of several replacement graphs leads to simpler graph structures and simpler equations, but also to an increase of the number of subproblems to be solved.

Example 7.12 *Let $U = \{u, v, w, x\}$ be a separating vertex set of the graph $G = (V, E)$ such that H and K are the split components with respect to U. Then one of the split components (say K) can be substituted by the five replacement graphs depictured in Figure 7.7 such that Equation (7.13) is satisfied. The bold drawn edges represent failure-free edges (edges of reliability 1). The resulting system of equations has a unique solution for the ten edge parameters and five reduction factors $\omega_1, ..., \omega_5$. The following table shows the induced partitions for each replacement graph:*

$H_1:$ $u/v/w/x,$ $u/vw/x,$ $u/v/wx,$ $u/vx/w,$ u/vwx
$H_2:$ $uv/w/x,$ $uvw/x,$ $uv/wx,$ $uvx/w,$ $uvwx$
$H_3:$ $uw/v/x,$ $uwx/v,$ $uw/vx,$ $uvwx$
$H_4:$ $ux/vw,$ $uvwx$
$H_5:$ $u/v/w/w,$ $ux/v/w$

Note that the five sets of induced partitions are not disjoint.

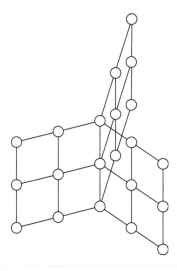

Figure 7.8: Wing graph

Exercise 7.1 *Let $G = (V, E)$ be a graph and $G^1, ..., G^r$ subgraphs of G such that $\bigcup_{i=1}^{r} G^i = G$ and $\bigcap_{i=1}^{r} G^i = (U, \emptyset)$. Find a generalized splitting formula for $R(G)$ as a function of all-terminal reliabilities of the r split components. How many different reliability polynomials have to be calculated in order to apply the generalized splitting formula to the wing graph depicted in Figure 7.8?*

Exercise 7.2 *Modify the splitting formula (7.7) for the case that either K^1 or K^2 is empty.*

Exercise 7.3 *Show by applying the splitting formula (7.8) that for a separating vertex pair $U = \{u, v\}$ the K-terminal reliability is given by*

$$R(G, K) = R(G^1_{uv})R(G^2_{\underline{u}/\underline{v}}) + R(G^1_{\underline{u}/v})R(G^2_{\underline{u}/v}) - R(G^1_{\underline{u}/v})R(G^2_{\underline{u}/\underline{v}})$$
$$+ R(G^1_{u/\underline{v}})R(G^2_{u/\underline{v}}) - R(G^1_{u/\underline{v}})R(G^2_{\underline{u}/v})$$
$$+ R(G^1_{\underline{u}/\underline{v}}) \left[R(G^1_{uv}) - R(G^2_{\underline{u}/v}) - R(G^1_{u/\underline{v}}) + R(G^2_{\underline{u}/v}) \right].$$

Exercise 7.4 *Show that for a set U of cardinality u the number of labeled partitions with exactly one labeled block is $|\mathbb{P}_1(U)| = B(u+1) - B(u)$, i.e. the difference of two successive Bell numbers.*

7.3   Planar and Symmetric Graphs

A graph is called *planar* if it can be drawn in the plane without any edge crossings, meaning that edges intersect only at their end points. A more precise definition requires some background from topology. The interested reader is referred to textbooks on topological graph theory such as [145] and [225]. Planar graphs have a lot of interesting properties and applications. For our purposes the most interesting facts are:

1. Planar graphs always have a "small" separating vertex set, more precisely, we can find in a planar graph of order n a separating vertex set of size at most $C\sqrt{n}$ such that each split component has at least $n/3$ vertices. Here $C \leq 3$ denotes a suitable small constant. This is the statement of the celebrated *planar separator theorem* by *Lipton* and *Tarjan* [203]. Moreover, there are efficient algorithms to find such a separator. This is indeed an advantage since we may not be able to find separating vertex sets of a size less than a constant multiple of n in general graphs.

2. If U is a separating vertex set of a planar graph G the split components induce only noncrossing partitions of $\mathbb{P}(U)$. Assume the vertices of U are ordered linearly. Let $u, v, w, x \in U$ be four vertices with $u < v < w < x$. Then we call a partition of U that has two different blocks X and Y with $u, w \in X$ and $v, x \in Y$ *crossing*. A partition without blocks of this kind is called *noncrossing*. As an example, a set of cardinality ten has $B(10) = 115975$ partitions but only 16796 noncrossing partitions. In general the number of noncrossing partitions of a set of size n is given by the *Catalan number* C_n, which is defined by

$$C_n = \frac{1}{n+1}\binom{2n}{n}.$$

3. A simple planar graph of order n ($n \geq 3$) has at most $3n - 6$ edges. In contrast, the number of edges of simple non-planar graphs may grow proportionally to n^2.

A measure for the symmetry of a graph is the order of its automorphism group. An *automorphism* of a graph $G = (V, E)$ is a bijective mapping $f : V \to V$ such that for all $u, v \in V$ the number of edges between u and v and the number of edges between $f(u)$ and $f(v)$ coincide. Let U be a separating vertex set of G. Then we can restrict our attention to automorphisms of G that map U onto U. This means that we consider the action of the symmetry group on U. In order to reduce the computational effort of splitting, certain edge reliabilities have to be equal in addition to the symmetry of the graph. Let G^i be a split component of G with respect to the separating vertex set U. We say that two partitions $\pi \in \mathbb{P}(U)$ and $\sigma \in \mathbb{P}(U)$ are *equivalent* in G^i if there exists a permutation $\alpha : U \to U$ such that $\sigma = \alpha(\pi)$ and $P(G^i, \pi) = P(G^i, \sigma)$.

Example 7.13 *Consider the graph H depicted in Figure 7.9. Let this graph be a split component of a graph with separating vertex set $U = \{1, 2, 3, 4\}$. We observe that the partition $13/24$ cannot be induced since the graph is planar.*

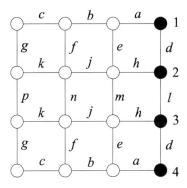

Figure 7.9: A symmetric graph

Assume that each two edges of H that are denoted by the same letter have equal failure probabilities. Then we obtain the following pairs of equivalent partitions:

$$(12/3/4, \ 1/2/34)$$
$$(13/2/4, \ 1/24/3)$$
$$(123/4, \ 1/234)$$
$$(124/3, \ 134/2)$$

We conclude that the application of the splitting formula for $4 \times n$ grid graph requires the computation of only 10 different partition probabilities (rather than 15 in a general graph with a separating vertex set of size four).

Exercise 7.5 *Assume that all edges of an $m \times n$ grid graph fail with identical probability $1-p$. We use the splitting formula (7.4) with respect to a separating vertex set U such that $G[U]$ is a path P_m as to compute the all-terminal reliability of the graph. How many different partition probabilities of the form $P(G^1, \pi)$ are to be calculated if planarity and symmetry of the grid graph is exploited?*

7.4   Splitting and Recurrent Structures

In this section, we investigate graphs with a regular structure, which are sometimes called *recurrent structures*. Let $G = (V, E, X, Y)$ be an undirected graph with vertex set V and edge set E together with two specified nonempty vertex subsets $X, Y \subseteq V$ with $|X| = |Y|$. We call G the *generator graph* (or short *generator*). A second graph $I = (W, F, U)$ with vertex set W and edge set F, called the *initiator graph*, has only one specified vertex subset $U \subseteq W$ with $|U| = |X|$. Now we define two bijections, $\alpha : U \to X$ and $\beta : Y \to X$. The graph $I *_\alpha G$ is obtained from the disjoint union of I and G by identifying

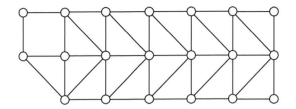

Figure 7.10: A recurrent graph

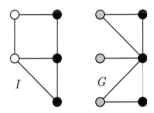

Figure 7.11: Initiator and generator

each vertex $u \in U$ of I with its image $\alpha(u) \in X$ of G. Analogously, we define $G *_\beta G$ as the graph obtained from the disjoint union of G with itself (i.e. from two disjoint copies of G) by identifying each vertex $y \in Y$ of the first copy of G with its image $\beta(y)$ of the second copy of G. We write short $I * G$ and $G * G$, assuming that the bijections α and β are known from the context. In addition, we define the power notation $G^0 = (X, \emptyset)$ and $G^n = G * G^{n-1}$ for $n \geq 1$. Figure 7.10 shows a recurrent graph that can be expressed as $I * G^5$ using the initiator and generator depictured in Figure 7.11. The vertex sets U of I and Y of G are blackened, whereas the grayly drawn vertices of G belong to X.

Now assume that the edges of I and G are weighted with reliabilities. How can we compute the all-terminal reliability of $I * G^k$ for given $k \in \mathbb{Z}^+$ by analyzing I and G separately? The splitting formula (7.5) applied to $I * G^k$ yields

$$R\left(I * G^k\right) = \sum_{\pi \in \mathbb{P}(X)} \sum_{\sigma \in \mathbb{P}(Y)} a^{-1}\left(\pi, \sigma\right) R\left(\left(I * G^{k-1}\right)_\sigma\right) R(G_\pi). \qquad (7.14)$$

At the first glance, we have a serious problem with the determination of $a^{-1}(\pi, \sigma)$, since $\pi \in \mathbb{P}(X)$ and $\sigma \in \mathbb{P}(Y)$ stem from different partition lattices. In order to resolve this problem, we extend the bijection $\alpha : Y \to X$ to $\mathbb{P}(Y)$ defining $\alpha(\sigma)$ as the partition of X obtained by replacing each element v of each block of σ by its image under α. Let $\alpha(\sigma) \in \mathbb{P}(X)$ be the so obtained partition of X. Then we can rewrite Equation (7.14) as

$$R\left(I * G^k\right) = \sum_{\pi \in \mathbb{P}(X)} \sum_{\sigma \in \mathbb{P}(Y)} a^{-1}\left(\pi, \alpha(\sigma)\right) R\left(\left(I * G^{k-1}\right)_\sigma\right) R\left(G_\pi\right). \quad (7.15)$$

Let $r = B(|X|)$ be the number of partitions of X and $f : \mathbb{P}(X) \to \{1, ..., r\}$ a bijective mapping such that $\sigma \leq \pi$ implies $f(\sigma) \leq f(\pi)$ for all $\sigma, \pi \in \mathbb{P}(X)$. If we index the partitions of X by their images under f, then we obtain

$$\mathbb{P}(X) = \left\{\hat{0} = \pi_1, \pi_2, ..., \pi_r = \hat{1}\right\}.$$

Let G_{ij} be the graph obtained from G by merging (the vertices of) all blocks of π_i and the vertices of Y that correspond to preimages of blocks of π_j with respect to α. Let $Q = (q_{ij})_{r,r}$ be a matrix with the entries $q_{ij} = R(G_{ij})$. The next matrix of interest, $A = (a_{ij})_{r,r}$, is defined in analogy to (7.2) by $a_{ij} = a(\pi_i, \pi_j)$, where $a(\pi_i, \pi_j) = 1$ if $\pi_i \vee \pi_j = \hat{1}$ and $a(\pi_i, \pi_j) = 0$, else. Let $\mathbf{s} = (R(I_{\pi_1}), ..., R(I_{\pi_r}))$ be the $1 \times r$ vector of all-terminal reliabilities of the modified initiator graphs, i.e. modified by merging blocks of π_i, $i = 1, ..., r$. Finally, let $\mathbf{e} = (1, 0, ..., 0)^\intercal$ be the first unit column vector of dimension r. Then we find

$$R\left(I * G\right) = \sum_{\pi \in \mathbb{P}(X)} \sum_{\sigma \in \mathbb{P}(Y)} a^{-1}\left(\pi, \alpha(\sigma)\right) R\left(\left(I * G^0\right)_\sigma\right) R\left(G_\pi\right)$$

$$= \sum_{\pi \in \mathbb{P}(X)} \sum_{\sigma \in \mathbb{P}(Y)} R\left(I_\sigma\right) a^{-1}\left(\alpha(\sigma), \pi\right) R\left(G_\pi\right)$$

$$= \mathbf{s} A^{-1} Q \mathbf{e},$$

which is just the matrix form of the splitting formula (7.5) applied to the graph $I \cup G$ with vertex separator X. For $k > 1$, we obtain from (7.15)

$$R\left(I * G^k\right) = \mathbf{s}(A^{-1}Q)^k \mathbf{e}. \quad (7.16)$$

Further generalizations and applications of this approach are presented in [143].

Exercise 7.6 *Define a proper initiator and generator graph for the "spider web" presented in Figure 7.12 and compute its all-terminal reliability assuming that all edge reliabilities are equal to p.*

Exercise 7.7 *How can we extend the recurrent splitting approach (7.16) so as to permit the computation of $R(G)$ for the graph G depicted in Figure 7.13?*

7.5   Approximate Splitting — Reliability Bounds

The computational effort for the application of splitting methods is exponentially increasing with the size of the separating vertex set. However, separating

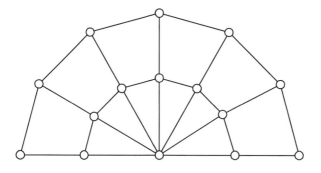

Figure 7.12: A spider web

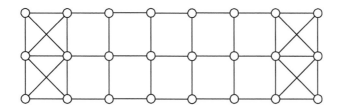

Figure 7.13: A modified $3 \times n$ grid graph

vertex sets can also be employed in order to find reliability bounds or approximate reliability values. In this case, the consideration of a subset of the partition lattice may be sufficient.

We assume again that $G^1 = (V^1, E^1)$ and $G^2 = (V^2, E^2)$ are two subgraphs of G such that $G^1 \cup G^2 = G$ and $G^1 \cap G^2 = (U, \emptyset)$. For any graph G and any vertex subset $X \subseteq V(G)$, let G_X denote the graph obtained from G by merging all vertices of X. Then we obtain by separation of the last term of the Splitting Formula (7.4)

$$R(G) = P\left(G^1, \hat{1}\right) R\left(G_{\hat{1}}^2\right) + \sum_{\pi < \hat{1}} P\left(G^1, \pi\right) R\left(G_{\pi}^2\right)$$

$$= R\left(G^1\right) R\left(G_U^2\right) + \sum_{\pi < \hat{1}} P\left(G^1, \pi\right) R\left(G_{\pi}^2\right). \qquad (7.17)$$

We can obviously exchange the roles of G^1 and G^2 in this equation, which yields the lower bound

$$R(G) \geq \max\left\{R\left(G^1\right) R\left(G_U^2\right), R\left(G^2\right) R\left(G_U^1\right)\right\}. \qquad (7.18)$$

The lower bound (7.18) can be improved in the following way. The sum of all partition probabilities of G^1 yields

$$\sum_{\pi \in \mathbb{P}(U)} P\left(G^1, \pi\right) = R(G^1_{\hat{1}}) = R(G^1_U)$$

and hence

$$\sum_{\pi < \hat{1}} P\left(G^1, \pi\right) = R(G^1_U) - R(G^1).$$

The inequality $R(G^i_\pi) \geq R(G^i)$ holds or all $\pi \in \mathbb{P}\left(U\right)$ and $i = 1, 2$, which follows from the monotonicity of the all-terminal reliability. Thus, replacing $R(G^2_\pi)$ by $R(G^2)$ in (7.17) yields

$$R\left(G\right) \geq R\left(G^1\right) R\left(G^2_U\right) + \sum_{\pi < \hat{1}} P\left(G^1, \pi\right) R\left(G^2\right)$$

$$= R\left(G^1\right) R\left(G^2_U\right) + R(G^1_U)R\left(G^2\right) - R(G^1)R\left(G^2\right). \tag{7.19}$$

In case of a separating vertex pair $U = \{u, v\}$, the last lower bound coincides with the exact formula given in Example 7.7.

A simple upper bound follows immediately from the observation that $R(G_U) \geq R(G)$. After merging U the graph G_U contains an articulation, which gives

$$R(G) \leq R(G_U) = R(G^1_U)R(G^2_U).$$

All bounds introduced so far can be modified in order to cover the K-terminal reliability, too. The 2-terminal reliability offers some additional possibilities to develop bounds. We presume that $s \in V^1$ and $t \in V^2$. We start from the splitting formula

$$R_{st}\left(G\right) = \sum_{\pi \in \mathbb{P}_1(U)} P\left(G^1, \underline{\pi}\right) R\left(G^2_{\underline{\sigma}}\right), \quad s \in V^1, t \in V^2. \tag{7.20}$$

The set $\mathbb{P}_1\left(U\right)$ consists of all labeled partitions with exactly one labeled block. Let $L \subseteq U$ be a non-empty subset of the separating vertex set U. We denote by π/L the extension of a partition $\pi \in \mathbb{P}(U \backslash L)$ with the block L. Remember that the only element of $\mathbb{P}(\emptyset)$ is the empty (blockless) partition. Let $P\left(G^1, \pi/L\right)$ be the probability that s is connected by an operating path with all vertices of L and that two vertices of U belong to the same component of G^1 if and only if they are contained in one block of π/L. Now we can reformulate the splitting formula as

$$R_{st}\left(G\right) = \sum_{\emptyset \neq L \subseteq U} \sum_{\pi \in \mathbb{P}(U \backslash L)} P\left(G^1, \pi/L\right) R_{Lt}(G^2_{\pi/L}). \tag{7.21}$$

Here $R_{Lt}(G^2_{\pi/L})$ denotes the probability that there is a path from L to t in the graph $G^2_{\pi/L}$ obtained from G^2 by merging L and all blocks of π. The probability that s is connected to all vertices from L but separated from all vertices of $U \setminus L$ in G^1 is

$$P_L^1 = \sum_{\pi \in \mathbb{P}(U \setminus L)} P\left(G^1, \pi/L\right).$$

In order to simplify the following presentation of some bounds, we introduce abbreviations for probabilities:

$$R_L^1 = R_{sL}(G_{\hat{0}/L}^1) = R_{sL}(G_L^1)$$

$$R_L^2 = R_{Lt}(G_{\hat{0}/L}^2) = R_{Lt}(G_L^2)$$

$$R_L^{1*} = R_{sL}(G_{\hat{1}/L}^1) = R_{sL}(G_{U \setminus L/L}^1)$$

$$R_L^{2*} = R_{Lt}(G_{\hat{1}/L}^2) = R_{Lt}(G_{U \setminus L/L}^2)$$

All these probabilities are two-terminal reliabilities; in R_L^i only the subset L is merged, whereas in R_L^{i*} the complement $U \setminus L$ is merged in addition.

Theorem 7.14 *Let $G^1 = (V^1, E^1)$ and $G^2 = (V^2, E^2)$ be split graphs of the graph G with respect to a separating vertex set U such that $s \in V^1$ and $t \in V^2$. Then the two-terminal reliability of G satisfies*

$$R_{st}(G) \geq \sum_{\emptyset \neq L \subseteq U} \sum_{J \supseteq U \setminus L} (-1)^{|U|+|L|+|J|+1} R_J^1 R_L^2$$

and

$$R_{st}(G) \leq \sum_{\emptyset \neq L \subseteq U} \sum_{J \supseteq U \setminus L} (-1)^{|U|+|L|+|J|+1} R_J^1 R_L^{2*}.$$

Proof. We prove only the first inequality since the second one can be obtained in the same way. We start with the following set of inequalities

$$R_L^2 = R_{Lt}(G_{\hat{0}/L}^2) \leq R_{Lt}(G_{\pi/L}^2) \leq R_{Lt}(G_{\hat{1}/L}^2) = R_L^{2*},$$

which are valid for all non-empty $L \subseteq U$ and all $\pi \in \mathbb{P}(U \setminus L)$. The inequalities give together with (7.21) the bounds

$$\sum_{\emptyset \neq L \subseteq U} P_L^1 R_L^2 \leq R_{st}(G) \leq \sum_{\emptyset \neq L \subseteq U} P_L^1 R_L^{2*}. \qquad (7.22)$$

In order to substitute P_L^1, we present R_M^1 as a sum of P_L^1-probabilities. We define $P_\emptyset^1 = 0$. The supervertex L is reached from s in G_L^1 if and only if there is an operating path from s to at least one vertex of L in G^1, which yields for all $M \subseteq U$ the following relation

$$R_M^1 = \sum_{L \cap M \neq \emptyset} P_L^1 = \sum_{L \subseteq U} P_L^1 - \sum_{L \cap M = \emptyset} P_L^1 = \sum_{L \subseteq U} P_L^1 - \sum_{L \subseteq U \setminus M} P_L^1$$

$$= R_U^1 - \sum_{L \subseteq U \setminus M} P_L^1.$$

We introduce the following notation for the last sum:

$$Q_{U\setminus M} = \sum_{L\subseteq U\setminus M} P_L^1 = R_U^1 - R_M^1$$

Consequently, we have also

$$\sum_{L\subseteq M} P_L^1 = Q_M = R_U^1 - R_{U\setminus M}^1,$$

which leads via Möbius inversion to

$$P_M^1 = \sum_{L\subseteq M} (-1)^{|M|-|L|} Q_L = \sum_{L\subseteq M} (-1)^{|M|-|L|} \left(R_U^1 - R_{U\setminus L}^1\right)$$

$$= (-1)^{|M|+1} \sum_{L\subseteq M} (-1)^{|L|} R_{U\setminus L}^1 = (-1)^{|U|+|M|+1} \sum_{L\supseteq U\setminus M} (-1)^{|L|} R_L^1.$$

The substitution of P_L^1 in (7.22) by the last presentation yields the first inequality of the theorem. ∎

7.6   Reliability Measures Based on Vertex Partitions

In this section, we present some generalizations of the K-terminal reliability that are defined via partitions of the vertex set. For a more detailed treatment of this subject the reader is referred to [293]. Let $G = (V, E)$ be a finite undirected graph. The edges of G may fail randomly and independently whereas the vertices are assumed to be perfectly reliable. Let $P(G, \pi)$ be the probability that G induces the partition $\pi \in \mathbb{P}(V)$. We denote by $E(G, \pi)$ the set of edges of G that link different blocks of π. The *partition probability* $P(G, \pi)$ is the probability that all blocks of π are separated and that each block forms a connected graph in G, i.e.

$$P(G, \pi) = \prod_{e\in E(G,\pi)} (1 - p_e) \prod_{X\in\pi} R(G[X]). \qquad (7.23)$$

The *partition reliability* of a graph G is defined by

$$R(G, \pi) = \sum_{\sigma\geq\pi} P(G, \sigma). \qquad (7.24)$$

Consequently, the partition reliability $R(G, \pi)$ is the probability that all vertices of any block of π are contained in one component of G. For the least and greatest element of $\mathbb{P}(V)$, we obtain

$$R(G, \hat{0}) = 1 = \sum_{\sigma\in\mathbb{P}(V)} P(G, \sigma),$$

$$R(G, \hat{1}) = P(G, \hat{1}) = R(G).$$

Let $K \subseteq V$ a given subset of terminal vertices of G. Then we define the partition $\pi(K) = K/\{\cdot\}/\{\cdot\}/.../\{\cdot\}$ as a partition of V consisting of the block K and singletons. Let $q_e = 1 - p_e$ be the edge failure probability for any $e \in E$. The application of (7.24) and (7.23) yields

$$R(G, K) = R(G, \pi(K))$$
$$= \sum_{\sigma \geq \pi(K)} \prod_{e \in E(G, \sigma)} q_e \prod_{X \in \sigma} R(G[X]),$$

which shows that the K-terminal reliability of G can be expressed as a sum over all-terminal reliabilities. The partition reliability can be considered as a kind of "at least connectivity" with respect to the given partition π. We require that at least each block of π taken by itself is connected. On the other hand, the demand that at most the blocks of π are connected leads us to the definition of the *separation probability*:

$$S(G, \pi) = \sum_{\sigma \leq \pi} P(G, \sigma)$$
$$= \prod_{e \in E(G, \pi)} q_e$$

By Möbius inversion, we obtain

$$P(G, \pi) = \sum_{\sigma \leq \pi} \mu(\sigma, \pi) \prod_{e \in E(G, \sigma)} q_e, \tag{7.25}$$

which gives together with (7.1)

$$R(G) = P(G, \hat{1}) = \sum_{\sigma \in \mathbb{P}(V)} (-1)^{|\sigma|-1} (|\sigma| - 1)! \prod_{e \in E(G, \sigma)} q_e. \tag{7.26}$$

This is the *node partition formula* first presented by *Buzacott* [77]. Let $\mathcal{C}(G)$ be the set of all cut sets of G that are minimal with respect to inclusion. If we separate the terms corresponding to one and two-block partitions of the sum in (7.26) then we obtain

$$R(G) = 1 - \sum_{C \in \mathcal{C}(G)} \prod_{e \in C} q_e + \sum_{\substack{\sigma \in \mathbb{P}(V) \\ |\sigma| \geq 3}} (-1)^{|\sigma|-1} (|\sigma| - 1)! \prod_{e \in E(G, \sigma)} q_e.$$

This equation suggests that $1 - \sum_{C \in \mathcal{C}(G)} \prod_{e \in C} q_e$ is a lower bound for the all-terminal reliability. This is correct, however, the expression is in some cases a very weak bound (or even completely worthless when lower than zero). A better lower bound is given for even k by

$$\sum_{\substack{\sigma \in \mathbb{P}(V) \\ |\sigma| \leq k}} (-1)^{|\sigma|-1} (|\sigma| - 1)! \prod_{e \in E(G, \sigma)} q_e + \frac{k!}{\left\{ {n \atop k+1} \right\}} \sum_{\substack{\sigma \in \mathbb{P}(V) \\ |\sigma| = k+1}} \prod_{e \in E(G, \sigma)} q_e,$$

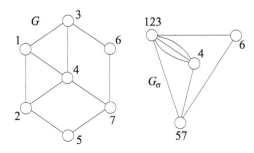

Figure 7.14: The merge operation for $\sigma = 123/4/57/6$

which is presented in [107].

Let G_σ be the graph obtained from G by merging the vertices of each block of $\sigma \in \mathbb{P}(V)$. Figure 7.14 shows an example graph G and the resulting graph G_σ. Notice that parallel edges are maintained in G_σ, whereas loops arising with the merge operation are removed. The graph G_σ may be connected even in case G is not connected. We can construct a graph with the same connectedness properties as G_σ by the following procedure. We introduce (failure-free) edges between all pairs of vertices of G that belong to the same block of σ. The resulting graph is denoted by G^σ. By Lemma 7.3 we obtain the following statement. If G induces the partition $\pi = \pi(G)$ then G^σ (and hence G_σ) induces the partition $\pi \vee \sigma$, which gives

$$R(G_\sigma, \tau) = \sum_{\pi:\pi\vee\sigma\geq\tau} P(G, \pi). \qquad (7.27)$$

An interesting special case is the all-terminal reliability:

$$R(G_\sigma) = \sum_{\pi:\pi\vee\sigma=\hat{1}} P(G, \pi) = \sum_{\pi\in\mathbb{P}(V)} a(\pi, \sigma) P(G, \pi),$$

where a denotes the a-function (7.2). By application of (7.3), we obtain

$$P(G, \pi) = \sum_{\sigma\in\mathbb{P}(V)} a^{-1}(\pi, \sigma) R(G_\sigma),$$

which yields via substitution in (7.24)

$$R(G, \tau) = \sum_{\pi\geq\tau} \sum_{\sigma\in\mathbb{P}(V)} a^{-1}(\pi, \sigma) R(G_\sigma).$$

The resilience of a graph, introduced in Section 4.4, can be represented as a sum over vertex partitions.

Theorem 7.15 *Let $G = (V, E)$ be a graph with independently failing edges. Let q_e be the failure probability of edge e for each $e \in E$. Then the resilience of G satisfies*

$$\text{Res}(G) = \sum_{\pi \in \mathbb{P}(V)} P(G, \pi) \sum_{X \in \pi} \binom{|X|}{2}$$

$$= \sum_{\pi \in \mathbb{P}(V)} \sum_{\sigma \leq \pi} \mu(\sigma, \pi) \prod_{e \in E(G, \sigma)} q_e \sum_{X \in \pi} \binom{|X|}{2}.$$

Proof. The second equality follows from Equation (7.25). To prove the first equality, we observe that the resilience is additive with respect to components. For a graph G consisting of k components $G_1, ..., G_k$, the equation

$$\text{Res}(G) = \sum_{i=1}^{k} \text{Res}(G_i)$$

follows from the fact that a communicating vertex pair has to be located within one component. The probability $P(G, \pi)$ corresponds to the random event that the operating edges of G induce a spanning subgraph whose components have the blocks of π as vertex sets. Now the number of communicating vertex pairs of G is the sum of communicating vertex pairs of its components. Clearly, a component of order $|X|$ leads to $\binom{|X|}{2}$ vertex pairs. ∎

We obtain a slightly different representation of $\text{Res}(G)$ if we sum over edge subsets first and then collect all edge subsets inducing the same partition of the vertex set:

$$\text{Res}(G) = \sum_{F \subseteq E} \prod_{e \in F} p_e \prod_{f \in E \setminus F} (1 - p_f) \sum_{X \in \pi(V, F)} \binom{|X|}{2}$$

Exercise 7.8 *Generalize the Node Partition Formula (7.26) as to compute the K-terminal reliability.*

Exercise 7.9 *Show that $R(G, K) = \sum_{X \supseteq K} R(G[X]) \prod_{e \in \partial(X)} q_e$, where $\partial(X)$ is the set of all edges of G that have exactly one of their end vertices in X.*

Exercise 7.10 * *Let λ be a (number) partition of the positive integer n, denoted by $\lambda \vdash n$. The representation $\lambda = 1^{k_1} \cdots n^{k_n}$ means that λ has k_j parts of size j for $j = 1, ..., n$. We define $w(\lambda) = \frac{1}{2}(n^2 - \sum_{i=1}^{r} \lambda_i^2)$, where $\lambda_1, ..., \lambda_r$ are the parts of λ. Prove by application of the node partition formula that the reliability polynomial of the complete graph is*

$$R(K_n, q) = n! \sum_{\substack{\lambda \vdash n \\ \lambda = 1^{k_1} \cdots n^{k_n}}} \prod_{j=1}^{n} \frac{1}{(j!)^{k_j} k_j!} (-1)^{|\lambda|-1} (|\lambda| - 1)! q^{w(\lambda)}.$$

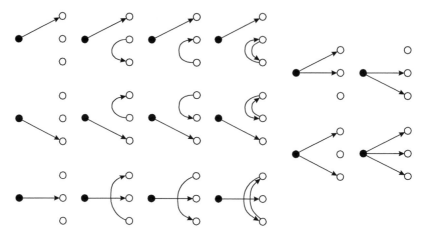

Figure 7.15: Different cases of connection

Exercise 7.11 *Show that for any fixed vertex $s \in V$ of a graph $G = (V, E)$, the resilience satisfies the recurrence equation*

$$\text{Res}(G) = \sum_{s \in X \subseteq V} R(G[X]) \prod_{e \in \partial X} (1 - p_e) \left[\binom{|X|}{2} + \text{Res}(G[V \setminus (X \cup \{s\})]) \right],$$

where ∂X is defined in Exercise 7.9.

7.7   Splitting in Directed Graphs

In this section, we investigate the splitting approach to the computation of the st-reachability in directed graphs. The edges of the digraph are supposed fail independently with given probabilities. Let $G = (V, E)$ be a directed graph and let $G^1 = (V^1, E^1)$, $G^2 = (V^2, E^2)$ be subgraphs of G such that $V^1 \cup V^2 = V$, $E^1 \cup E^2 = E$, $V^1 \cap V^2 = U \neq \emptyset$, and $E^1 \cap E^2 = \emptyset$. We assume that $s \in V^1$ and $t \in V^2$. Paths between s and t may cross the separating vertex set several times. In contrast to the undirected case, we have to distinguish directions of edges now, which leads to a dramatic increase of the number of terms of a splitting formula. We write in the following $u \underset{G}{\rightsquigarrow} v$ if there is a directed path from u to v in the digraph G.

Figure 7.15 shows all possibilities of connection within G^1, where here a separating vertex set U with three vertices is assumed. The (blackened) vertex s can be connected via directed paths to one, two, or three vertices of U. In addition, we have to distinguish if the remaining vertices of U are connected and in which direction they are reachable from each other. We call these connection possibilities *configurations* of G^1 and G^2, respectively. If $F \subseteq E$ is a set of operating edges of G such that (V, F) contains an st-path, then $F \cap E^1$ induces exactly one of the configurations of G^1.

Lemma 7.16 *Let G be a digraph with a separating vertex set of size u inducing the splitting of G into the subgraphs G^1 and G^2. Then the number of configurations of G^1 (and G^2) is*

$$c_u = \sum_{k=1}^{u} \binom{u}{k} \sum_{j=0}^{u-k} \left\{ \begin{matrix} u-k \\ j \end{matrix} \right\} A_j, \qquad (7.28)$$

where A_j denotes the number of acyclic transitive digraphs on j vertices.

Proof. First we select a subset of the separating vertex set U of vertices reachable from s in G^1. There are $\binom{u}{k}$ such selections of size k. Then we partition the remaining set X of cardinality $u - k$ into j blocks. The corresponding number of choices is given by the Stirling number of the second kind. Let π be a given partition of X. Each block of the partition π represents a strongly connected component of G^1. If we remove all vertices that can be reached from s and contract all blocks of the partition π, then an acyclic graph G' remains. We replace each path that connects blocks of π by a single directed edge between these blocks (vertices). We obtain an acyclic graph H with j vertices. It suffices to count acyclic transitive digraphs (or posets), since taking the transitive closure does not change the reachability of vertices. ∎

The numbers A_j have been computed by *Brinkmann* and *McKay* [70] up to $n = 16$. The numbers c_u grow rapidly with increasing size u of the separating vertex set. The first eight values are

$$1, \ 3, \ 16, \ 145, \ 2111, \ 47624, \ 1626003, \ 82564031.$$

Consequently, splitting seems suitable for digraphs only in case of very small vertex separators. However, there are special cases where splitting in digraphs is even simpler than in undirected graphs. Suppose all vertices of U have outdegree zero in G^1 (or no incoming edges in G^2). This condition is satisfied if G is acyclic, but also in more general situations. In this case, any path from s to t can pass the separating vertex set U only once. Let $P(G^1, s, X)$ be the probability that in G^1 all vertices of X, $X \subseteq U$ are reachable from s but no vertex of $U \setminus X$. We conclude that $\sum_{X \subseteq U} P(G^1, s, X) = 1$. For any given subset $X \subseteq U$, let G^2/X be the graph obtained by merging all vertices of X in G^2. We denote, in this context only, the vertex obtained by merging X again with X, which will not lead to any confusion. Then $R(G^2/X, X, t)$ equals the probability that there exists in G^2 an operating path from at least one vertex of X to t, which yields

$$R_{st}(G) = \sum_{\emptyset \neq X \subseteq U} P(G^1, s, X) R(G^2/X, X, t). \qquad (7.29)$$

This splitting formula has $2^{|U|} - 1$ terms, which is a considerable advantage in comparison with the splitting in general digraphs as given in Equation (7.28). The main problem when applying (7.29) is the computation of the probabilities $P(G^1, s, X)$. In order to surmount this obstacle, we relate the probabilities $P(G^1, s, X)$ to reachabilities, which provides the following theorem.

Theorem 7.17 *Let G be a digraph with a separating vertex set U such that $G^1 = (V^1, E^1)$ and $G^2 = (V^2, E^2)$ are the two split components of G. Assume further that all vertices of U have outdegree zero in G^1 or all vertices of U have indegree zero in G^2. Let s and t be vertices with $s \in V^1$ and $t \in V^2$. Then*

$$R_{st}(G) = \sum_{X \subseteq U} \sum_{Y \subseteq U : Y \supseteq U \setminus X} (-1)^{|X|+|Y|-|U|+1} R(G^1/Y, s, Y) R(G^2/X, X, t),$$

where $R(G^1/\emptyset, s, \emptyset) = R(G^2/\emptyset, \emptyset, t) = 0$.

Proof. First we observe that there exists a path from s to X in G^1/X if and only if there is a path from s to at least one vertex of X in G^1, which gives for any $X \subseteq U$,

$$R(G^1/X, s, X) = \sum_{\substack{Y \subseteq U \\ X \cap \bar{Y} \neq \emptyset}} P(G^1, s, Y)$$

$$= 1 - \sum_{Y \subseteq U \setminus X} P(G^1, s, Y).$$

Defining $f(X) = 1 - R(G^1/(U \setminus X), s, U \setminus X)$, we obtain

$$f(X) = \sum_{Y \subseteq X} P(G^1, s, Y),$$

which yields via Möbius inversion

$$P(G^1, s, X) = \sum_{Y \subseteq X} (-1)^{|X|-|Y|} f(Y)$$

$$= \sum_{Y \subseteq X} (-1)^{|X|-|Y|} \left[1 - R(G^1/(U \setminus Y), s, U \setminus Y) \right]$$

$$= \sum_{Y \subseteq X} (-1)^{|X|-|Y|+1} R(G^1/(U \setminus Y), s, U \setminus Y)$$

$$= \sum_{U \setminus Y \subseteq X} (-1)^{|X|+|Y|-|U|+1} R(G^1/Y, s, Y)$$

$$= \sum_{Y \supseteq U \setminus X} (-1)^{|X|+|Y|-|U|+1} R(G^1/Y, s, Y).$$

Substituting $P(G^1, s, X)$ in (7.29) by the last sum yields the statement of the theorem. ∎

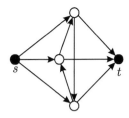

Figure 7.16: Example graph

Exercise 7.12 *Compute the st-reachability for the graph shown in Figure 7.16 by splitting according to Theorem 7.17. Assume that all edge reliabilities are equal (say p) and use the symmetry of the graph.*

Exercise 7.13 *Show that Theorem 7.17 remains valid for digraphs with a separating vertex set U of cardinality one or two even if we do not impose any indegree or outdegree requirements on vertices of U.*

Exercise 7.14 *Let G be a digraph with a separating vertex set U such that $G^1 = (V^1, E^1)$, $G^2 = (V^2, E^2)$ are the two split components of G, $s \in V^1$, $T \subseteq V^2$, and $d^-_{G^2}(u) = 0$ for all $u \in U$. Prove the following splitting formula:*

$$R(G, s, T) = \sum_{\emptyset \neq L \subseteq U} \sum_{L \subseteq M \subseteq U} (-1)^{|M|-|L|} R(G^1, s, M) R(G^2_L, L, T)$$

Chapter 8

Algorithmic Aspects of Network Reliability

An explicit formula is from the mathematical point of view the most satisfying solution of a problem. As an example, the reliability polynomial of a cycle, $R(C_n, p) = p^n + np^{n-1}(1 - p)$, is such a nice result. However, we may find rarely any problems in engineering applications that exhibit such closed-form solutions. Often we look for an algorithm that solves a general class of problems. Clearly we expect in the first place that the algorithm works correctly for any intended input. The input may be a graph from a well-prescript class of graphs (e.g. an undirected connected finite graph) and a set of parameters such as edge reliabilities. An answer can be a reliability polynomial or a numerical value for the desired reliability measure. In the latter case, we enter first difficulties in deciding what "a correct answer" means. Processing numerical data with computers always produces errors. If these errors are small enough then we will be willing to accept them. Numerical errors are usually not the hardest obstacle in reliability analysis, since in most cases we do not need a great number of iterations.

There is a second problem that causes more trouble. If we design an algorithm that computes the reliability of a network with 20 vertices within 40 seconds, but for a network with 30 vertices the computation needs several years then the user is perhaps not particularly happy. The investigation of the complexity of reliability problems is the topic of the first section in this chapter. Often we can find special classes of graphs that permit an efficient reliability computation, even if the same problem is intractable for general graphs. In further sections, we present special graph classes and corresponding algorithms. If the networks under consideration do not belong to a graph class permitting efficient computation then we are forced to search for bounds, approximation and simulation methods.

8.1 Complexity of Network Reliability Problems

We assume that the reader is familiar with basic notions of complexity theory, otherwise we recommend classic texts, such as [133] and [246]. The first property that our problem should possess is an input size of a network reliability problem that is polynomially bounded by the number of vertices of the network. This property excludes lists of all cut sets or all path sets as input. On the other hand, we should not accept real numbers as parameters for edge or vertex reliabilities. Hence we assume that our algorithms deal with rational numbers exclusively, which does not lead to any restrictions for practical applications.

Valiant [300] developed a complexity theory for counting problems in combinatorics that is similar to the well-known theory of **NP**-completeness. The hardest problems among the class **#P**, which consists of counting problems that can be solved in polynomial time by a non-deterministic Turing machine equipped with a counting mechanism, are called **#P**-*complete* problems. *Rosenthal* [261] and *Ball* [10] showed that almost all reliability problems considered here belong to the class of **#P**-complete problems. This means that it is extremely unlikely that we can find a polynomial algorithm for these problems. The general strategy to prove that a given counting problem is **#P**-complete is similar to the **NP**-completeness proofs. First we have to show that there is a **#P**-complete problem, i.e. we need an equivalent to Cook's theorem, then we use polynomial-time reductions, see [305].

In order to simplify the presentation, we prove here only that a reliability problem is **NP**-hard rather than it is **#P**-complete. To show that a reliability problem Q is **NP**-hard it suffices to find an **NP**-complete problem Q' such that any instance of Q' can be transformed in polynomial time into an instance of Q and the corresponding solution of Q can be converted efficiently into an answer for Q'.

Example 8.1 *A well-known* **NP**-*complete problem is the Steiner tree problem in graphs, which is defined as follows.*

Input: *An undirected graph $G = (V, E)$, a subset $K \subseteq V$ of terminal vertices, and a positive integer l.*

Problem: *Is there a Steiner tree in G with respect to K with at most l edges?*

A Steiner tree is a tree $T = (W, F)$ in G such that $K \subseteq W$ and if v is a degree-1-vertex (a leaf) of T then $v \in K$. Now assume we have computed the K-terminal reliability $R(G, K, p)$ of G as a polynomial in p, where p denotes the common edge reliability of all edges. Remember that the coefficients of $R(G, K, p)$ correspond to the number of path sets of a given size in G. But a path set of minimum size is a Steiner tree, because it must be a connected graph without any cycles containing all terminal vertices. Consequently, the coefficient of the smallest power in p in $R(G, K, p)$ gives the number of minimum Steiner trees in G. If the minimum degree of the polynomial $R(G, K, p)$

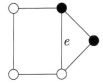

Figure 8.1: Generation of irrelevant edges

*is at most l then we know that G has a Steiner tree with at most l edges. Hence the K-terminal reliability problem is at least as hard as the Steiner tree problem in graphs, which means it is **NP**-hard.*

The numerical version of any reliability problem is not harder than the computation of a corresponding reliability polynomial since we can easily evaluate a polynomial at any rational point. Here we mean with "reliability polynomial" not only the all-terminal reliability polynomial but any univariate polynomial that solves a reliability problem for identical edges or vertex reliabilities. Also, the reverse statement is true: If we are able to solve a reliability problem numerically, then we can easily determine the reliability polynomial, since any polynomial of degree n is uniquely reconstructible from its values at $n + 1$ points by solving a system of linear equations for the coefficients.

8.2 Decomposition–Reduction Algorithms for Network Reliability

The simplest algorithm for all kinds of reliability problems consists of enumerating all path sets and adding up the corresponding probabilities, which means essentially that we have to list all subsets of the edge set of a graph. We can do slightly better by listing only minimum path sets or cut sets. In case of the all-terminal reliability, the path sets are the edge sets of spanning trees. There are different algorithms for the generation of all spanning trees of an undirected graph available, see e.g. [279]. However, even if we have the set of all spanning trees or cuts of the graph, we need additional procedures like inclusion-exclusion or disjoint sum methods as explained in the first part of this book.

A more direct approach to K-terminal reliability is the application of edge decomposition formulae in combination with reduction techniques. Algorithm 8.1 shows the framework of decomposition-reduction methods.

Clearly, the implementation of Algorithm 8.1 requires some care. We can avoid the test for K-connectedness if we ensure that the input graph is K-connected and that we choose an edge e for the decomposition such that $G - e$ remains K-connected. The proper choice of the decomposition edge is also essential in order to avoid the creation of irrelevant edges. Figure 8.1 shows

---

Algorithm 8.1 KREL(G)

Input: G,K {graph, set of terminal vertices}, **p** {edge reliabilities}
Output: $R(G)$ {K-terminal reliability of the graph}
  if G is not K-connected **then**
    return 0
  end if
  Perform all possible reductions.
  if G is one single terminal vertex **then**
    return 1
  else
    Choose an edge e.
    return $p_e \cdot \text{KREL}(G/e) + (1 - p_e) \cdot \text{KREL}(G - e)$
  end if

---

a graph with two terminal vertices and an edge e whose contraction produces irrelevant edges.

The step *"Perform all possible reductions"* obtains a clear meaning if we define in advance a list of reductions to be performed. In addition, we need subprocedures that find reducible configurations such as parallel edges, degree-2-vertices, or triangles. In some cases, a special order of reductions is desirable. For instance, the combination of a degree-2-reduction and a parallel reduction (if possible) may reduce the effort of searching parallel edges. Another item to be considered is the handling of reduction factors that arise in bridge reductions or polygon-to-chain reductions. Assume Ω is the product of all reduction factors obtained during one reduction phase of the algorithm, i.e. before reaching a return statement. Then the return value has to be multiplied by Ω.

The Algorithm 8.1 is quite flexible with respect to change of the reliability measure or the introduction of additional parameters. Assume, for instance, that we want to include vertex reliabilities as introduced in Section 5.3. Then we can include two additional steps, namely the choice of an unreliable vertex and a return statement representing the vertex decomposition formula (5.12). Furthermore, we may want to extend the algorithm by insertion of special reductions respecting vertex failures, which is also easily done by placing them in a list of allowed reductions and designing the corresponding subroutines. Even completely different reliability measures, like the *diameter constrained network reliability* [80] or the residual network reliability (see Section 5.4) can be computed using the decomposition-reduction framework.

The set of reductions used within Algorithm 8.1 is essential for the complexity. Assume we apply exclusively bridge reductions for the computation of the all-terminal reliability of a graph G, which means essentially that we do not produce any disconnected graphs while applying the edge decomposition. The number of leaves in the decomposition tree is in this case equal to the number of spanning trees of G. When we enrich our algorithm with degree-2 and parallel reductions, then the number of leaves of the decomposition tree drops down to the domination $D(G)$, which is a result of *Satyanarayana* and *Chang* [269].

Example 8.2 *Let G be a 3×3 grid graph consisting of nine vertices and twelve edges. Then an edge decomposition algorithm for the computation of the all-terminal reliability without any reductions has to solve $2^{12} = 4096$ subproblems. An algorithm with bridge reductions produces $t(G) = 192$ leaves in the decomposition tree. The application of bridge and parallel reductions leads to 79 subproblems (leaves). The additional application of degree-2-reductions reduces the total computational effort to only three subproblems.*

Algorithm 8.2 $\mathrm{Res}(G)$

Input: (G, ϕ) {a vertex labeled graph}, **p** {edge reliabilities}
Output: $\mathrm{Res}(G)$ {resilience of the graph}
  Perform all parallel and degree-1 reductions.
  if G is edgeless **then**
    return $\sum_{v \in V(G)} \binom{\phi(v)}{2}$
  else
      Choose an edge e.
      return $p_e \cdot \mathrm{Res}(G/e) + (1 - p_e) \cdot \mathrm{Res}(G - e)$
  end if

In Section 4.4, we introduced the resilience $\mathrm{Res}(G)$ as the expected number of vertex pairs that can communicate via paths of operating edges in G. The computation of $\mathrm{Res}(G)$ by a decomposition algorithm requires a vertex labeling $\phi : V \to \mathbb{Z}^+$. The proposed algorithm takes as input a graph G together with its edge reliabilities. In addition, we assign an initial label value of 1 to each vertex of G. The vertex labels are modified with each edge contraction. If edge $e = \{u, v\}$ is contracted, then the new label for the vertex obtained from merging u and v is the sum of the vertex labels of u and v. We maintain all other vertex labels of G. We know that $\mathrm{Res}(G) = 0$ if G consists of a single vertex, which we use as initial condition for the algorithm. However, if G consists of a single v vertex with a vertex label $\phi(v) = k > 1$, then this vertex represents a connected component of k vertices merged into one vertex through a sequence of edge contractions. In this case, the correct value of the resilience is $\binom{k}{2}$ giving the number of vertex pairs of a graph of order k. Algorithm 8.2 shows the main steps of the computation of $\mathrm{Res}(G)$. Note that the edge contraction step involves the modification of vertex labels. There are two reductions applied in this algorithm. The parallel reduction is used in the same manner as for the computation of the all-terminal reliability. The degree-1 reduction is more interesting. This reduction can be applied if G contains a vertex v of degree 1 with the label $\phi(v) = 1$. If the single edge $e = \{v, w\}$ that is incident to v fails then the component $(v, \emptyset)$ does not contribute to $\text{Res}(G)$. Hence we can contract e using p_e as reduction factor. The new label of w (the only adjacent vertex of v) is $\phi(w) + 1$.

Example 8.3 *Figure 8.2 shows the application of Algorithm 8.2 to a graph of order four. The abbreviations G1 and PA stand for degree-1 and parallel reduction, respectively. The return values are given in boxes at the leaves of the decomposition tree. The final result for this example graph is*

$$\text{Res}(G) = 2ae + 2bd + 4abe + 4ace + 4ade + 6bce + 4bde + 4abd + 6acd + 4bcd$$
$$- 10abce - 14abde - 10acde - 10bcde - 10abcd + 20abcde.$$

Exercise 8.1 *Design a decomposition-reduction algorithm for the computation of the st-reachability in directed graphs (see Section 5.5). This task requires the definition of a suitable set of reductions for digraphs and a clear strategy for the choice of the decomposition edge.*

Exercise 8.2 *Extend the degree-1 reduction for the resilience in order to cover the case where the vertex weight $\phi(v)$ of degree-1 vertex v is greater than 1.*

8.3 Algorithms for Special Graph Classes

Almost all network reliability problems considered in this book are **NP**-hard for general graphs. For special graph classes, like trees, cycles, wheels, or complete graphs, we can often find polynomial algorithms. In this section, we investigate more general graph classes that are still accessible by efficient algorithms. First, we define a measure of graph complexity that we usually call *width* of the graph. Then we restrict the allowed input graphs to those having width at most k for a given positive integer k. Assume we have an algorithm A for a certain network reliability problem. If there exist a constant a and a function f such that the running time of the algorithm A is bounded by $f(k) \cdot n^a$, $n = |V(G)|$, then we call the problem *fixed-parameter tractable* – a concept introduced by *Downey* and *Fellows* [110].

8.3.1 Graphs of Bounded Tree-Width

The notion of tree-width is one of the most important advances in graph theory during the last three decades. It was introduced by *Robertson* and *Seymour* [255]. Tree-width measures in a certain sense the tree-likeness of a graph. It is also a measure of global decomposability or of complexity of a graph. Besides its numerous theoretical consequences and applications, graphs of bounded tree-width often permit an easy (polynomial time) solution of otherwise computationally hard problems. On the other hand, graphs of bounded tree-width form a very general graph class (even if some important graph classes, like planar graphs, may have arbitrarily high tree-width).

We start here with a concise introduction into tree and path decompositions of graphs (for a deeper introduction, see [56]) and proceed with applications in network reliability analysis.

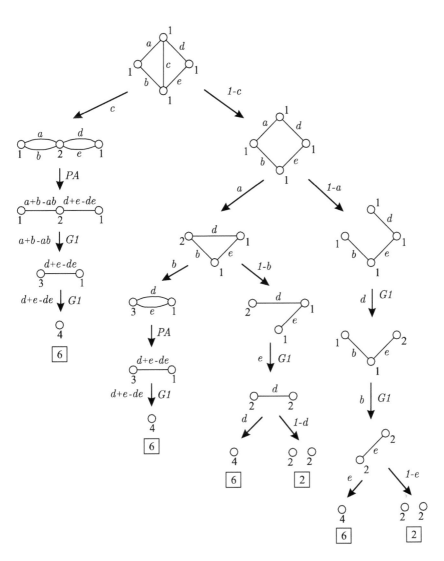

Figure 8.2: A decomposition algorithm for resilience

Tree Decompositions

Let $G = (V, E)$ be an undirected graph. A *tree decomposition* of G is a pair (T, ϕ) where $T = (W, F)$ is a tree and $\phi : W \to 2^V$ is a mapping that assigns a subset of V to each tree *node* (vertex of the tree) such that the following conditions are satisfied:

1. The union of all sets $\phi(w)$ for $w \in W$ equals V.

2. For each edge $\{u, v\} \in E$, there exists a node $w \in W$ with $\{u, v\} \subseteq \phi(w)$.

3. If $u, v, w \in W$ and v is on the path from u to w in T then $\phi(u) \cap \phi(w) \subseteq \phi(v)$.

Remark 8.4 *Here and in the following we use the term* node *for a vertex of the tree* T *in order to distinguish it easily from the vertices of the graph* G.

Often we identify the nodes of T with the assigned subsets of V. Thus we say also a vertex $v \in V$ "is contained" in a node $w \in W$. The third condition implies that all nodes of T that contain a given vertex $v \in V$ induce a subtree of T. The *width* of a tree decomposition $((W, F), \phi)$ is max $\{|\phi(w)| : w \in W\} - 1$. The *tree-width* $tw(G)$ of a graph G is the minimum width of a tree decomposition of G. Any graph G has a trivial tree decomposition where T consists of one node w with $\phi(w) = V$ and no edge.

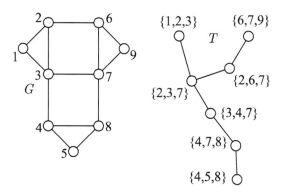

Figure 8.3: A graph and its tree decomposition

Example 8.5 *Figure 8.3 shows a graph together with a tree decomposition of width 2 for this graph. One can easily show that any tree decomposition of a graph G must contain each clique of G completely in one node. Since our example graph possesses cliques of size 3, the tree decomposition presented is of minimum width. Consequently, the graph has tree-width 2.*

A tree decomposition (T, ϕ) of G is called *nice* if the tree T is rooted such that each node has at most two children, and if v is the son of u in T then $\phi(v) \subseteq \phi(u)$ or $\phi(u) \subseteq \phi(v)$ such that $\phi(u)$ and $\phi(v)$ differ in exactly one vertex. It can be shown that each tree decomposition can be transformed easily into a nice tree decomposition.

Remark 8.6 *A nice tree decomposition is defined differently by other authors, see [56].*

A *path decomposition* of a graph G is a tree decomposition (T, ϕ) of G such that T is a path. The *path-width* $pw(G)$ is the minimum width of all path decompositions of G. There is an easier way to define a path decomposition. A path decomposition of a graph $G = (V, E)$ is a sequence $X = (X_1, ..., X_r)$ of subsets of V such that

- $\bigcup\limits_{i=1}^{r} X_i = V,$

- for any edge $\{u, v\} \in E$, there is an $i \in \{1, ..., r\}$ such that $\{u, v\} \subseteq X_i$,

- for all $i, j, k \in \{1, ..., r\}$, if $i < j < k$ then $X_i \cap X_k \subseteq X_j$.

Example 8.7 *(continued) The following sequence is a path decomposition for the graph G of Figure 8.3:*

$$(\{1, 2, 3, 6\}, \{3, 6, 7, 9\}, \{3, 4, 7, 8\}, \{4, 5, 8\})$$

It can be shown that this path decomposition has minimum width, i.e. $pw(G) = 3$.

A path decomposition is called *nice* if the first and last set of the sequence X is empty and if each two successive sets differ in exactly one vertex. Any path decomposition can be transformed into a nice one. There is a concise form to store a nice path decomposition as a sequence $\mathbf{x} = (x_1, ..., x_{2n})$ of signed vertices of G. We define for $i = 1, ..., 2n$,

$$x_i = \begin{cases} +v \text{ if } X_{i+1} \setminus X_i = \{v\} \\ -v \text{ if } X_i \setminus X_{i+1} = \{v\}. \end{cases}$$

Each vertex of G appears exactly twice in a nice path decomposition — once with positive and once with negative sign.

A *composition order*, a notion introduced in [251], of an undirected graph $G = (V, E)$ is a sequence $\mathbf{y} = (y_1, ..., y_{2n+m})$ of signed vertices and edges of G such that

1. the removal of all edges from $\mathbf{y}$ results in a nice path decomposition of G,

2. each edge $e \in E$ appears exactly once in $\mathbf{y}$,

3. if $y_k = \{u, v\}$ is an edge of G then there exist four different indices i, j, p, q satisfying $i < k$, $j < k$, $p > k$, $q > k$ such that $y_i = u$, $y_j = v$, $y_p = -u$, and $y_q = -v$.

Any valid composition order *defines* a graph. Hence, we may consider a composition order as a way to store a graph. Each graph $G = (V, E)$ with $V = \{v_1, ..., v_n\}$ and $E = \{e_1, ..., e_m\}$ has the trivial composition order, namely $(v_1, ..., v_n, e_1, ..., e_m, -v_1, ..., -v_n)$.

Example 8.8 *(continued) A nice path decomposition for the graph of Figure 8.3 is*

$$\mathbf{x} = (1, 2, 3, -1, 6, -2, 7, 9, -6, -9, 4, -3, 8, -7, 5, -4, -8, -5).$$

As a composition order, we obtain

$$\mathbf{y} = (1, 2, \{1, 2\}, 3, \{1, 3\}, \{2, 3\}, -1, 6, \{2, 6\}, -2, 7, \{6, 7\}, \{3, 7\}, 9,$$
$$\{6, 9\}, \{7, 9\}, -6, -9, 4, \{3, 4\}, -3, 8, \{7, 8\}, \{4, 8\}, -7, 5, \{4, 5\},$$
$$\{5, 8\}, -4, -8, -5).$$

Note that this composition order is not unique, even with respect to the given path decomposition.

For algorithmic purposes, a path (or tree) decomposition of minimum width is highly desirable, because the computational effort depends heavily on the width of the decomposition. Unfortunately, the determination of an optimal path or tree decomposition, i.e. one of minimal width, is itself an **NP**-hard problem. However, there exist different heuristic methods to find a "good" path or tree decompositions and methods to determine bounds for the tree-width [57, 58].

All-Terminal Reliability in Graphs of Bounded Path-Width

The basic idea of a path decomposition algorithm is the sequential reconstruction of the given graph using a given composition order as a "construction manual." We distinguish three main steps: *introducing a vertex, removing a vertex* and *processing an edge*. We denote these basic steps (according to the composition order) by $+v$, $-v$ and $\{u, v\}$, respectively. Each step corresponds to a *transformation* of a *set of states*. A *state* is a pair (π, f) consisting of an *index* π and a *value*. Let $\mathbf{y} = (y_1, ..., y_r)$ be the given composition order of the graph $G = (V, E)$. We refer to the number i in y_i as the *step* of the composition order. For any given step $i \in \{1, ..., r\}$, we define

$$A_i = (V \cap \{y_1, ..., y_i\}) \setminus (V \cap \{-y_1, ..., -y_i\})$$

and call A_i the *active vertex set* in step i. For many enumeration and reliability problems on graphs, the index is a subset, a set partition, or a labeled set partition of the active vertex set. For the all-terminal reliability problem, we use a partition $\pi \in \mathbb{P}(A_i)$ as index for any state in step i. The value corresponds to the objective of our calculation. For the all-terminal reliability, we expect a value in form of a rational number (a probability) or a polynomial with integer coefficients (the reliability polynomial of G).

Assume we have individual reliabilities p_e ($e \in E$) for the edges as input for the problem, in addition to the composition order $\mathbf{y}$. Then the desired output of the algorithm is a numerical value for the all-terminal reliability. Thus the value of a state is also a numerical value in $[0, 1]$. The state set is initialized by $Z_0 = \{(\emptyset, 1)\}$. Here $\emptyset$ stands for the empty set partition. Each step in the algorithm performs a transformation $Z_{i-1} \longmapsto Z_i$ of the state set. The transformations are the following:

1. **Introducing a vertex** $(+v)$: For a given partition $\pi \in \mathbb{P}(A_{i-1})$, let π/v be the partition obtained from π by extending π with the singleton $\{v\}$. The the new state set is defined by

$$Z_i = \{(\pi/v, f) \mid (\pi, f) \in Z_{i-1}\}.$$

Observe that the value of all states remains unchanged while performing this transformation.

2. **Removing a vertex** $(-v)$: We construct the new state set Z_i in three phases. First we remove all states from Z_{i-1} that have an index containing $\{v\}$ as a singleton. We call the modified state set Z_i^*. The reason for this reduction is that after the removal of v no edge can be attached to v, hence those states do not lead to connected graphs. During the second phase we delete v from the index of any state in Z_i^*. The result is a state set Z_i^+. Finally, we reduce the number of states of Z_i^+ by combining states with the same index. Let (π, f_1) and (π, f_2) be two states in Z_i^+ with coinciding index. Then we replace these states by one new state, which is $(\pi, f_1 + f_2)$. This last phase produces the new state set Z_i, in which any two states have different indices.

3. **Processing an edge** (e): Let $e = \{u, v\}$ be the edge to be processed. We denote by $\pi \vee e$ the partition obtained from $\pi \in \mathbb{P}(A_i)$ by merging the blocks of π that contain u and v. If both vertices u and v are contained in the same block of π then $\pi \vee e = \pi$. Each state $(\pi, f) \in Z_{i-1}$ is replaced by two states $(\pi, (1 - p_e) f)$ and $(\pi \vee e, p_e f)$. Finally, in order to obtain Z_i, we replace any two states (π, f_1) and (π, f_2) with a coinciding index by one new state $(\pi, f_1 + f_2)$.

If in step i the state $(\hat{1}, f)$ is a member of Z_i then f is the all-terminal reliability of the graph that consists of all vertices introduced up to step i and all edges processed so far. For a given graph $G = (V, E)$ with a composition order $\mathbf{y} = (y_1, ..., y_r)$, we define $V_i = V \cap \{y_1, ..., y_i\}$ and $E_i = E \cap \{y_1, ..., y_i\}$. Then the graph $G_i = (V_i, E_i)$ is the subgraph of G that is composed until step i. The value f is for each state $(\pi, f) \in Z_i$ the probability that the components of G_i induce the partition $\pi \in \mathbb{P}(A_i)$ *and* that all vertices of $V_i \setminus A_i$ are connected by a path to a vertex in A_i.

Algorithm 8.3 All-terminal reliability via path decomposition

---

Input: $\mathbf{z}$ {composition order}, $\mathbf{p}$ {vector of edge reliabilities}
Output: $R(G)$ {all-terminal reliability of the graph defined by $\mathbf{z}$}
   $Z := \{(\emptyset, 1)\}$ {initialization of the state set}
   for $i = 1$ to $2n + m$ **do**
      if $v := z_i$ is a vertex **then**
         IncludeVertex(v)
      else if $v := -z_i$ is a vertex **then**
         RemoveVertex(v)
      else
         ProcessEdge(z_i)
      end if
   end for
   return *value(Z)*

Algorithm 8.3 shows the computation of the all-terminal reliability based on a given composition order of the graph. The number of transformations of the state set is given by the length of the composition order, which is $2n + m$. The number of states in step i is given by $|\mathbb{P}(A_i)| = B(|A_i|)$, i.e. the number of partitions of A_i. Hence, for a path decomposition of width k, the maximum number of states in one transformation step is $B(k)$. The time complexity of Algorithm 8.3 is linear in $m + n$ but exponential in k. Consequently, to find a path decomposition of small width is crucial to the practical application of this algorithm.

Algorithm 8.3 can easily be modified in order to determine the reliability polynomial of a graph. All we have to do is to replace the edge reliabilities p_e by a formal variable p and to define the value of a state as a polynomial.

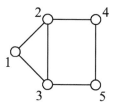

Figure 8.4: Example graph

Example 8.9 *The graph depicted in Figure 8.4 can be represented by the*

composition order

$$[1, 2, 12, 3, 13, -1, 23, 4, 24, -2, 5, 35, -3, 45, -4, -5],$$

where we used uv as a short notation for the edge $\{u, v\}$. The Table 8.1 shows the computation of the reliability polynomial of the example graph with Algorithm 8.3. The first column indicates the transformation step. The second and third row of the table shows the index and value of a state, respectively. The value of the last row yields the desired reliability polynomial.

All-Terminal Reliability in Graphs of Bounded TreeWidth

Let $G = (V, E)$ be an edge-weighted graph with a given nice tree decomposition (T, ϕ). In this case, $T = (W, F)$ is a rooted tree in which each node has at most two children. The algorithm for the computation of the all-terminal reliability of G is quite similar to the above described path decomposition algorithm. The algorithm again uses a state set that is transformed into a new one in each step. A step is an introduction of a vertex, the removal of a vertex, processing an edge, or processing a branch node of the tree. The set of states assigned to a node v of the decomposition tree is denoted by Z_v. If $z = (\pi, f)$ is a state then we write $\pi = \text{ind } z$ and $f = \text{val } z$ in order to indicate that π is the index and f the value of z. At the beginning, we label all edges of G as "unprocessed."

We start the process by choosing a leaf x of the tree T. The nice tree decomposition ensures that the leaf x contains exactly one vertex v of G. We assign the state set $Z_x = \{(v, 1)\}$ to the leaf, corresponding to the start partition consisting of the single block $\{v\}$ and the probability 1 as value. Assume that a certain node w of the tree is reached within the last step; then we climb up the tree. If the father u of the presently reached node w contains one vertex (say v) more than w then we perform the introduction step $(+v)$, which results in a transformed state set. All unprocessed edges that link the introduced vertex with already activated vertices are processed immediately after the vertex introduction. An important modification with respect to the path decomposition algorithm is that we label each processed edge of the graph, in order to avoid that an edge is processed twice in different branches of the tree. In case the father node u contains one vertex less than w, we apply the vertex removal $(-v)$.

Something new happens when we reach a branching node y while climbing up the tree. In this case, we stop after introducing the new vertex and processing the edges. If we reach the node y the first time then we store the corresponding state set as Z_y^1 and restart the process with a leaf node not considered so far. Otherwise the node y has been encountered once already, which means there is a state set Z_y^1 stored for this node. Now we have generated a second state set Z_y^2 by entering y again. The final state set Z_y is generated

transformation step	index	value
init	$\emptyset$	1
+1	1	1
+2	1/2	1
$\{1,2\}$	1/2	$1-p$
	12	p
+3	1/2/3	$1-p$
	12/3	p
$\{1,3\}$	1/2/3	$(1-p)^2$
	13/2	$(1-p)\,p$
	12/3	$(1-p)\,p$
	123	p^2
-1	2/3	$2(1-p)p$
	23	p^2
$\{2,3\}$	2/3	$2(1-p)p$
	23	$p^2+2(1-p)p^2$
+4	2/3/4	$2p-4p^2+2p^3$
	23/4	$3p^2-2p^3$
$\{2,4\}$	2/3/4	$2p-6p^2+6p^3-2p^4$
	24/3	$2p^2-4p^3+2p^4$
	23/4	$3p^2-5p^3+2p^4$
	234	$3p^3-2p^4$
-2	3/4	$5p^2-9p^3+4p^4$
	34	$3p^3-2p^4$
+5	3/4/5	$5p^2-9p^3+4p^4$
	34/5	$3p^3-2p^4$
$\{3,5\}$	3/4/5	$5p^2-14p^3+13p^4-4p^5$
	35/4	$5p^3-9p^4+4p^5$
	34/5	$3p^3-5p^4+2p^5$
	345	$3p^4-2p^5$
-3	4/5	$8p^3-14p^4+6p^5$
	45	$3p^4-2p^5$
$\{4,5\}$	4/5	$8p^3-22p^4+20p^5-6p^6$
	45	$11p^4-16p^5+6p^6$
-4	5	$11p^4-16p^5+6p^6$
-5	$\emptyset$	$11p^4-16p^5+6p^6=R(G,p)$

Table 8.1: Computation of the reliability polynomial via path decomposition

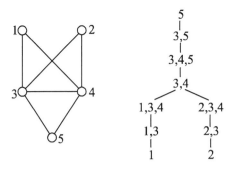

Figure 8.5: A graph and a nice tree decomposition

out of Z_y^1 and Z_y^2 as follows. Let π be a partition of the active vertex set (i.e. the set of those vertices of G that appear in the indices of states) and define

$$A(\pi) = \left\{ (a, b) \mid a \in Z_y^1,\ b \in Z_y^2,\ \text{ind } a \vee \text{ind } b = \pi \right\}.$$

The new state set is then defined by

$$Z_y = \left\{ \left(\pi, \sum_{(a,b) \in A(\pi)} (\text{val } a)\,(\text{val } b) \right) : A(\pi) \neq \emptyset \right\}.$$

The algorithm terminates as soon as the there is a state set Z_r assigned to the root r of the decomposition tree. We can easily verify that Z_r has exactly one state. The value of this state is the all-terminal reliability of G.

Example 8.10 *Figure 8.5 shows an example graph G together with a nice tree decomposition for this graph. We start the computation of the reliability polynomial (all-terminal reliability) at node 1. In Figure 8.5, the nodes of the tree are denoted by the list of vertices of G contained in the node. The following table shows the first part of the state transformations:*

Transformation step	Index	Value
+1	1	1
+3, $\{1,3\}$	1/3	$1 - p$
	13	p
+4, $\{1,4\}$, $\{3,4\}$	1/3/4	$1 - 3p + 3p^2 - p^3$
	13/4	$p - 2p^2 + p^3$
	14/3	$p - 2p^2 + p^3$
	34/1	$p - 2p^2 + p^3$
	134	$3p^2 - 2p^3$
-1	3/4	$2p - 4p^2 + 2p^3$
	34	$3p^2 - 2p^3$

Now the calculation stops, because we have reached a branching node x. The last two rows of the table display the state set Z_x^1. The process restarts with the leaf denoted by 2:

Transformation step	Index	Value
$+2$	2	1
$+3, \{2,3\}$	2/3	$1 - p$
	23	p
$+4, \{2,4\}$	2/3/4	$1 - 2p + p^2$
	23/4	$p - p^2$
	24/3	$p - p^2$
	234	p^2
-2	3/4	$2p - 2p^2$
	34	p^2

Here the last two rows correspond to Z_x^2. Observe that the edge $\{3,4\}$ does not appear in this table. This edge has been labeled as "processed" within the first phase of the calculation. Now the branching step starts:

Transformation step	Index	Value
branching	3/4	$4p^2 - 12p^3 + 12p^4 - 4p^5$
	34	$8p^3 - 11p^4 + 4p^5$
$+5, \{3,5\}, \{4,5\}$	3/4/5	$4p^2 - 20p^3 + 40p^4 - 40p^5 + 20p^6 - 4p^7$
	34/5	$8p^3 - 27p^4 + 34p^5 - 19p^6 + 4p^7$
	35/4	$4p^3 - 16p^4 + 24p^5 - 16p^6 + 4p^7$
	45/3	$4p^3 - 16p^4 + 24p^5 - 16p^6 + 4p^7$
	345	$20p^4 - 42p^5 + 31p^6 - 8p^7$
-3	4/5	$12p^3 - 43p^4 + 58p^5 - 35p^6 + 8p^7$
	45	$20p^4 - 42p^5 + 31p^6 - 8p^7$
-4	5	$20p^4 - 42p^5 + 31p^6 - 8p^7$

An alternative approach to the computation of network reliability in graphs of bounded tree-width is given in [308].

8.3.2   Domination Reliability of Cographs

The investigation of the following graph class requires the notion of a complement of a graph. Let $G = (V, E)$ be a simple graph. Then the *complement* of G is a simple graph $\bar{G}$ with vertex set V in which two vertices are adjacent if and only if these vertices are non-adjacent in G. A *cograph* is a graph that can be obtained from graphs $K_1 = (\{v\}, \emptyset)$, i.e. from single vertices, by repeated operations of the form disjoint union and join of graphs.

More precisely, a cograph can be defined recursively as follows:

- The complete graph K_1, consisting of one vertex and no edges, is a cograph.

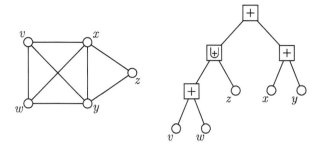

Figure 8.6: A cograph and it cotree

- If G and H are cographs, then their disjoint union $G \uplus H$ is a cograph, too.

- If G and H are cographs, then their join $G + H$ is a cograph.

Cographs have some interesting properties. They can be defined, alternatively, as graphs that can be constructed from single vertices by recursive application of disjoint union and forming the complement of the graph. This construction follows from the fact that a join of two graphs G and H can be realized by taking the complement of G and H separately, forming the disjoint union of these two complements and taking the complement of the whole graph. Cographs are exactly those graphs not containing a P_4 (a path with four vertices) as vertex induced subgraph [67]. Figure 8.6 shows a cograph with five vertices and its *cotree*. The cotree T that is assigned to a cograph $G = (V, E)$ is a tree whose leaves are the vertices of V. The internal nodes of T correspond to the two operations disjoint union ($\uplus$) and join ($+$).

Now let $G = (V, E)$ be a cograph and T a cotree assigned to G. Assume further that for each vertex $v \in V$ a vertex availability p_v is given. Our aim is the computation of the domination reliability $D(G)$ of G. We know from Section 5.6, Lemma 5.23 and Theorem 5.28 that the domination reliability of a disconnected graph is the product of the domination reliabilities of its components. Lemma 5.24 provides a formula for the join of two graphs G and H:

$$
D(G + H) = \left(1 - \prod_{v \in V(G)} (1 - p_v) \right) \left(1 - \prod_{w \in V(H)} (1 - p_w) \right) \tag{8.1}
$$
$$
+ D(G) \prod_{w \in V(H)} (1 - p_w) + D(H) \prod_{v \in V(G)} (1 - p_v)
$$

Example 8.11 *The application of the two elementary transformations to the* *graph given in Figure 8.6 yields*

$$D(G) = p_x + p_y - p_x p_y + p_v p_z + p_w p_z - p_v p_w p_z - p_v p_y p_z$$
$$- p_w p_y p_z - p_v p_x p_z - p_w p_x p_z + p_v p_w p_x p_z$$
$$+ p_v p_w p_y p_z + p_v p_x p_y p_z + p_w p_x p_y p_z - p_v p_w p_x p_y p_z.$$

8.3.3   Graphs of Bounded Clique-Width

Graphs of bounded clique-width form a graph class that generalizes simultaneously the classes of bounded tree-width graphs and cographs. Consequently, graphs of high edge-density like complete graphs and complete bipartite graphs as well as sparse graphs like trees or series-parallel graphs have small clique-width. In order to define the clique-width of a graph, a concept introduced in [98], we need labeled graphs.

Let k be a positive integer. A *labeled graph* or a *k-graph* $G = (V, E, \alpha)$ is a simple graph (V, E) together with a labeling function $\alpha : V \to \{1, ..., k\}$ that assigns a positive integer not greater than k to each vertex of G. We denote a labeled graph consisting of a single vertex with label i by $[i]$. We introduce the following operation for k-graphs:

(1) The *disjoint union* $G \oplus H$ of two k-graphs G and H corresponds to the disjoint union of the underlying graphs where all the labels are preserved. Observe that also $G \oplus G$ is a well-defined labeled graph with twice as many vertices as G.

(2) Let $G = (V, E, \alpha)$ be a k-graph and $i, j \in \{1, ..., k\}$ with $i \neq j$. The *relabeling operation* $\rho_{i \to j}(G) = (V, E, \beta)$ produces a labeled graph based on the same graph (V, E) but with labeling function β, defined by

$$\beta(v) = \begin{cases} \alpha(v) \text{ if } \alpha(v) \neq i, \\ j \text{ if } \alpha(v) = i. \end{cases}$$

Consequently, each vertex carrying label i is relabeled into j.

(3) Let $G = (V, E, \alpha)$ be a k-graph and $i, j \in \{1, ..., k\}$ with $i \neq j$. Then the *i, j-join* $\eta_{i,j}(G)$ is a k-graph with vertex set V and edge set

$$E \cup \{\{u, v\} \mid u, v \in V, \alpha(u) = i, \alpha(v) = j\}.$$

This operation does not change any vertex labels.

A well-formed expression using only the symbols $[i]$, $\eta_{i,j}$, $\rho_{i \to j}$, $($, $)$, $\oplus$ with $i, j \in \{1, ..., k\}$ is called a *k-expression*. The *clique-width* of an (unlabeled) graph G is the minimum k such that there exist a labeling function α and a k-expression for (G, α).

Example 8.12 *A valid 3-expression for the graph presented in Figure 8.7 is*

$$\eta_{1,3}(\rho_{1 \to 2}(\eta_{2,3}(\eta_{1,2}([1] \oplus [2] \oplus [2]) \oplus [3] \oplus [3])) \oplus [1]). \tag{8.2}$$

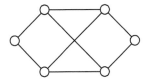

Figure 8.7: Example graph

Graphs of clique-width 1 are edgeless graphs. A graph G with at least one edge has clique-width 2 if and only if G is a cograph [98]. Each graph with n vertices possesses a trivial n-expression. First we generate the labeled one-vertex graphs $[1]$ until $[n]$, then we build their disjoint union and apply join operators $\eta_{i,j}$ for all edges $\{i, j\}$ of G. However, for algorithmic purposes, expression of minimum width are highly desirable as the computational effort increases exponentially with the number of labels. The clique-width of the complement of a graph G is at most $2 \cdot \text{cwd}(G)$ [99]. If $\mathcal{G}$ is a graph class of bounded tree-width then $\mathcal{G}$ is also class of graphs of bounded clique-width, more precisely we have for each graph G the relation $\text{cwd}(G) \le 3 \cdot 2^{\text{tw}(G)-1}$ [97].

Now let us assume that we have a graph G together with a k-expression for G. How can we use these data for reliability computations? The idea is to find transformations of the reliability measure in question that correspond to the elementary operations for labeled graphs. We consider again the computation of the domination reliability. In order to simplify the presentation of the algorithm, we assume that all vertex probabilities are equal, say p. A graph consisting of a single vertex and no edge has domination probability p, i.e. $R_d([i]) = p$. By Lemma 5.23, we obtain

$$D(G \oplus H) = D(G)D(H). \tag{8.3}$$

Let $A, B \subseteq V$ and $R_d(G, A, B)$ be the probability that in G all vertices of B are covered (dominated) by vertices of A. Thus, we have $R_d(G) = R_d(G, V, V)$. In Section 5.6 we introduced $R_d(G, A, B)$ as the domination reliability of G with respect to A and B.

Disjoint union. It is not too hard to verify that Equation (8.3) generalizes to the domination reliability:

$$R_d(G \oplus H, A, B) = R_d(G, A \cap V(G), B \cap V(G))R_d(H, A \cap V(H), B \cap V(H))$$

Relabeling. The relabeling of vertices does not affect the domination reliability, which yields

$$R_d(\rho_{i \to j}(G), A, B) = R_d(G, A, B)$$

for any $i, j \in \{1, ..., k\}$.

Join. The translation of the last operation, the i,j-join, is more difficult. We denote by V_i, $i \in \{1, ..., k\}$, the set of all vertices of G with label i. First we count all dominating sets of G that does not include any vertex from V_i but at least one vertex of V_j. The probability of operation of such a set is

$$R_1 = R_d(G, A \setminus V_i, B \setminus V_i) - R_d(G, A \setminus (V_i \cup V_j), B \setminus V_i). \qquad (8.4)$$

Assume that a partial dominating set X in G exists that does not include any vertex from V_i but at least one vertex of V_j and covers all vertices of $B \setminus V_i$. Then X covers in $\eta_{i,j}(G)$ all vertices of B. Exchanging the roles of i and j gives

$$R_2 = R_d(G, A \setminus V_j, B \setminus V_j) - R_d(G, A \setminus (V_i \cup V_j), B \setminus V_j) \qquad (8.5)$$

as the probability for a dominating set that does not include any vertex from V_j but at least one vertex of V_i. The probability that all vertices of B in $\eta_{i,j}(G)$ are covered by a subset of A that includes a vertex from V_i and a vertex from V_j is

$$R_3 = R_d(G, A, B \setminus (V_i \cup V_j)) - R_d(G, A \setminus V_i, B \setminus (V_i \cup V_j)) - $$
$$R_d(G, A \setminus V_j, B \setminus (V_i \cup V_j)) + R_d(G, A \setminus (V_i \cup V_j), B \setminus (V_i \cup V_j)). \quad (8.6)$$

Here we reduce B by $V_i \cup V_j$, since we count only partial dominating sets containing a vertex of V_i and a vertex of V_j. Those partial dominating sets cover all vertices of B in $\eta_{i,j}(G)$ as each vertex of V_i is adjacent to all vertices of V_j — and vice versa. The four terms of Equation (8.6) result from the principle of inclusion-exclusion. Finally we count all dominating sets that do not include any vertices of V_i or V_j:

$$R_4 = R_d(G, A \setminus (V_i \cup V_j), B) \qquad (8.7)$$

The random events whose probabilities are given by (8.4) – (8.7) are disjoint. Consequently, we obtain

$$R_d(\eta_{i,j}(G), A, B) = R_1 + R_2 + R_3 + R_4.$$

Clearly, an obstacle remains for the application of this procedure. We presuppose the existence of a k-expression for the input graph. Finding the optimal k-expression (of minimum width) is **NP**-hard [121].

Remark 8.13 *The Tutte polynomial of a graph with bounded clique-width can be computed in subexponential time [139]. Hence the reliability polynomial can also be evaluated by this method. Is there perhaps a polynomial-time algorithm for the reliability polynomial of graphs of bounded clique-width?*

Exercise 8.3 *Let $G = (V, E)$ be a connected graph with a connected complement $\bar{G}$. Show that G is not a cograph. Develop an algorithm to find the cotree for a given cograph.*

Exercise 8.4 *Prove Equation (8.1).*

Exercise 8.5 *Verify the result given in Example 8.11.*

Exercise 8.6 *Find a 3-expression for the cycle C_6. Show that $\mathrm{cwd}(C_6) = 3$.*

Exercise 8.7 *Calculate the domination reliability of the graph shown in Figure 8.7 by using the 3-expression (8.2).*

Exercise 8.8 *Analyze the complexity of algorithms for the computation of the all-terminal reliability of graphs of bounded path-width and bounded tree-width, respectively.*

8.4   Simulation and Probabilistic Algorithms

Simulation has become a standard technique for all kinds of complex calculations in engineering and science. It is often the fastest and easiest way to obtain results. In network reliability we deal with random edge and/or vertex failures. Consequently we can employ random number generators in order to mimic the probabilistic behavior of the network components. We will not discuss here any issues concerning the generation of (pseudo-) random numbers with a computer. Instead, we assume that there is a well-tested random number generator that produces a rational number with sufficiently many decimal places to give a good approximation to a uniformly distributed real number in $[0, 1]$. However, the reader should know that there are some traps in dealing with pseudorandom numbers; we recommend to consult a corresponding book such as [182]. We presuppose in the following that the instruction "`random(0,1)`" of an algorithm returns (a rational approximation to) a real number that is uniformly distributed in $[0, 1]$.

8.4.1   Testing of States

The next basic element of our simulation algorithm is a method (a subprocedure) that is able to test whether a given (deterministic) network is in operating state. Thus we assume that the function call "`operating(G)`" returns *true* if the graph G is operating with respect to the given reliability measure, otherwise it returns *false*. As an example, we consider the K-terminal reliability problem. Given the graph $G = (V, E)$ and the set K of terminal vertices, the function "`operating(G, K)`," presented in Algorithm 8.4, returns *true* if and only if all vertices of K belong to one and the same component of G. Here we use an additional parameter K within the function "`operating`."

The test for K-connectedness of a graph $G = (V, E)$ can be performed in $O(|E|)$ steps, assuming a suitable implementation. In a similar manner we can design algorithms for testing the operation of the graph with respect to other reliability measures.

Algorithm 8.4 Test for K-connectivity

---

procedure operating(G, K)

Input: G {graph}, K {terminal vertex set}
Output: *true* {if G is K-connected}

  procedure DFS(v)
    $L := L \cup \{v\}$
    for $w \in N(v)$ **do**
      if $w \notin L$ **then**
        DFS(w)
      end if
    end for
  end procedure

  begin
    $L := \emptyset$
    Choose $v \in K$
    DFS(v)
    if $K \subseteq L$ **then**
      return *true*
    else
      return *false*
    end if
  end

8.4.2 Basic Monte Carlo Sampling

The main procedure of a simulation is quite easy. We agree on a number k of simulation steps. Then we start a k-step run through a loop in which we first produce a copy of the input graph, remove edges and/or vertices according to the given reliability distribution and the realization of the random number generator, test the operation as described above, and increment a success counter in case of an operating structure. Finally we return the ratio (success counter)/k as an approximation for the desired reliability measure.

Algorithm 8.5 shows the principle of a simulation procedure. The algorithm can be easily extended in order to cover vertex failures or other problem-specific needs.

We expect that a suitable high number of simulation runs (say one thousand or even one million) should produce an accurate estimation for the value of the reliability measure. However, there is one exception. In case we deal with high-reliable systems with extremely small edge or vertex failure probabilities, the simulation may always produce operating networks, which gives an estimated reliability of one hundred percent.

Algorithm 8.5 Simulation

Input: $G = (V, E)$ {graph}, **p** {reliabilities and additional parameters}, k
{number of simulation steps}
Output: $\tilde{R}(G)$ {estimation for the reliability measure}
  $succ := 0$ {success counter}
  for $i = 1$ **to** k **do**
    $H := \text{copy}(G)$
    for $e \in E$ **do**
      if $random(0, 1) \geq p_e$ **then**
        $RemoveEdge(e)$
      end if
    end for
    if $operating(G)$ **then**
      $succ := succ + 1$
    end if
  end for
  return $\dfrac{succ}{k}$

8.4.3 Refined Sampling Methods

A simple method that works in case of identical edge failure probability $q = 1 - p$ is *biased sampling*. First we compute the size c of a minimum cut in the given graph $G = (V, E)$ with m edges. Clearly, any failure set of edges must comprise at least c edges. Hence, any edge failure event with less than c edges is worthless for our simulation. The binomial distribution tells us that the probability of a failure with exactly k edges is $P_k = \binom{m}{k} q^k p^{m-k}$. Let S_k be the probability that the graph is connected if exactly k random chosen edges fail. The probability S_k can be estimated by a simulation using random k-samples out of an m-set. We denote by $\hat{S}_k$ the estimated value of S_k. Then we obtain

$$\hat{R}(G) = \sum_{k=c}^{m} \binom{m}{k} q^k p^{m-k} \hat{S}_k$$

as an estimation for the all-terminal reliability of G. Since small cuts are more likely to cause the failure, the simulation should estimate S_k for values of k close to c more precisely than those for higher values.

Remark 8.14 *The method of biased sampling can be generalized to non-identical edge reliabilities if the edge set can be decomposed in to a small number of classes such that all edges belonging to one class have identical reliability.*

David Karger [170, 171] observed that for networks with high edge reliability the simulation of edge failures can be restricted to minimum cuts (or near-minimum cuts). Moreover, he proposed a *randomized fully polynomial time approximation scheme* (FPRAS), which means essentially that we can obtain an estimation of the all-terminal reliability (more precisely of $1 - R(G)$) that has with a probability of $1 - \left(\frac{1}{2}\right)^k$ a relative error of $1 \pm \varepsilon$ and can be computed in time polynomial in n, k and $\frac{1}{\varepsilon}$.

Karger used a result from *Karp* and *Luby* [174] to show that in case of $1 - R(G) > n^{-4}$ the probability $1 - R(G)$ can be estimated within $1 + \varepsilon$ in a time proportional to $\frac{mn^4}{\varepsilon^2}$ via standard Monte Carlo simulation as presented above. If $R(G)$ is very close to 1, i.e. $R(G) > 1 - n^{-4}$ then with high probability only minimum (or near minimum) cuts cause the network failure. An *α-minimum cut* is a cut in a graph with at most α times the number of edges of a minimum cut. Let $q = 1 - p$ the identical failure probability of all edges of a given graph $G = (V, E)$. We denote by c the cardinality of a minimum cut of G. A particular α-minimum cut fails with probability $q^{\alpha c}$. If $q^c < n^{-4}$ then clearly $q^{\alpha c} < n^{-4\alpha}$. It can be shown that there are at most $n^{2\alpha}$ α-minimum cuts in G, which yields that the probability any α-minimum cut fails is less than $n^{-2\alpha}$. Which means that essentially only minimum cuts contribute to a network failure.

We give only an outline of the algorithm; for details, the reader is referred to [170, 172, 174]. First we enumerate all minimum cuts. This can be done by a randomized algorithm as explained in standard textbooks on randomized algorithms such as [224] or [230]. Then we can define a Boolean formula in disjunctive normal form (DNF) that yields one if and only if at least one minimum cut fails. Each term represents one (α-) minimum cut, such that there are less than $Cn^{2\alpha}$ terms in the DNF. Now we can use the FPRAS given in [175] in order to estimate the probability that the DNF becomes true. Karger shows that a choice of $\alpha = 2 - \ln\left(\frac{\varepsilon}{2}\right)(2 \ln n)^{-1}$ is sufficient to guarantee the error bounds.

There are many many additional methods for the estimation of network reliability by Monte Carlo methods. You will find terms like *variance reduction methods*, *importance sampling*, *dagger sampling*, *sequential destruction* within the huge literature to this topic, see e.g. [113, 112, 11, 78, 79, 127, 189]. We present here only some basic ideas; the interested reader is refered to the excellent book by *Gertsbakh* and *Shpungin* [135].

The idea of dagger sampling is to reuse random numbers in order to generate different samples. The samples produced in this way are no longer independent but are negatively correlated. Let us consider a dagger sampling for the estimation of the all-terminal reliability of a graph. If $e \in E$ is an edge with reliability p_e, then we generate $k_e = \lfloor 1/(1 - p_e) \rfloor$ sample graphs $\{H_1, ..., H_{k_e}\}$ from one random number. A sample graph is a spanning subgraph of G obtained by removing edges from G. Let r be an uniform random number out of $[0, 1]$. We remove edge e in sample graph H_i ($i = 1, ..., k_e$) if and only if $(i - 1)/(1 - p_e) \leq r < i/(1 - p_e)$. This step can be performed independently for all edges of G. The only problem is that the numbers k_e differ from edge to edge, which can be overcome by setting $k = \text{lcm}\{k_e \mid e \in E\}$

and repeating the sampling procedure k/k_e times for edge e [126]. Dagger sampling saves not only the amount of random numbers needed to obtain a huge number of samples but also reduces the variance due to the negative correlation of edge samples [189].

Lower and upper bounds for network reliability provide the next useful tool in order to reduce the variance of Monte Carlo methods. We start with Equation (5.1) for the all-terminal network reliability:

$$R(G) = \sum_{F \subseteq E} \psi(F) \prod_{e \in F} p_e \prod_{f \in E \setminus F} q_f$$

$$= \sum_{F \subseteq E} \psi(F) P(F)$$

$$= \mathbb{E}\psi$$

Here

$$\psi(F) = \begin{cases} 1 \text{ if } (V, F) \text{ is connected,} \\ 0 \text{ else.} \end{cases}$$

is an indicator function for the connectedness of the spanning subgraph (V, F) of $G = (V, E)$ and

$$P(F) = \prod_{e \in F} p_e \prod_{f \in E \setminus F} q_f$$

is the probability that exactly the subgraph (V, F) is induced by operating edges of G. Let $T = \{T_1, ..., T_k\}$ be a collection of edge sets of edge-disjoint spanning trees of G. We define two set systems:

$$\mathcal{E}_1 = \{F \mid F \subseteq E \text{ and } \exists T_i \in T : T_i \subseteq F\},$$
$$\mathcal{E}_2 = \{F \mid F \subseteq E \text{ and } \nexists T_i \in T : T_i \subseteq F\} = 2^E \setminus \mathcal{E}_1$$

Consequently, $\mathcal{E}_1$ is the set of all edge subsets of G that contain all edges of at least one spanning tree from T. All remaining subsets are pooled in $\mathcal{E}_2$. The graph G is connected if at least one subset of $\mathcal{E}_1$ is operating, which means that

$$R_L(G) = 1 - \prod_{i=1}^{k}(1 - P(T_i))$$

yields a lower bound for the all-terminal reliability of G (see also Section 6.4). We obtain

$$R(G) = \sum_{F \in \mathcal{E}_1} \psi(F) P(F) + \sum_{F \in \mathcal{E}_2} \psi(F) P(F)$$

$$= R_L(G) + \sum_{F \in \mathcal{E}_2} \psi(F) P(F). \tag{8.8}$$

The sampling space is now reduced to $\mathcal{E}_2$. Moreover the part of the result given by the lower bound R_L is exact.

Upper bounds for the all-terminal reliability result from cut sets (cf. Section 6.4). Let $\mathcal{C} = \{C_1, ..., C_l\}$ be a collection of edge-disjoint cut sets of G. If all edges of a cut fail then G is disconnected. Hence,

$$R_U(G) = \prod_{i=1}^{l} \left(1 - \prod_{e \in C_i} (1 - p_i) \right)$$

is an upper bound for $R(G)$. We define

$$\mathcal{F}_1 = \{F \mid F \subseteq E \text{ and } \exists C_i \in \mathcal{C} : C_i \subseteq E \setminus F\},$$
$$\mathcal{F}_2 = \{F \mid F \subseteq E \text{ and } \nexists C_i \in \mathcal{C} : C_i \subseteq E \setminus F\} = 2^E \setminus \mathcal{F}_1.$$

Consequently, if $F \in \mathcal{F}_1$ then (V, F) is disconnected, which yields

$$R(G) = \sum_{F \in \mathcal{F}_1} \psi(F) P(F) + \sum_{F \in \mathcal{F}_2} \psi(F) P(F)$$

$$= \sum_{F \in \mathcal{F}_2} \psi(F) P(F). \tag{8.9}$$

The combination of (8.8) and (8.9) gives

$$R(G) = R_L(G) + \sum_{F \in \mathcal{E}_2 \cap \mathcal{F}_2} \psi(F) P(F).$$

The remaining sample space $\mathcal{E}_2 \cap \mathcal{F}_2$ has the probability measure $R_U(G) - R_L(G)$ which leads to a considerable variance reduction. However, sampling from $\mathcal{E}_2 \cap \mathcal{F}_2$ requires more effort than sampling from the whole space. The simplest idea is to sample subsets from 2^E and to reject them in case they do not belong to $\mathcal{E}_2 \cap \mathcal{F}_2$.

Remark 8.15 *The application of bounds for sampling methods that we demonstrated here for the all-terminal reliability can easily be adopted to other reliability measures such as the K-terminal reliability or reachability problems in directed graphs.*

Markov chain sampling methods use a Markov chain in order to traverse the state space, for instance the set of all spanning subgraphs of a given graph. The advantage of Markov chain simulation is the fast and efficient generation of new states by small perturbations of existing ones. Standard Monte Carlo simulation for the all-terminal reliability generates states, i.e. spanning subgraphs, by complete enumeration of the edge set and setting or removing each single edge depending on the outcome of a random experiment. In a Markov chain simulation a sequence $\{H_i\}_{i \in \mathbb{Z}^+}$ is generated. Let H_0 be the initial state, which may be created at random. The subgraph H_{i+1} is obtained from H_i by inserting or removing a single edge. The choice of the transition probabilities for the Markov chain should ensure that the stationary distribution coincides with the desired distribution of our sample space [73]. A complete overview about this topic is beyond the scope of this book — we refer the interested reader to the literature.

Part III

Maintenance Models

Chapter 9

Random Point Processes in System Replacement

9.1 Basic Concepts

Point processes A *point process* is a sequence $\{t_1, t_2, \cdots\}$ of real numbers with

$$t_1 < t_2 < \cdots \quad \text{and} \quad \lim_{i \to \infty} t_i = \infty.$$

That means, a point process is a strictly increasing sequence of real numbers, which does not have a finite limit point. Generally, the t_i are called *event times* or, more exactly, *event time points*. In reliability and maintenance theory, point processes are mainly generated by the failure time points of systems or by the time points at which repaired or renewed systems resume their work, given that the life and repair periods continue unlimitedly. Strictly speaking, since every repair-failure process will be stopped after a finite time, the generated point processes have to be considered finite samples from a point process.

Random point processes Usually, the event times are random variables. A sequence of random variables $\{T_1, T_2, \cdots\}$ with

$$0 < T_1 < T_2 < \cdots \quad \text{and} \quad \Pr\left(\lim_{i \to \infty} T_i = \infty\right) = 1$$

is called a *random point process (rpp)*. By introducing the random *interevent (interarrival) times*

$$\{Y_1, Y_2, \cdots\} \text{ with } Y_i = T_i - T_{i-1}, \ i = 1, 2, ..., \ T_0 = 0,$$

an *rpp* can equivalently be defined as a sequence of positive random variables $\{Y_1, Y_2, \cdots\}$ with property

$$\Pr\left(\lim_{n \to \infty} \sum_{i=1}^{n} Y_i = \infty\right) = 1.$$

An *rpp* is called *simple* if, with probability 1, at any time point t not more than one event can occur.

An *rpp* is said to be *recurrent* if its corresponding sequence of interevent times $\{Y_1, Y_2, \cdots\}$ is a sequence of independent, identically distributed random variables. The most popular recurrent point processes are homogeneous Poisson processes and renewal processes.

Sometimes it is more convenient or even necessary to define an *rpp* as a doubly infinite sequence, which tends to $-\infty$ to the left and to $+\infty$ to the right, each with probability 1:

$$\cdots - T_2, -T_1, T_0 = 0, +T_1, +T_2, \cdots,$$

Random counting processes   Frequently, the event times are of less interest than the number $N(t)$ of events, which occur in $(0, t]$:

$$N(t) = \begin{cases} \max\{n, T_n \leq t\}, & n = 1, 2, \ldots \\ 0, & t < T_1 \end{cases}. \tag{9.1}$$

$\{N(t), \ t \geq 0\}$ is said to be the *random counting process* (*rcp*) belonging to the *rpp* $\{T_1, T_2, \cdots\}$. This stochastic process has parameter space $\mathbf{T} = [0, \infty)$ and state space $\mathbf{Z} = \{0, 1, \ldots\}$. Further properties are:

1) $N(0) = 0$.

2) $N(s) \leq N(t)$ for $s \leq t$.

3) For any s, t with $0 \leq s < t$, the increment $N(s, t) = N(t) - N(s)$ is equal to the number of events, which occur in $(s, t]$.

Conversely, every stochastic process in continuous time, which has these three properties, is the *rcp* of exactly one *rpp* $\{T_1, T_2, \ \cdots\}$. Thus, from the statistical point of view, the stochastic processes $\{T_1, T_2, \ \cdots\}$, $\{Y_1, Y_2, \cdots\}$, and $\{N(t), \ t \geq 0\}$ are equivalent.

Obviously, "$N(t) \geq n$" = "$T_n \leq n$", so that

$$F_{T_n}(t) = P(T_n \leq t) = P(N(t) \geq n) \tag{9.2}$$

and, letting $F_{T_0}(t) \equiv 1$,

$$p_n(t) = \Pr(N(t) = n) = F_{T_n}(t) - F_{T_{n+1}}(t), \ n = 0, 1, \ldots \tag{9.3}$$

Therefore, the trend function of the *rcp* $\{N(t), \ t \geq 0\}$ is

$$m(t) = \mathbb{E}(N(t)) = \sum_{n=1}^{\infty} n \, p_n(t),$$

or, by making use of $\mathbb{E}(N(t)) = \sum_{n=1}^{\infty} \Pr(N(t) \geq n)$ and (9.3),

$$m(t) = \sum_{n=1}^{\infty} F_{T_n}(t). \tag{9.4}$$

Definition 9.1 *A rpp is called* stationary *if its sequence of interarrival times is strongly stationary, that is, if for any sequence of integers* $i_1, i_2, \cdots, i_k$ *with* $i_1 < i_2 < \cdots < i_k$, $k = 1, 2, ...,$ *and for any* $\tau = 0, 1, 2, ...,$ *the two random vectors*

$$\{Y_{i_1}, Y_{i_2}, \cdots, Y_{i_k}\} \text{ and } \{Y_{i_1+\tau}, Y_{i_2+\tau}, \cdots, Y_{i_k+\tau}\}$$

have the same probability distribution.

It is easy to show that if a *rpp* is stationary, the corresponding counting process $\{N(t), t \geq 0\}$ has homogeneous increments and vice versa. This implies the following corollary:

Corollary 9.2 *A rpp is stationary if and only if its corresponding counting process has homogeneous increments.*

Note that a stochastic process $\{X(t), t \in \mathbf{T}\}$ is said to have *homogeneous increments* if, for an arbitrary but fixed $s \geq 0$, the probability distribution of its increments $X(s + t) - X(s)$ depends only on t, $t \geq 0$. Thus, the *rcp* belonging to a stationary *rpp* has property

$$p_n(t) = \Pr(N(t) = n) = \Pr(N(s, s + t) = n), \ n = 0, 1, ..., \quad (9.5)$$

where $N(s, s + t) = N(s + t) - N(s)$.

For stationary *rpp'es*, the expected number of events occurring in $[0, 1]$ (or in any other interval of length 1) is called the *intensity* of the process and will be denoted as λ. Analytically, λ is given by

$$\lambda = m(1) = \sum_{n=1}^{\infty} n \, p_n(1).$$

In case of a nonstationary point process, the role of λ is taken over by an *intensity function* $\lambda(t)$. This function gives us the trend function of the *rcp* and its expected increments:

$$m(t) = \int_0^t \lambda(x) dx, \ \ m(s, t) = m(t) - m(s) = \int_s^t \lambda(x) dx, \quad (9.6)$$

where $0 \leq s < t$. Thus, for $\Delta t \to 0$,

$$\Delta m(t) = m(t + \Delta t) - m(t) = \lambda(t) \Delta t + o(\Delta t). \quad (9.7)$$

Hence, for small Δt the product $\lambda(t) \Delta t$ is approximately equal to the expected number of events occurring in $(t, t+\Delta t]$. For this reason, the intensity function $\lambda(t)$ is also called the *arrival rate* of events at time t. In case of a simple *rpp*, (9.7) can be equivalently interpreted as follows: If Δt is sufficiently small, then $\lambda(t) \Delta t$ is approximately the probability of the occurrence of an event in the interval $(t, t + \Delta t]$.

Random Marked Point Processes Frequently, in addition to their arrival times, events come with another piece of information. In maintenance, the following situations are common: If T_i is the time point of the ith failure of a system, then the time (or cost) necessary for removing the failure may be assigned to T_i. Or if at time T_i a repaired system resumes its work, then the random profit arising from operating the system to its next failure, may be assigned to T_i. Generally, a decision of any kind may be assigned to T_i (decision processes). This leads to the concept of a marked point process: Let $\{T_1, T_2, \cdots\}$ be a *rpp* with random marks M_i out of a mark space $\mathbf{M}$ assigned to T_i. Then the sequence $\{(T_1, M_1), (T_2, M_2), \cdots\}$ is called a *random marked point process (rmpp)*.

Random marked point processes have been dealt with in full generality in [217]. For other high-level mathematical treatments, see [185] or [286]. *Beichelt and Franken* [42] were probably the first who fully made use of the theory of stationary and synchronous *rmpp'es* in the reliability analysis of complex systems.

Compound Stochastic Processes Let $\{(T_1, M_1), (T_2, M_2), \cdots\}$ be a random marked point process and $\{N(t), \ t \geq 0\}$ be the *rcp* which belongs to the *rpp* $\{T_1, T_2, \cdots\}$. The stochastic process $\{C(t), \ t \geq 0\}$ defined by

$$C(t) = \sum_{i=0}^{N(t)} M_i, \quad M_0 = 0, \tag{9.8}$$

is called a *compound (cumulative, aggregate) stochastic process*.

For instance, if T_i is the time point of the ith breakdown of a system, and M_i is the corresponding repair cost, then $C(t)$ is the total repair cost in $(0, t]$.

9.2   Renewal Processes

9.2.1   Renewal Function

The motivation for this section within the context of this book is a simple maintenance policy: A system is replaced after every failure by another one, which is equivalent to the failed one with regard to its lifetime distribution. Principally, such an equivalence may also be achieved by a *perfect repair*. In this section, we usually use the terminology *renewed system* and *renewal* instead of *replaced system* and *replacement*, respectively.

Definition 9.3 *An* (ordinary) renewal process *is a sequence of nonnegative iid random variables* $\{L_1, L_2, \cdots\}$.

Situations which can be modeled by renewal processes, do not only occur in engineering, but also in the natural, economic, and social sciences. They are a basic stochastic tool e.g. for particle counting, population development, and arrivals of customers at a service station. In the latter context, L_n is the time between the arrival of the $(n-1)^{th}$ and the nth customer. Renewal processes are particularly important in actuarial risk analysis, namely for modeling the arrival of claims at an insurance company. With regard to maintenance applications, a renewal process specifies the times between two neighboring renewals given that the renewal times are negligibly small or are not taken into account. Hence,

$$T_n = \sum_{i=1}^{n} L_i, \; n = 1, 2, ... \tag{9.9}$$

is the time point at which the nth renewal of a failed system occurs. Obviously, the sequences $\{L_1, L_2, ...\}$ and $\{T_1, T_2, ...\}$ are statistically equivalent, so that $\{T_1, T_2, ...\}$ is called *renewal process* as well. The time intervals between neighboring renewals are called *renewal cycles*, and the difference $L_n = T_n - T_{n-1}$ is the *length* of the nth renewal cycle, $n = 1, 2, ..., T_0 = 0$. In what follows, it will be assumed that the L_i are identically distributed as L with distribution function $F(t) = \Pr(L \le t)$ and density $f(t) = dF(t)/dt$ if the latter exists. L is called the *typical cycle length* of the renewal process. Since the $L_1, L_2, ...$ are independent, $T_2 = L_1 + L_2$ has distribution function

$$F_{T_2}(t) = \int_0^t F(t-x)dF(x).$$

Thus, $F_{T_2}(t)$ is the *convolution* of F with itself. The distribution function of T_n is the *convolution* of $F_{T_{n-1}}$ and F, i.e. the n-fold convolution of F with itself: $F_{T_{n-1}} * F = F^{*(n)}$, where

$$F_{T_n}(t) = F^{*(n)}(t) = \int_0^t F_{T_{n-1}}(t-x)dF(x), \; n = 3, 4, ... \tag{9.10}$$

Generally, the *convolution* of two integrable functions g_1 and g_2 with $g_i(t) \equiv 0$ for $t < 0$, $i = 1, 2$, in particular probability densities of nonnegative random variables, is defined as

$$(g_1 * g_2)(t) = (g_2 * g_1)(t) = \int_0^t g_1(t-x)g_2(x)dx. \tag{9.11}$$

Thus, by differentiating (9.10) on both sides with regard to t (mind the variable integration bound), the density of T_n is seen to be equal to the n-fold convolution of f with itself, $n = 2, 3.... :$

$$f_{T_n}(t) = f^{*(n)}(t) = (f_{T_{n-1}} * f)(t) = \int_0^t f_{T_{n-1}}(t-x)f(x)dx. \tag{9.12}$$

The number $N(t)$ of renewals occuring in $(0, t]$ as given by (9.1) defines the *renewal counting process* $\{N(t), \; t \ge 0\}$. (Thus, $N(t)$ does not count the renewal at time $t = 0$). The expected value of the number of renewals in $(0, t]$,

$$H(t) = \mathbb{E}(N(t)), \; t > 0,$$

is called the *renewal function*. By (9.4) and (9.10), this function can be represented as

$$H(t) = \sum_{n=1}^{\infty} F^{*(n)}(t). \tag{9.13}$$

Given that the first renewal occurs at time $T_1 = x$, $0 < x \le t$, there are on average $1 + H(t - x)$ renewals in $(0, t]$. Since $F(t) = \Pr(T_1 \le t)$, the expected number of renewals in $(0, t]$ is $\int_0^t [1 + H(t - x)] \, dF(x)$. Hence the renewal function satisfies the integral equation

$$H(t) = F(t) + \int_0^t H(t - x) dF(x). \tag{9.14}$$

(A rigorous derivation of (9.14) can be done by using (9.10) and (9.13) (exercise 9.12)). If the density $f(t) = dF(t)/dt$ exists, then the *renewal density* $h(t) = dH(t)/dt$ exists as well. By differentiating equation (9.14) on both sides, $h(t)$ is seen to satisfy the integral equation

$$h(t) = f(t) + \int_0^t h(t - x) f(x) dx. \tag{9.15}$$

(9.14) and (9.15) are special *integral equations of renewal type*.

Example 9.4 *Let the typical cycle length L be exponentially distributed with parameter λ: $F(t) = 1 - e^{-\lambda t}$, $\lambda > 0$, $t \ge 0$. The corresponding renewal process is called* the homogeneous Poisson process. *In this special case, T_n is a sum of independent, identically exponentially distributed random variables, and has, therefore, an* Erlang-distribution *with parameters λ and n:*

$$F_{T_n}(t) = 1 - e^{-\lambda t} \sum_{i=0}^{n-1} \frac{(\lambda t)^i}{i!}; \quad n = 1, 2, \dots \tag{9.16}$$

From (9.3),

$$\Pr(N(t) = n) = \frac{(\lambda t)^i}{i!} e^{-\lambda t}; \quad i = 0, 1, \dots \tag{9.17}$$

The expected number of renewals in $(0, t]$ is a linear function in t:

$$H(t) = \lambda t. \tag{9.18}$$

Hence, the parameter λ, for being the expected number of renewals per unit time, is the intensity of the homogeneous Poisson process. (Note that the homogeneous Poisson process has homogeneous increments so that the underlying renewal process, namely $\{L_1, L_2, \dots\}$, is stationary.)

Generally, integral equations of the renewal type need to be solved by numerical methods. However, explicit solutions exist in the space of the Laplace transforms if the functions involved satisfy conditions 1 and 2 listed in the following definition.

Definition 9.5 *Let $g(t)$ be any function defined on $[0, \infty)$, which has the following properties: 1) $g(t)$ is piecewise continuous, and 2) there exist real constants a and b so that $g(t) \leq ae^{-bt}$ for all $t \geq 0$. Then the Laplace transform of $g(t)$, denoted as $L(g)$ or as $\widehat{g}(s)$, is defined as the integral*

$$\widehat{g}(s) = \int_0^\infty e^{-st} g(t) dt \tag{9.19}$$

for any complex s with $\operatorname{Re}(s) > b$.

Thus, the Laplace transformation transforms a real function of a real variable t into a complex function of a complex variable s. We will need the following three properties of the Laplace transformation: Let $g_1(t)$ and $g_2(t)$ satisfy conditions 1 and 2 in definition 9.5. Then,

$$L(g_1 + g_2) = L(g_1) + L(g_2), \tag{9.20}$$

$$L\left(\frac{dg(t)}{dt}\right) = s\widehat{g}(s) - g(0), \quad L\left(\int_0^t g(x)dx\right) = \frac{1}{s}\widehat{g}(s), \tag{9.21}$$

$$L(g_1 * g_2) = \widehat{g}_1(s) \cdot \widehat{g}_2(s). \tag{9.22}$$

Let $\widehat{F}(s)$, $\widehat{f}(s)$, $\widehat{H}(s)$, and $\widehat{h}(s)$ be in this order the Laplace transforms of $F(t)$, $f(t)$, $H(t)$, and $h(t)$. Applying the Laplace transformation on both sides of (9.15), making use of its properties (9.20) and (9.22), transforms this integral equation for $h(t)$ into a simple linear algebraic equation for $\widehat{h}(s)$:

$$\widehat{h}(s) = \widehat{f}(s) + \widehat{f}(s) \cdot \widehat{h}(s).$$

Hence and by (9.21),

$$\widehat{h}(s) = \frac{\widehat{f}(s)}{1 - \widehat{f}(s)}, \quad \widehat{H}(s) = \frac{\widehat{f}(s)}{s(1 - \widehat{f}(s))}. \tag{9.23}$$

Retransformation of $\widehat{h}(s)$ yields $h(t)$. Very helpful for retransformation are *contingency tables*, which show standard functions (*pre-images*) and their corresponding Laplace-transforms (*images*) and vice versa. But the first step when doing retransformation should be to simplify $\widehat{h}(s)$ by representing it as a sum of "simple" terms, which in turn should be represented as products of "simple" factors. Then properties (9.20) and (9.22) can be used for finding its pre-image. For instance, if $\widehat{h}(s)$ is a rational function, decomposing it into partial fractions will facilitate its retransformation. This is illustrated by the following example.

Example 9.6 *1. Let $F(t) = 1 - e^{-\lambda t}$, $\lambda > 0$, $t \geq 0$. Then $f(t) = \lambda e^{-\lambda t}$ and $\widehat{f}(s) = \lambda/(s + \lambda)$ so that*

$$\widehat{h}(s) = \lambda/s, \quad h(t) = \lambda, \quad H(t) = \lambda t.$$

2. Let $F(t) = (1 - e^{-\lambda t})^2$, $\lambda > 0$, $t \geq 0$. Then,

$$f(t) = 2\lambda e^{-\lambda t}(1 - e^{-\lambda t}), \quad \widehat{f}(s) = 2\lambda^2/(s + \lambda)(s + 2\lambda),$$

and, by (9.23),

$$\widehat{h}(s) = \left(2\lambda^2 \cdot \frac{1}{s} \cdot \frac{1}{s + 3\lambda}\right).$$

The pre-image of $\widehat{g}_1(s) = 1/s$ is $g_1(t) \equiv 1$ and that of $\widehat{g}_2(s) = 1/(s + 3\lambda)$ is $g_2(t) = e^{-3\lambda t}$. Multiplying the convolution of g_1 and g_2 by $2\lambda^2$ gives $h(t)$, integrating $h(t)$ yields $H(t)$:

$$h(t) = \tfrac{2}{3}\lambda \left(1 - e^{-3\lambda t}\right), \quad H(t) = \tfrac{2}{3}\left[\lambda t + \tfrac{1}{3}\left(e^{-3\lambda t} - 1\right)\right]. \tag{9.24}$$

We next summarize some results concerning the asymptotic behavior of $H(t)$ if $t \to \infty$. Letting $\mu = \mathbb{E}(L) < \infty$, we have:

1. *Elementary renewal theorem:*

$$\lim_{t \to \infty} \frac{H(t)}{t} = \frac{1}{\mu}. \tag{9.25}$$

2. *Key renewal theorem:* If $F(t)$ is nonarithmetic and $g(t)$ an integrable function on $[0, \infty)$, then

$$\lim_{t \to \infty} \int_0^\infty g(t - x)dH(x) = \frac{1}{\mu} \int_0^\infty g(x)dx. \tag{9.26}$$

3. *Blackwell's renewal theorem:* If $F(t)$ is nonarithmetic, then, for any $h > 0$,

$$\lim_{t \to \infty} [H(t + h) - H(t)] = \frac{h}{\mu}.$$

4. If $F(t)$ is nonarithmetic and $\sigma^2 = Var(Y) < \infty$, then

$$\lim_{t \to \infty} \left(H(t) - \frac{t}{\mu}\right) = \frac{\sigma^2}{2\mu^2} - \frac{1}{2}.$$

Note that $F(t)$ or L are called *arithmetic* if there is a constant d so that L has range $\{0, d, 2d, ...\}$, i.e. L is a special discrete random variable, which can only assume values which are multiples of d. Further asymptotic results are:

5.

$$\Pr\left(\lim_{t \to \infty} \frac{N(t)}{t} = \frac{1}{\mu}\right) = 1.$$

6.

$$\lim_{t \to \infty} \Pr\left(\frac{N(t) - t/\mu}{\sigma\sqrt{t\mu^{-3}}} \leq x\right) = \Phi(x), \tag{9.27}$$

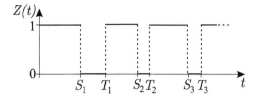

Figure 9.1: Alternating renewal process

where $\Phi(x)$ is the distribution function of the standardized normal distribution. From (9.27), for t being sufficiently large, we get for any $0 < \varepsilon < 1$,

$$\Pr \frac{t}{\mu} - z_{\alpha/2}\sigma\sqrt{t\mu^{-3}} \le N(t) \le \frac{t}{\mu} + z_{\alpha/2}\sigma\sqrt{t\mu^{-3}} \approx 1 - \alpha, \qquad (9.28)$$

where z_ε is the $(1 - \varepsilon)$-percentile of the standardized normal distribution, $0 < \varepsilon < 1$, and

$$\Pr\left(\frac{N(t) - t/\mu}{\sigma\sqrt{t\mu^{-3}}} \le z_\alpha\right) \approx 1 - \alpha.$$

Hence, the smallest number of spare systems, which are necessary to maintain the renewal process in the intervall $(0, t]$ with probability $1 - \alpha$, is approximately

$$n_{\min} \approx \frac{t}{\mu} + z_{\alpha/2}\sigma\sqrt{t\mu^{-3}}.$$

9.2.2 Alternating Renewal Processes

Alternating renewal processes take explicitly into account nonnegligible renewal times: A system starts working at time $t = 0$. On a failure at time L_1 it will be renewed. The renewal requires the time R_1. The system immediately starts operating as soon as the renewal is completed. Proceeding in this way, a marked point process $\{(L_i, R_i), \ i = 1, 2, ...\}$ is generated.

Definition 9.7 *If $\{L_1, L_2, ...\}$ and $\{R_1, R_2, ...\}$ are sequences of independent, nonnegative random variables, then the marked point process $\{(L_i, R_i), \ i = 1, 2, ...\}$ with L_i and R_i being independent, $i = 1, 2, ...$, is called an* alternating renewal process.

The random variables

$$S_1 = L_1, \quad S_n = \sum_{i=1}^{n-1}(L_i + R_i) + L_n, \quad n = 2, 3, ...$$

are the time points at which failures occur, and the random variables

$$T_0 = 0, \quad T_n = \sum_{i=1}^{n}(L_i + R_i), \quad n = 1, 2, ...$$

are the time points at which the system resumes working. A "1" is assigned to the time intervals in which the (renewed) system is working, and a "0" to the time intervals in between. Then a binary indicator variable of the state the alternating renewal process is at, is

$$Z(t) = \begin{cases} 0 & \text{if } t \in [S_n, T_n), & n = 1, 2, \ldots \\ 1 & \text{if } t \in [T_n, S_{n+1}), & n = 0, 1, \ldots \end{cases} . \qquad (9.29)$$

An alternating renewal process can equivalently be defined by the stochastic process in continuous time $\{Z(t), \ t \geq 0\}$, $Z(+0) = 1$ (Figure 9.1). In what follows, all L_i and R_i are assumed to be distributed as L and R with distribution functions $F_L(t) = \Pr(L \leq t)$ and $F_R(t) = \Pr(R \leq t)$, respectively. By agreement, $\Pr(Z(+0) = 1) = 1$.

Let $H_r(t)$ be the expected number of renewals in $(0, t]$. Then $H_r(t)$ is the renewal function of a renewal process with *typical cycle length* $L + R$. Thus, $H_r(t)$ satisfies (9.14) with F replaced by $F_L * F_R$. Now, let

$$A_x(t) = \Pr(Z(t) = 1, \ L_t > x)$$

be the probability that a system is operating at time t and that its residual lifetime L_t is greater than x. $A_x(t)$ is called *interval reliability* of the system with regard to the interval $[t, \ t + x]$. Then

$$A_x(t) = \overline{F}_L(t + x) + \int_0^t \overline{F}_L(t + x - u) dH_r(u), \qquad (9.30)$$

since, by (9.7) and the subsequent comment, $dH_r(u)$ is the "probability" that a (renewed) system starts operating at time u.

Let $A(t)$ be the probability that a system is available (operating) at time t:

$$A(t) = \mathbb{E}(Z(t)) = \Pr(Z(t) = 1). \qquad (9.31)$$

This important characteristic of an alternating renewal process is called *availability* (of the system) or, more exactly, *point availability*, since it refers to a fixed, but arbitrary, time point t. It is given by equation (9.30) with $x = 0$:

$$A(t) = \overline{F}_L(t) + \int_0^t \overline{F}_L(t - u) dH_r(u). \qquad (9.32)$$

The *average availability* of the (renewed) system $\overline{A}(t)$ and its *random total operating time* $U(t)$ in $[0, t)$ are

$$\overline{A}(t) = \frac{1}{t} \int_0^t A(u) du \text{ and } U(t) = \int_0^t Z(u) du,$$

respectively. By changing the order of integration,

$$\mathbb{E}(U(t)) = \int_0^t \mathbb{E}(Z(u)) du = \int_0^t A(u) du = t \, \overline{A}(t).$$

As an immediate corollary from the fundamental renewal theory, we get the following limits as $t \to \infty$ for $A_x(t)$ and $A(t)$:

$$A_x = \lim_{t \to \infty} A_x(t) = \frac{1}{\mathbb{E}(L) + \mathbb{E}(R)} \int_x^\infty \overline{F}_L(u)du, \qquad (9.33)$$

$$A = \lim_{t \to \infty} A(t) = \frac{\mathbb{E}(L)}{\mathbb{E}(L) + \mathbb{E}(R)}. \qquad (9.34)$$

A_x is the *stationary* or *long-run interval availability (reliability)* with regard to an interval of length x, and A is the *stationary* or *long run availability* of the underlying system. It is important that Equation (9.34) is also true if within the renewal cycles the L_i and R_i are dependent random variables. (Analogously to renewal processes, a *renewal cycle* is defined as the interval between two neighboring time points, at which the renewed system resumes its work.) Generally, numerical methods have to be applied to determine interval- and point availabilities via the formulas (9.30) and (9.32), respectively. Explicit results can be obtained in the following example.

Example 9.8 *Let life- and renewal times have exponential distributions:*

$$f_L(t) = \lambda e^{-\lambda t}, \quad f_R(t) = \rho e^{-\rho t}; \quad \lambda, \rho > 0, \ t \geq 0.$$

Letting $dH_r(u)/du = h_r(u)$, application of the Laplace transformation to (9.32) yields

$$\widehat{A}(s) = \widehat{F}_L(s) + \widehat{F}_L(s) \cdot \widehat{h}_r(s) = \frac{1}{s + \lambda}\left[1 + \widehat{h}_r(s)\right]. \qquad (9.35)$$

To determine $\widehat{h}_r(s)$, note that

$$L\{f_L * f_R\} = \frac{\lambda\rho}{(s + \lambda)(s + \rho)}.$$

Hence, from (9.23)

$$\widehat{h}_r(s) = \frac{\lambda\rho}{s(s + \lambda + \rho)}.$$

By inserting $\widehat{h}_r(s)$ into (9.35) and decomposing $\widehat{A}(s)$ into partial fractions,

$$\widehat{A}(s) = \frac{1}{s + \lambda} + \frac{\lambda}{s(s + \lambda)} - \frac{\lambda}{s(s + \lambda + \rho)}.$$

Retransformation yields the point availability

$$A(t) = \frac{\rho}{\lambda + \rho} + \frac{\lambda}{\lambda + \rho}e^{-(\lambda + \rho)t}, \ t \geq 0.$$

The corresponding stationary availability is $A = \rho/(\lambda + \rho)$.

9.2.3 Compound Renewal Processes

Let $\{T_1, T_2, ...\}$ be a renewal process with the corresponding renewal counting process $\{N(t), \ t \geq 0\}$. A renewal at time T_i comes with a random *mark* M_i. This generates a marked point process

$$\{(T_1, M_1), (T_2, M_2), ...\},$$

which in turn determines the *compound renewal process* $\{C(t), \ t \geq 0\}$ with

$$C(t) = \sum_{i=0}^{N(t)} M_i, \quad M_0 = 0. \tag{9.36}$$

   Assumption *The sequences $\{L_1, L_2, ...\}$ and $\{M_1, M_2, ...\}$ are indepen- dent of each other and each consists of independent, nonnegative random vari- ables, which are identically distributed as L and M, respectively, with $\mathbb{E}(L)$ and $\mathbb{E}(M)$ finite. However, L_i and M_j may depend on each other if $i = j$, i.e. if they refer to the same renewal cycle.*
   Under this assumption, the trend function of $\{C(t), \ t \geq 0\}$ is

$$\mathbb{E}(C(t)) = \mathbb{E}(M)H(t), \quad t \geq 0, \tag{9.37}$$

where $H(t)$ is the renewal function of the underlying renewal process, namely $\{L_1, L_2, ...\}$. From (9.25),

$$\lim_{t \to \infty} \frac{C(t)}{t} = \frac{\mathbb{E}(M)}{\mathbb{E}(L)}. \tag{9.38}$$

This asymptotic relationship is frequently called the *renewal reward theorem*, in paricular if M is a profit arising from a renewal. If M is a loss, then (9.38) should be called the *renewal loss theorem*.

Distribution of C(t) If M has distribution function $G(x)$, then $C(t)$ has the conditional distribution function $\Pr(C(t) \leq x \,|\, N(t) = n) = G^{*(n)}(x)$, where $G^{*(n)}(x)$ is the n-fold convolution of G with itself, and $G^{*(0)}(x) \equiv 1$, $G^{*(1)}(x) = G(x)$. Hence, by the total probability rule,

$$F_{C(t)}(x) = P(C(t) \leq x) = \sum_{n=1}^{\infty} G^{*(n)}(x)P(N(t) = n). \tag{9.39}$$

1. If $\{N(t), \ t \geq 0\}$ is a homogeneous Poisson process with intensity λ, then, for $x > 0$, $t > 0$,

$$F_{C(t)}(x) = e^{-\lambda t} \sum_{n=0}^{\infty} G^{*(n)}(x) \frac{(\lambda t)^n}{n!}.$$

2. If $M = N(\mu, \sigma^2)$ with $\mu > 3\sigma^2$, then, for $x > 0$, $t > 0$,

$$F_{C(t)}(x) = e^{-\lambda t} \left[1 + \sum_{n=1}^{\infty} \Phi\left(\frac{x - \mu}{\sigma \sqrt{n}}\right) \frac{(\lambda t)^n}{n!} \right].$$

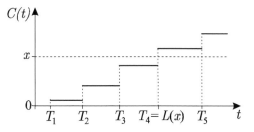

Figure 9.2: First passage time

In most cases, tractable formulas for $F_{C(t)}$ cannot be obtained via (9.39). Hence, much effort has been put into constructing bounds and asymptotic expansions for $F_{C(t)}$ (for surveys see [307], [259]). The following result of *Gut* [150] is particularly useful: Let

$$\gamma^2 = Var\left(\mathbb{E}(L)M - \mathbb{E}(M)L\right) > 0. \tag{9.40}$$

Then

$$\lim_{t \to \infty} Pr\left(\frac{C(t) - \frac{\mathbb{E}(M)}{\mathbb{E}(L)}t}{[\mathbb{E}(L)]^{-1.5}\gamma\sqrt{t}} \leq x\right) = \Phi(x). \tag{9.41}$$

Thus, for t sufficiently large,

$$C(t) \approx N\left(\frac{\mathbb{E}(M)}{\mathbb{E}(L)}t, [\mathbb{E}(L)]^{-3}\gamma^2 t\right). \tag{9.42}$$

The condition (9.40) only excludes the case $\gamma = 0$, i.e. linear dependency between L and M.

First Passage Time of C(t)   Let $L(x)$ be the time when the compound stochastic process $\{C(t),\ t \geq 0\}$ for the first time hits or exceeds a given level x:

$$L(x) = \inf_t\{t,\ C(t) \geq x\}. \tag{9.43}$$

If, for instance, x is the critical mechanical wear limit of a component, then crossing level x is referred to as a *drift failure* of the component. Since under our assumptions the process $\{C(t),\ t \geq 0\}$ has nondecreasing sample paths, there is the following obvious relationship between the distribution of $C(t)$ and $L(x)$ (Figure 9.2):

$$Pr(L(x) \leq t) = Pr(C(t) \geq x), \tag{9.44}$$

or

$$F_{L(x)}(t) = 1 - F_{C(t)}(x). \tag{9.45}$$

Hence, if $\{N(t),\ t \geq 0\}$ is the homogeneous Poisson process, then, from (9.39),

$$F_{L(x)}(t) = 1 - e^{-\lambda t} \sum_{n=0}^{\infty} G^{*(n)}(x) \frac{(\lambda t)^n}{n!}, \quad t > 0,\ x > 0.$$

In addition to (9.42), with γ again given by (9.40), *Gut* [150] proved that

$$\lim_{x \to \infty} \Pr\left(\frac{L(x) - \frac{\mathbb{E}(L)}{\mathbb{E}(M)} x}{[\mathbb{E}(M)]^{-1.5}\, \gamma \sqrt{x}} \leq t \right) = \Phi(t).$$

Equivalently, for x sufficiently large,

$$L(x) \approx N\left(\frac{\mathbb{E}(L)}{\mathbb{E}(M)} x,\ [\mathbb{E}(M)]^{-3} \gamma^2 x \right). \tag{9.46}$$

The distribution on the right-hand side of (9.42) and (9.46), respectively, is the *Birnbaum–Saunders distribution*.

Example 9.9 *Mechanical wear of an item is caused by shocks, the interarrival times of which are distributed as L with parameters $\mathbb{E}(L) = 6\ [h]$ and $\sqrt{Var(L)} = 2\ [h]$. The random wear M caused by a shock has parameters*

$$\mathbb{E}(M) = 9.2\ [10^{-4} mm] and \sqrt{Var(M)} = 2.8\ [10^{-4} mm].$$

What is the probability that the total wear occuring in $[0, 600\ h]$ does not exceed $0.1\ [mm]$? To answer this question, (9.46) is applicable, because in this case level x is large enough compared to $\mathbb{E}(M)$. Since $\gamma = 0.0024914$,

$$\Pr\left(L(0.1) > 600\right) = 1 - \Phi\left(\frac{600 - \frac{6}{9.2} 10^3}{(9.2)^{-1.5} \cdot 2491.4 \cdot \sqrt{0.1}} \right)$$

$$= 1 - \Phi(-1.848) = 0.967.$$

It is interesting to compare this numerical value to the one obtained by making use of (9.42) and (9.44):

$$\Pr\left(L(0.1) > 600\right) = \Phi\left(\frac{0.1 - \frac{9.2 \cdot 10^{-4}}{6} 600}{6^{-1.5} 0.0024914 \sqrt{600}} \right)$$

$$= 0.973.$$

9.3   Minimal Repair Processes

9.3.1   Pure Minimal Repair Policy and Nonhomogeneous Poisson Processes

Renewals of systems after failure as discussed in Section 9.2 are an extreme form of maintenance actions. They are usually exercised by replacements of failed systems by a new one, which, with regard to its lifetime distribution, is equivalent to the previous one. The other extreme case is the minimal repair. The system lifetime L is assumed to have density $f(t) = F'(t)$, so that the failure rate $\lambda(t) = f(t)/\overline{F}(t)$ exists. Then, by (1.5), $F(t)$ has structure

$$F(t) = \Pr(L \le t) = 1 - e^{-\Lambda(t)}, \text{ with } \Lambda(t) = \int_0^t \lambda(x)dx, \ t \ge 0. \qquad (9.47)$$

Definition 9.10 *A minimal repair performed after a failure enables the system to continue its work, but does not affect the failure rate of the system, i.e. the failure rate immediately before the failure is equal to the failure rate of the system immediately after the completion of its repair.*

Policy 1 After every failure the system is minimally repaired. The repair time is negligibly small, and the system resumes its work immediately after a minimal repair. This process continues to infinity.

We now consider a special *rcp*, namely the *minimal repair process (mrp)* $\{N(t), \ t \ge 0\}$, where $N(t)$ denotes the random number of minimal repairs in $(0, t]$, which occur when following policy 1. Let $0 \le t_1 < t_2 < t_3$. Then the development in time of $\{N(t), \ t \ge 0\}$ in $(t_2, t_3]$ depends only on $\lambda(t_2)$, irrespective of its development in $(t_1, t_2]$, which in turn is fully determined by $\lambda(t_1)$. Hence, $\{N(t), \ t \ge 0\}$ has independent increments, i.e. for any $0 \le t_1 < t_2 < t_3 < \cdots < t_{n-1} < t_n$, the increments $N(t_1, t_2), N(t_2, t_3), \cdots,$ $N(t_{n-1}, t_n)$ are independent random variables.

Since, by (9.47), $F(+0) = 0$, the *mrp* $\{N(t), \ t \ge 0\}$ is simple, i.e. it has property $\Pr(N(t, t + \Delta t) \ge 2) = o(\Delta t)$ for $\Delta t \to 0$.

Finally, from (9.7), $\Pr(N(t, t + \Delta t) = 1) = \lambda(t)\Delta t + o(\Delta t)$ for $\Delta t \to 0$.

Thus, we have verified that the *mrp* $\{N(t), \ t \ge 0\}$ is a nonhomogeneous Poisson process with intensity function $\lambda(t)$:

Definition 9.11 *A rcp $\{N(t), \ t \ge 0\}$ is called a nonhomogeneous Poisson process (nhPp) with intensity function $\lambda(t)$ if it has the following properties:*
1. $\{N(t), \ t \ge 0\}$ has independent increments.
2. $\{N(t), \ t \ge 0\}$ is a simple rcp.
3. $\Pr(N(t, t + \Delta t) = 1) = \lambda(t)\Delta t + o(\Delta t)$.

In what follows, we continue referring to $\{N(t), \ t \ge 0\}$ as the mrp. Following [38], results on *mrp*'es are summarized, which we will need later.

Distribution of increments   The increments $N(s,t) = N(t) - N(s)$, $s < t$, of the *mrp* have a Poisson distribution with expected value

$$m(s,t) = \int_s^t \lambda(x)dx = \Lambda(t) - \Lambda(s), \tag{9.48}$$

i.e., for $n = 0, 1, \ldots$

$$p_n(s,t) = \Pr(N(s,t) = n) = \frac{(\Lambda(t) - \Lambda(s))^n}{n!} e^{-(\Lambda(t) - \Lambda(s))}. \tag{9.49}$$

In particular, the absolute state probabilities of the *mrp* are

$$p_n(t) = \Pr(N(t) = n) = \frac{\Lambda(t)^n}{n!} e^{-\Lambda(t)}, \quad n = 0, 1, \ldots \tag{9.50}$$

Hence, its trend function is

$$m(t) = \Lambda(t), \quad t \geq 0. \tag{9.51}$$

Distribution of the time points of minimal repairs   Let $\{T_1, T_2, \ldots\}$ be the point process which belongs to the *mrp* $\{N(t), \ t \geq 0\}$. Since $T_1 = L$, the distribution function of T_1 is given by (9.47). Generally, the distribution function of T_n, the time point of the nth minimal repair or failure, respectively, is

$$F_{T_n}(t) = e^{-\Lambda(t)} \sum_{i=n}^{\infty} \frac{(\Lambda(t))^n}{n!}, \quad t \geq 0, \ n = 1, 2, \ldots, \tag{9.52}$$

which follows from (9.2) and (9.50). Differentiation with respect to t yields the probability density of T_n (all terms but one cancel each other):

$$f_{T_n}(t) = \frac{(\Lambda(t))^{n-1}}{(n-1)!} \lambda(t) e^{-\Lambda(t)}, \quad t \geq 0, \ n = 1, 2, \ldots, \tag{9.53}$$

or, equivalently,

$$f_{T_n}(t) = \frac{(\Lambda(t))^{n-1}}{(n-1)!} f_{T_1}(t), \quad t \geq 0, \ n = 1, 2, \ldots \tag{9.54}$$

By (1.2) and (9.54),

$$\mathbb{E}(T_n) = \int_0^{\infty} e^{-\Lambda(t)} \left(\sum_{i=0}^{n-1} \frac{(\Lambda(t))^i}{i!} \right) dt, \quad n = 1, 2, \ldots \tag{9.55}$$

Hence, the expected time between the $(n-1)^{th}$ and the nth minimal repair $\mathbb{E}(L_n) = \mathbb{E}(T_n) - \mathbb{E}(T_{n-1})$ is

$$\mathbb{E}(L_n) = \frac{1}{(n-1)!} \int_0^{\infty} (\Lambda(t))^i e^{-\Lambda(t)} dt, \quad n = 1, 2, \ldots \tag{9.56}$$

Letting $\lambda(t) \equiv \lambda$ and $\Lambda(t) = \lambda t$ yields the corresponding results for the homogeneous Poisson process. In this extreme case, every minimal repair is at the same time a complete renewal of the failed system.

Joint density of $(T_1, T_2, ..., T_n)$ The conditional probability

$$F_{T_2}(t_2 \,|t_1) = \Pr(T_2 \leq t_2 \,|T_1 = t_1)$$

is the probability that at least one minimal repair occurs in $(t_1, t_2]$, $t_1 < t_2$. Thus, by (9.49),

$$F_{T_2}(t_2 \,|t_1) = 1 - P(N(t_1, t_2) = 0) = 1 - e^{-[\Lambda(t_2) - \Lambda(t_1)]}.$$

By differentiation with regard to t_2, the corresponding conditional density is seen to be

$$f_{T_2}(t_2 \,|t_1) = \lambda(t_2)e^{-[\Lambda(t_2) - \Lambda(t_1)]}, \; 0 \leq t_1 < t_2.$$

Hence, since

$$f_{(T_1, T_2)}(t_1, t_2) = f_{T_2}(t_2 \,|t_1)\, f_{T_1}(t_1),$$

we have

$$f_{(T_1, T_2)}(t_1, t_2) = \begin{cases} \lambda(t_1)\lambda(t_2)e^{-\Lambda(t_2)}, \; 0 \leq t_1 < t_2 \\ 0, \; \text{elsewhere} \end{cases}.$$

Starting with $f_{(T_1, T_2)}(t_1, t_2)$, one inductively gets the joint density of the random vector $\mathbf{T} = (T_1, T_2, ..., T_n)$: Letting $\mathbf{t} = (t_1, t_2, ..., t_n)$,

$$f_{\mathbf{T}}(\mathbf{t}) = \begin{cases} \lambda(t_1)\lambda(t_2)\cdots\lambda(t_n)e^{-\Lambda(t_n)}, \; t_1 < t_2 < \cdots < t_n \\ 0, \; \text{elsewhere} \end{cases}. \qquad (9.57)$$

9.3.2   Minimal Repair Process with Embedded Renewals

To restore the ability of a failed system to resume its work only by minimal repairs is certainly not always possible and definitely not possible over a sufficiently long time period. Wear out, accidents and other causes will sooner rather than later lead to system failures, which require major maintenance actions. As a first step in modeling situations like that we assume that a system failure occurring at time t (with regard to the time point of the start of its work) will (and can be) removed by a minimal repair with probability $1 - p(t)$, and will be removed by a renewal (replacement) of the system with probability $p(t)$.

Policy 2 A system failure occuring at time t is with probability $1 - p(t)$ of type 1 and with probability $p(t)$ of type 2. Type 1-failures and type 2-failures are removed by minimal repairs and renewals, respectively. Repair- and renewal times are assumed to be negligibly small. Repaired or renewed systems immediately resume their work. This process continues unlimitedly.

This policy can be described by a marked point process

$$\{(T_1, B_1), \; (T_2, B_2), \cdots\},$$

where T_i is the time point, at which the ith failure occurs, and B_i is the binary indicator variable of the type of the ith failure. The $B_1, B_2, \cdots$ are assumed to be independent.

Time between renewals   Under Policy 2, the time intervals between two neighboring renewals, called *renewal cycles*, generate a recurrent point process. For our subsequent applications, we only need to analyze one renewal cycle, since they are all statistically equivalent. Let the random length Y of a renewal cycle have distribution function $G(t) = \Pr(Y \le t)$. Since type 1- (type 2- failures) occur with intensities $(1 - p(t))\lambda(t)$ and $p(t)\lambda(t)$, respectively, we have from formula (1.6),

$$\Pr(t < Y \le t + \Delta t \,|\, L > t) = p(t)\lambda(t)\Delta t + o(\Delta t),$$

so that, from (1.5), $p(t)\lambda(t) = G'(t)/\overline{G}(t)$. By integration,

$$\overline{G}(t) = e^{-\int_0^t p(x)\lambda(x)dx}, \ t \ge 0. \tag{9.58}$$

If $p(t) \equiv 1$, then $\overline{G}(t)$ is the survival function of the system.

Number of minimal repairs in a renewal cycle   Let Z be the random number of minimal repairs in a renewal cycle, and

$$f(t) = f_{T_1}(t) = \lambda(t)e^{-\Lambda(t)}, \ t \ge 0.$$

Then,

$$\Pr(Z = 0) = \int_0^\infty p(t)f(t)dt.$$

From (9.57) with $\overline{p}(t) = 1 - p(t)$, $n \ge 1$,

$$\Pr(Z = n) = \int_0^\infty \int_0^{t_{n+1}} \int_0^{t_n} \cdots \int_0^{t_2} \prod_{i=1}^{n} [\overline{p}(t_i)\lambda(t_i)dt_i]p(t_{n+1})f(t_{n+1})dt_{n+1}.$$

By making use of the formula, $n = 2, 3, ...,$

$$\int_0^t \int_0^{x_n} \cdots \int_0^{x_3} \int_0^{x_2} \prod_{i=1}^{n} g(x_i)dx_1 dx_2 \cdots dx_n = \frac{1}{n!}\left[\int_0^t g(x)dx\right]^n, \tag{9.59}$$

the probability $\Pr(Z = n)$ becomes

$$\Pr(Z = n) = \frac{1}{n!}\int_0^\infty \left(\int_0^t \overline{p}(x)\lambda(x)dx\right)^n p(t)f(t)dt. \tag{9.60}$$

After some algebra

$$\mathbb{E}(Z) = \sum_{n=1}^\infty n\Pr(Z = n) = \int_0^\infty \Lambda(t)dG(t) - 1. \tag{9.61}$$

If $p(t) \equiv p > 0$, then Z has a geometric distribution with parameter p so that

$$\mathbb{E}(Z) = \frac{1-p}{p}, \quad \overline{G}(t) = \left[\overline{F}_L(t)\right]^p, \ t \ge 0. \tag{9.62}$$

Now, let Z_t be the random number of minimal repairs in the interval $(0, \min\{Y, t\})$ and

$$r_t(n) = \Pr(Z_t = n \mid Y = t), \quad n = 0, 1, \dots$$

Then,

$$r_t(n) = \lim_{\Delta t \to 0} \frac{\Pr(Z_t = n \cap t \leq Y \leq t + \Delta t)}{\Pr(t \leq Y \leq t + \Delta t)}$$

or, equivalently,

$$r_t(n) = \lim_{\Delta t \to 0} \frac{\Pr(t \leq Y = T_{n+1} \leq t + \Delta t)}{G(t + \Delta t) - G(t)}. \tag{9.63}$$

Using (9.57) and (9.59), the numerator in (9.63) becomes

$$\Pr(t \leq Y = T_{n+1} \leq t + \Delta t)$$

$$= \int_t^{t+\Delta t} \int_0^{t_{n+1}} \int_0^{t_n} \cdots \int_0^{t_2} \prod_{i=1}^n \left[\overline{p}(t_i)\lambda(t_i)dt_i \right] p(t_{n+1})f(t_{n+1})dt_{n+1}$$

$$= \frac{1}{n!} \int_t^{t+\Delta t} \left(\int_0^y \overline{p}(x)\lambda(x)dx \right)^n p(y)f(y)dy.$$

Dividing numerator and denominator in (9.63) by Δt and letting $\Delta t \to 0$, yields

$$r_t(n) = \frac{1}{n!} \left(\int_0^t \overline{p}(x)\lambda(x)dx \right)^n e^{-\int_0^t \overline{p}(x)\lambda(x)dx}. \tag{9.64}$$

Hence, on condition $Y = t$, the random variable Z_t has a Poisson distribution with expected value

$$\mathbb{E}(Z_t \mid Y = t) = \int_0^t \overline{p}(x)\lambda(x)dx. \tag{9.65}$$

From

$$\mathbb{E}(Z_t \mid Y < t) = \frac{1}{G(t)} \int_0^t \mathbb{E}(Z_x \mid Y = x) \, dG(x)$$

we obtain

$$\mathbb{E}(Z_t \mid Y < t) = \frac{1}{G(t)} \int_0^t \left(\int_0^x \overline{p}(y)\lambda(y)dy \right) dG(x). \tag{9.66}$$

Moreover,

$$\mathbb{E}(Z_t \mid Y \geq t) = \mathbb{E}(Z_t \mid Y = t) = \int_0^t \overline{p}(x)\lambda(x)dx. \tag{9.67}$$

From $\mathbb{E}(Z_t) = \mathbb{E}(Z_t \mid Y < t) \, G(t) + \mathbb{E}(Z_t \mid Y \geq t) \, \overline{G}(t)$, making use of (9.66) and (9.67):

$$\mathbb{E}(Z_t) = \int_0^t \overline{G}(x)\lambda(x)dx - G(t). \tag{9.68}$$

For $t \to \infty$, $\mathbb{E}(Z_t)$ tends to $\mathbb{E}(Z)$ given that there exists a p_0 with $\lim_{t\to\infty} p(t) = p_0 > 0$. But according to its meaning, $p(t)$ can be expected to be an increasing function in t anyway.

Remark 9.12 *To be mathematically correct, for proving (9.64) it has to be shown that for every positive ε, however small, there exists a positive δ so that*

$$\left| r_t(n) - \frac{\Pr(t \leq Y = T_{n+1} \leq t + \Delta t)}{G(t + \Delta t) - G(t)} \right| < \varepsilon$$

whenever $|\Delta t| < \delta$. But this is a routine procedure.

The results of this section were published in *Beichelt* [26], see also *Block, Borges, and Savits* [55]. Extended versions are given in the monographs [28] and [42]. *Kapur, Garg and Kumar* [169] (see also the literature cited there) were likely the first to apply these results to software reliability analysis: Software failures occur according to a nonhomogeneous Poisson point process. A failure occuring at time t is removed from the software with probability $p(t)$ and is not removed with probability $1 - p(t)$ ("imperfect debugging").

9.4   Exercises

Exercise 9.1 *The number of catastrophic accidents at Sosal & Sons can be described by a homogeneous Poisson process with intensity $\lambda = 3$ a year.*
(1) What is the probability $p_{\geq 2}$ that at least two catastrophic accidents will occur in the second half of the current year?
(2) Determine the same probability given that two catastrophic accidents have occurred in the first half of the current year.

Exercise 9.2 *The number of cars, which pass a certain intersection daily between 12:00 and 14:00, follows a homogeneous Poisson process with intensity $\lambda = 40$ per hour. Among these there are 0.8% which disregard the STOP-sign. What is the probability that at least one car disregards the STOP-sign between 12:00 and 13:00?*

Exercise 9.3 *A Geiger counter is struck by radioactive particles according to a homogeneous Poisson process with intensity $\lambda = 1$ per 12 seconds. On average, the Geiger counter only records 4 out of 5 particles.*
(1) What is the probability that the Geiger counter records at least 2 particles a minute?
(2) What are the expected value μ [min] and variance σ^2 [min^2] of the random time Y between the occurrence of two successively recorded particles?

Exercise 9.4 *An electronic system is subject to two types of shocks which arrive independently of each other according to homogeneous Poisson processes with intensities $\lambda_1 = 0.002$ and $\lambda_2 = 0.01$ per hour, respectively. A shock of* type 1 always causes a system failure, whereas a shock of type 2 causes a system failure with probability 0.4.

What is the probability of the event A that the system fails within 24 hours due to a shock?

Exercise 9.5 *Consider two independent homogeneous Poisson processes 1 and 2 with respective intensities λ_1 and λ_2. Determine the expected value of the random number of events of process 2, which occur between any two successive events of process 1.*

Exercise 9.6 *By making use of the independence and homogeneity of the increments of a homogeneous Poisson process $\{N(t),\ t \geq 0\}$ with intensity λ show that its covariance function $C(s,t) = \mathbb{E}[N(s) - \mathbb{E}N(s)][N(t) - \mathbb{E}N(t)]$ is given by $C(s,t) = \lambda \min(s,t)$.*

Exercise 9.7 *Statistical evaluation of a large sample justifies modeling the number of cars which arrive daily for petrol between 1:00 p.m. and 4:00 p.m. at a particular filling station by a nonhomogeneous Poisson process with intensity function*

$$\lambda(t) = 8 - 4(t-1) + 3(t-1)^2 \ [h^{-1}], \quad 1 \leq t \leq 4.$$

(1) How many cars arrive on average between 1:00 p.m. and 4:00 p.m.?
(2) What is the probability that at least 30 cars arrive between 2:00 and 4.00 a.m.?

Exercise 9.8 *A system starts working at time $t = 0$. Its lifetime has approximately a normal distribution with expected value $\mu = 120$ and standard deviation $\sigma = 24$ [hours]. After a failure, the system is replaced by an equivalent new one in negligible time and immediately resumes its work. How many spare systems must be available in order to be able maintain the replacement process over an interval of length 10000 hours (1) with probability 0.90 and (2) with probability 0.99?*

Exercise 9.9 *An ordinary renewal process has the renewal function $H(t) = 10$. Determine the probability $\Pr(N(10) \geq 2)$.*

Exercise 9.10 *By means of the Laplace transformation find the renewal function $H(t)$ of an ordinary renewal process whose cycle lengths have an Erlang distribution with parameters $n = 2$ and λ.*

Exercise 9.11 *The probability density function of the cycle lengths of an ordinary renewal process is the mixture of two exponential distributions:*

$$f(t) = p\lambda_1 e^{-\lambda_1 t} + (1-p)\lambda_2 e^{-\lambda_2 t}, \ 0 \leq p \leq 1, \ t \geq 0, \ \lambda_1, \lambda_2 > 0.$$

By making use of the Laplace transformation, determine the associated renewal function.

Exercise 9.12 *Prove integral equation (9.14) by means of formulas (9.10) and (9.13).*

Exercise 9.13 *Verify that the probability $p(t) = \Pr(N(t) = odd)$ satisfies the integral equation*

$$p(t) = F(t) - \int_0^t p(t - x)dF(x).$$

Determine $p(t)$ if the cycle lengths are exponentially distributed with parameter λ.

Exercise 9.14 *The cycle length Y of an ordinary renewal process is a discrete random variable with probability distribution $p_k = \Pr(Y = k)$, $k = 0, 1, ...$*
(1) Show that the corresponding renewal function $H(n)$ satisfies

$$H(n) = q_n + H(0)p_n + H(1)p_{n-1} + \cdots + H(n)p_0$$

with $q_n = \Pr(Y \le n) = p_0 + p_1 + \cdots + p_n$, $n = 0, 1, ...$
(2) Consider the special cycle length distribution

$$\Pr(Y = 0) = p, \ \ \Pr(Y = 1) = 1 - p$$

and determine the corresponding renewal function. (This special renewal process is sometimes referred to as the negative binomial process.*)*

Exercise 9.15 *Let (Y, Z) be the typical cycle of an alternating renewal process, where Y and Z have an Erlang distribution with joint parameter λ and parameters $n = 2$ and $n = 1$, respectively.*
For $t \to \infty$ determine the probability that the system is in state 1 at time t and that it stays in this state over the entire interval $[t, t + x]$.

Exercise 9.16 *The time intervals between successive repairs of a system generate an ordinary renewal process $\{Y_1, Y_2, ...\}$ with typical cycle length Y. The costs of repairs are mutually independent, independent of $\{Y_1, Y_2, ...\}$ and identically distributed as M. The random variables Y and M have parameters*

$$\mathbb{E}(Y) = 180 \ [days], \ \ \sqrt{Var(Y)} = 30,$$
$$\mathbb{E}(M) = \$200, \ \ \sqrt{Var(M)} = 40.$$

Determine approximately the probabilities that
(1) the total repair cost arising in $[0, \ 3600 \ days]$ does not exceed \$4500,
(2) a total repair cost of \$3000 is not exceeded before 2200 days.

Exercise 9.17 *Prove formula (9.64) according to remark 9.12.*

Chapter 10

Time-Based System Replacement

The choice of maintenance policies dealt with in this book is governed by the interests of the authors. The focus is on basic models, not on the mathematically more sophisticated ones. In applications, maintenance models need to be tailored to the specific practical situation. Hence, the maintenance policies analyzed in what follows can only provide the general framework. For more comprehensive recent summaries, including such important aspects as warranty and optimal maintenance of multi-unit systems, see e.g. Gertsbakh [136], *Pham* [249], *Wang* and *Pham* [303], *Nakagawa* [237, 238], *Sarka, Panja,* and *Sarka* [268]. However, some of the maintenance models considered in what follows, although formally designed for simple systems, are actually applicable to the cost-optimal maintenance of complex systems.

Notation and Assumptions

If not stated otherwise, in Part 3 we will assume that all repair and replacement (renewal) times are negligibly small, and that the lifetime distribution of the replaced systems are the same as those of the previous ones, i.e. replaced systems are *as good as new*. Distribution function, density, failure rate, integrated failure rate, expected value of the system lifetime L, and residual lifetime of a system which has survived $[0, x]$, are $F(t)$, $f(t)$, $\lambda(t)$, $\Lambda(t)$, μ, and L_x, respectively. The respective costs of emergency replacements, preventive replacements, and minimal repairs are assumed to be constant and equal to

$$c_e, \ c_p, \ \text{and} \ c_m \ \text{with} \ 0 < c_m < \ c_p < c_e.$$

Preventive maintenance only makes sense if the system is wearing out (aging), at least from a certain time point $t_0 \geq 0$. We usually assume that $F(t)$ is IFR in $[0, \infty)$. The maintenance process is carried out unlimitedly.

10.1   Age and Block Replacement

The two basic replacement policies (strategies) considered in this section we need for the sake of references and comparisons.

Policy 3 (*age replacement policy*)   A system is replaced on failure by an *emergency replacement* or by a *preventive replacement* τ time units after the last replacement (emergency or preventive), whichever occurs first [280].

The random maintenance cost C within a renewal cycle (time between to neighboring replacements of any kind) and the random length X of a renewal cycle are

$$C = \begin{cases} c_e \text{ with probability } F(\tau) \\ c_p \text{ with probability } \overline{F}(\tau) \end{cases}, \qquad X = \begin{cases} L \text{ if } L \leq \tau \\ \tau \text{ if } L > \tau \end{cases}.$$

Hence, by the renewal reward theorem (9.38), the *long-run* or *stationary maintenance cost per unit time* (in what follows shortly referred to as *cost rate*) is

$$K_3(\tau) = \frac{\mathbb{E}(C)}{\mathbb{E}(X)} = \frac{c_e F(\tau) + c_p \overline{F}(\tau)}{\int_0^\tau \overline{F}(t)dt}. \tag{10.1}$$

The optimum $\tau = \tau^*$ we get from $dK_3(\tau)/d\tau = 0$ as solution of the equation

$$\lambda(\tau) \int_0^\tau \overline{F}(t)dt - F(t) = \frac{c_p}{c_e - c_p}. \tag{10.2}$$

A sufficient condition for the existence of a unique solution τ^* of equation (10.2) is that $\lambda(t)$ is strictly increasing and $\lambda(\infty) > c_e/\mu(c_e - c_p)$ (Exercise 10.2). The latter condition is always fulfilled if $\lambda(t) \to \infty$ for $t \to \infty$. By combining formulas (10.1) and (10.2),

$$K_3(\tau^*) = (c_e - c_p)\lambda(\tau^*).$$

Age Replacement Policy for a Two-Unit System   It is interesting that equation (10.2) for an optimal preventive maintenance interval not only appears in conjunction with Policy 3, but also in quite another situation, namely in conjunction with the optimal maintenance of a two-unit-system with regard to its expected lifetime. Here a standard situation is considered (see [20] and references cited there): A system consists of two identical elements. Initially, one of them is in cold standby. If the active unit fails, it immediately undergoes maintenance, and the standby element becomes active. Different from the previous policies, the times required for maintenance (preventive or emergency replacements) are not negligibly small. The active unit is principally subject to an age replacement policy with interval τ. But, in order to avoid system failures (none of the elements is active), there is a modification of Policy 3: If at the time of the planned replacement of the active element the other one is still under maintenance, then its preventive replacement is postponed till the replacement of the other element is finished. As soon as an active unit fails while the other unit is under maintenance, the system fails (system lifetime L_S).

Let $F(t)$ be the c.d.f. of the lifetime L of an active unit and $F_e(t)$ and $F_p(t)$ be the c.d.f.'s of the random times L_e and L_p to do an emergency replacement and a preventive replacement, respectively, with positive and finite expected values $\mu = \mathbb{E}(L)$, $\mu_e = \mathbb{E}(L_e)$, and $\mu_p = \mathbb{E}(L_p)$. Let further

$$\alpha = \int_0^\infty \overline{F}_e(t)dF(t) \quad \text{and} \quad \beta = \int_0^\infty \overline{F}_p(t)dF(t)$$

be the probabilities of the events that the operating unit fails during an ongoing emergency replacement and that the operating unit fails during an ongoing preventive replacement, respectively.

To justify preventive replacements, we will assume that $\alpha > \beta$. If the random times involved are independent and have finite expected values, then the expected system lifetime is

$$\mathbb{E}(L_S) = \int_0^\tau \overline{F}(t)dt + \frac{\left(\mu - \int_\tau^\infty \overline{F}(t)F_e(t)dt\right)\left[\beta F(\tau) + \int_0^\tau F_p(t)dF(t)\right]}{\alpha\beta + \alpha\int_0^\tau F_p(t)dF(t) + \beta\int_\tau^\infty F_e(t)dF(t)}$$

$$+ \frac{\left(\mu - \int_\tau^\infty \overline{F}(t)F_p(t)dt\right)\left[\alpha\overline{F}(\tau) + \int_\tau^\infty F_e(t)dF(t)\right]}{\alpha\beta + \alpha\int_0^\tau F_p(t)dF(t) + \beta\int_\tau^\infty F_e(t)dF(t)}.$$

To simplify this formula, let us make some assumptions, which normally are fulfilled in practice: $\mu_p \ll \mu$, and $\mu_e \ll \mu$. If, in addition, $\mu \lesssim \tau$, then $F_p(\tau) \gtrsim 1$ and $F_e(\tau) \gtrsim 1$ so that

$$\alpha \approx \int_0^\tau \overline{F}_e(t)dF(t), \quad \beta \approx \int_0^\tau \overline{F}_p(t)dF(t).$$

These assumptions lead to a rather simple approximative formula for the expected system lifetime:

$$\mathbb{E}(L_S) \approx \widetilde{\mathbb{E}}(L_S) = (1+\alpha)\frac{\int_0^\tau \overline{F}(t)dt}{\alpha F(\tau) + \beta\overline{F}(\tau)}.$$

A replacement interval $\tau = \tau^*$ being optimal with regard to $\widetilde{\mathbb{E}}(L_S)$ satisfies Equation (10.2) derived for policy 3 with $\alpha = c_e$ and $\beta = c_p$.

Policy 4 (*block replacement policy*) On failure, the system is replaced by a new one (emergency replacement). At fixed time points τ, 2τ, ..., the system is preventively replaced [17].

The time points τ, 2τ, ... partition the positive real axis into renewal cycles of constant length τ. Hence, by the renewal reward theorem (9.38), the corresponding cost rate is

$$K_4(\tau) = \frac{c_p + c_e H(\tau)}{\tau}, \qquad (10.3)$$

where $H(t)$ is the renewal function which belongs to the lifetime distribution function $F(t)$. Provided its existence, an optimum $\tau = \tau^*$ satisfies the equation $dK_4(\tau)/d\tau = 0$ or

$$\tau h(\tau) - H(\tau) = c_p/c_e. \qquad (10.4)$$

Example 10.1 *Let*

$$F(t) = (1 - e^{-\lambda t})^2, \ \lambda > 0, \ t \geq 0.$$

By (9.24), the corresponding renewal density $h(t)$ and the renewal function $H(t)$ are

$$h(t) = \tfrac{2}{3}\lambda \left(1 - e^{-3\lambda t}\right), \ H(t) = \tfrac{2}{3}\left[\lambda t + \tfrac{1}{3}\left(e^{-3\lambda t} - 1\right)\right],$$

so that Equation (10.4) becomes

$$(1 + 3\lambda\tau)e^{-3\lambda\tau} = \frac{2c_e - 9c_p}{2c_e}.$$

A unique solution $\tau = \tau^*$ exists if $c_e > 4.5c_p$.

There are numerous modifications and generalizations of policies 3 and 4, see for instance, *Cox* [100], *Beichelt* [34] *Wang* and *Pham* [303], and the following sections.

From the point of view of the structure of its cost rate, Policy 4 is a special case of a simpler, but nevertheless more general replacement policy:

Policy 5 (*economic lifetime*)   A system is preventively replaced at time points τ, 2τ, ... A deterministic loss $V(t)$ arises from the time point of a replacement $n\tau$ till $n\tau + t$, $t < \tau$, $n = 0, 1, ...$ [210].

$V(t)$ is caused by maintenance actions (repairs, inspections, staff and material costs etc.). The corresponding cost rate is

$$K_5(\tau) = \frac{c_p + V(\tau)}{\tau}. \qquad (10.5)$$

A replacement interval $\tau = \tau^*$ minimizing $K_5(\tau)$ is called *economic lifetime* [91]. Letting $v(t) = dV(t)/dt$, an optimal replacement interval τ^* satisfies the equation

$$K_5(\tau) = v(\tau). \qquad (10.6)$$

10.2   Replacement and Minimal Repair

In this section, after a series of minimal repairs, replacements will be scheduled preventively or on failures.

Policy 6 (*Block replacement with minimal repair*)   At fixed time points $\tau, 2\tau, ...$, the system is preventively replaced. In between, failures are removed by minimal repairs [17].

This policy reflects the common approach of preventively overhauling complicated systems at the end of fixed time periods, while in between only the absolutely necessary repairs are done.

Let c_m be the cost of a minimal repair. Then the cost rate is

$$K_6(\tau) = \frac{c_p + c_m \Lambda(\tau)}{\tau}, \tag{10.7}$$

since, by (9.51), $\Lambda(\tau)$ is the expected number of minimal repairs in a renewal cycle of length τ. In case of its existence, an optimal replacement interval $\tau = \tau^*$ is solution of the equation

$$\tau \lambda(\tau) - \Lambda(\tau) = c_p/c_m. \tag{10.8}$$

A unique solution exists if $\lambda(t)$ is strictly increasing in t. By combining formulas (10.7) and (10.8), the minimal maintenance cost rate is seen to be

$$K_6(\tau^*) = c_m \lambda(\tau^*)$$

Just as Policy 4, from the point of view of the structure of its cost rate, Policy 6 is a special case of Policy 5, too.

Policy 7   A system is replaced at the first failure which occurs <u>after</u> a fixed time τ. Failures occuring between replacements are removed by minimal repairs [229].

This policy fully makes use of the system lifetime so that, from this point of view, it is preferable to Policy 6. However, the partial unertainty about the times of replacements leads to higher replacement costs than with Policy 6. Thus, in practice the maintenance cost rate of Policy 7 may actually exceed the one of Policy 6.

By (1.2) and (1.10), the expected residual lifetime L_τ of the system when having survived interval $[0, \tau]$ is

$$\mu(\tau) = \mathbb{E}(L_\tau) = e^{\Lambda(\tau)} \int_\tau^\infty e^{-\Lambda(t)} dt.$$

Strictly speaking, a replacement is now an emergency replacement. Hence, the corresponding cost rate is

$$K_7(\tau) = \frac{c_e + c_m \Lambda(\tau)}{\tau + \mu(\tau)}.$$

Given its existence, an optimal $\tau = \tau^*$ satisfies the equation

$$\left[\Lambda(\tau) + \frac{c_e}{c_m} - 1\right]\mu(\tau) = \tau.$$

A related policy is the following one [234]: If the system fails at age t with $t < \tau$, then a minimal repair is done. But if it fails at an age t with $t \geq \tau$, then it is replaced with a new one. However, a more general version had already been proposed and analyzed by *Makabe and Morimura* [213]: Given a sequence of positive, increasing time points $\tau_1, \tau_2, ...$, the system is minimally repaired if $T_i < \tau_i$, and replaced if $T_i \geq \tau_i$, where T_i is the time point of the ith failure (with regard to the time point of the last replacement), $i = 1, 2, ...$

Policy 8　A system is replaced at the first failure which occurs after a fixed time τ_1. However, if there is no failure in $[\tau_1, \tau_2]$, $0 < \tau_1 < \tau_2$, then a preventive replacement takes place at time τ_2. Failures in $(0, \tau_1)$ are removed by minimal repairs [292].

Under this policy, the replacement cycle length is

$$X = \tau_1 + \min\{L_{\tau_1}, \tau_2 - \tau_1\}.$$

Hence, the expected cycle length is

$$\mathbb{E}(X) = \tau_1 + \mu(\tau_1, \tau_2) \quad \text{with } \mu(\tau_1, \tau_2) = \int_0^{\tau_1 - \tau_2} \overline{F}_{\tau_1}(t)dt.$$

The cost rate is

$$K_8(\tau_1, \tau_2) = \frac{c_m \Lambda(\tau_1) + c_e F_{\tau_1}(\tau_2 - \tau_1) + c_p \overline{F}_{\tau_1}(\tau_2 - \tau_1)}{\tau_1 + \mu(\tau_1, \tau_2)}.$$

The optimal vector (τ_1^*, τ_2^*) satisfies $\partial K_8(\tau_1, \tau_2)/\partial \tau_i = 0$, $i = 1, 2$:

$$\lambda(\tau_2)\mu(\tau_1, \tau_2) + \overline{F}_{\tau_1}(\tau_2 - \tau_1) - c_m/(c_e - c_p) = 0,$$

$$(c_e - c_p)\tau_1 \lambda(\tau_2) - c_m \Lambda(\tau_1) = c_e - c_m.$$

As Tahara and Nishida have shown, a unique solution (τ_1^*, τ_2^*) with $0 \leq \tau_1^* < \tau_2^*$ exists if $\lambda(t)$ strictly increases to ∞ and if $0 < c_e - c_p < c_m < c_e < \infty$. In this case, the minimal maintenance cost rate is

$$K_8(\tau_1^*, \tau_2^*) = (c_e - c_p)\lambda(\tau_2^*).$$

Policy 9 The first $n-1$ failures are removed by minimal repairs. At the time point of the nth failure, an (emergency) replacement is carried out [213, 212, 214].

In this case, the replacement cycle length is T_n. Hence, the cost rate is

$$K_9(n) = \frac{c_e + (n-1)c_m}{\mathbb{E}(T_n)}, \tag{10.9}$$

where $\mathbb{E}(T_n)$ is given by (9.55). By analyzing the behavior of the difference $K_9(n) - K_9(n-1)$, an optimal $n = n^*$ is seen to be the smallest integer n satisfying

$$\mathbb{E}(T_n) - [n - 1 + c_e/c_m]\mathbb{E}(L_{n+1}) \geq 0, \tag{10.10}$$

where $\mathbb{E}(L_{n+1})$, the expected time between the nth and the $(n+1)^{th}$ minimal repair, is given by (9.56).

Example 10.2 *Let the system lifetime L have a Weibull distribution:*

$$\overline{F}(t) = e^{-(t/\theta)^\beta}, \quad \Lambda(t) = (t/\theta)^\beta, \quad \lambda(t) = \frac{\beta}{\theta}(t/\theta)^{\beta-1}, \quad \beta > 1. \tag{10.11}$$

Then condition (10.10) simplifies to $\beta n - [n - 1 + c_e/c_m] \geq 0$. Hence, for $c_e > c_m$,

$$n^* = \left\| \frac{1}{\beta - 1}\left(\frac{c_e}{c_m} - 1\right) \right\| + 1,$$

where $\|x\|$ denotes the largest integer which is less than or equal to x.

10.3 Replacement Policies Based on the Failure Type

To be able to remove every failure by a minimal repair, as tacitly assumed in Policies 6 to 8, is generally technically not feasible and/or economically not advantageous. Hence, in this section we consider the case that every failure comes with the information about whether the system operation needs to be restored by a minimal repair or by a replacement. Thus, we have to distinguish between two failure types:

   Type 1-failure: These failures are (*and can be*) removed by minimal repairs.
   Type 2-failure: These failures are removed by replacements.
   Thus, type 1-failures are minor ones, which can be removed without much effort, where type 2-failures may be complete system breakdowns, e.g. if a car is reduced to scrap after a serious traffic accident. However, from a technological point of view, a failure may be a minor one, but identifying the cause of the failure can be a huge problem. A failure happening at time t is with probability $1 - p(t)$ of type 1 and of type 2 with probability $p(t)$. The types of the failures are assumed to occur independently of each other. Thus, the system failures generate the marked point process, which we have already introduced and investigated in Section 9.3.2.

Policy 10 (*generalized age replacement*) The system is maintained according to the failure type, i.e. according to Policy 2. In addition, preventive replacements are carried out τ time units after the previous replacement ([26]).

If Y denotes the random time between two neighboring type 2-failures, the random length of a replacement cycle under Policy 10 is

$$X(\tau) = \min(Y, \tau) \quad \text{with} \quad \mathbb{E}(X(\tau)) = \int_0^\tau \overline{G}(t)dt, \tag{10.12}$$

where $G(t) = \Pr(Y \le t)$ is given by (9.58):

$$\overline{G}(t) = e^{-\int_0^t p(x)\lambda(x)dx}, \ t \ge 0.$$

Hence, the corresponding cost rate has structure

$$K_{10}(\tau) = \frac{c_m \mathbb{E}(Z_\tau) + c_e G(\tau) + c_p \overline{G}(\tau)}{\mathbb{E}(X(\tau))},$$

where the expected number of minimal repairs in a replacement cycle $\mathbb{E}(Z_\tau)$ is given by (9.68). Thus,

$$K_{10}(\tau) = \frac{c_m \left[\int_0^\tau \overline{G}(t)\lambda(t)dt - G(\tau)\right] + c_e G(\tau) + c_p \overline{G}(\tau)}{\int_0^\tau \overline{G}(t)dt}. \tag{10.13}$$

An optimal preventive replacement interval satisfies

$$(p(\tau) + c)\lambda(\tau)\int_0^\tau \overline{G}(t)dt - c\int_0^\tau \overline{G}(t)\lambda(t)dt - G(\tau) = \frac{cc_p}{c_m} \tag{10.14}$$

with $c = c_m(c_e - c_p - c_m)$. As shown in (*Beichelt* [26]), there is a unique solution $\tau = \tau^*$ of this equation if $\lambda(t)$ is strictly increasing to ∞ and if $c > 0$, i.e. if $c_e > c_p + c_m$.

In particular, for $p(t) \equiv p$, (10.13) and (10.14) simplify to

$$K_{10}(\tau) = \frac{\{c_e + c_m\left[(1-p)/p\right]\} G(\tau) + c_p \overline{G}(\tau)}{\int_0^\tau \overline{G}(t)dt}, \tag{10.15}$$

$$p\lambda(\tau)\int_0^\tau \overline{G}(t)dt - G(\tau) = \frac{pc_p}{p(c_e - c_p) + (1-p)c_m}. \tag{10.16}$$

Obviously, the functional structures of the Equations (10.2) and (10.16) for the respective optimal replacement intervals of Policies 3 and 11 on condition $p(t) \equiv p$ are identical, since $p\lambda(t)$ is the failure rate belonging to $G(t)$. Thus, together with the two-unit-system considered in Section 10.1, we have a third example for the appearance of the "classical" (age replacement) Equation (10.2). These three examples have been discussed in [20], but Equation (10.16) can already be found as a special case in [26].

If $\tau = \infty$, then there is no preventive maintenance, and (10.13) becomes the maintenance cost rate of Policy 2 (= maintenance according to the failure type):

$$K_2 = \frac{c_m \left[\int_0^\infty \overline{G}(t)\lambda(t)dt - 1\right] + c_e}{\int_0^\infty \overline{G}(t)dt},$$ (10.17)

and, for $p(t) \equiv p$,

$$K_2 = \frac{c_e + c_m \left[(1-p)/p\right]}{\int_0^\infty \overline{G}(t)dt}.$$ (10.18)

For $p(t) \equiv 1$ and $p(t) \equiv 0$, Policy 10 contains as special cases policies 3 and 6, respectively.

So far we have assumed that $p(t)$ is determined by the failure behavior of the system. But simply by letting

$$p(t) = \begin{cases} 0, & 0 \le t \le \tau \\ 1, & t > \tau \end{cases}$$

we see that Policy 7 is a special case of Policy 10 as well.

Policy 11 (*generalized block replacement*) At fixed time points $\tau, 2\tau, ...$, the system is preventively replaced. In between, the system is maintained according to the failure type [31].

Let $H_m(t)$ and $H_e(t)$ be the expected numbers of minimal repairs and emergency replacements, respectively, which occur in $(0, t]$, $0 < t \le \tau$. Then the cost rate is

$$K_{11}(\tau) = \frac{1}{\tau} \left[c_m H_m(\tau) + c_e H_e(\tau) + c_p\right].$$ (10.19)

The time points of the occurrence of type 2-failures generate a renewal process with renewal function $H_e(t)$. Hence, by (9.14), $H_e(t)$ is solution of the integral equation, $t \le \tau$,

$$H_e(t) = G(t) + \int_0^t H_e(t-x)dG(x).$$

$H_m(t)$ satisfies the following integral equation, $t \le \tau$:

$$H_m(t) = \mathbb{E}(Z_t \,|\, Y \ge t)\overline{G}(t)$$
$$+ \int_0^t \left[\mathbb{E}(Z_x \,|\, Y = x) + H_m(t-x)\right]dG(x),$$

which, with $\mathbb{E}(Z_t)$ given by (9.68), becomes

$$H_m(t) = \mathbb{E}(Z_t) + \int_0^t H_m(t-x)dG(x).$$

With $h_m(t) = dH_m(t)/dt$ and $h_e(t) = dH_e(t)/dt$ an optimal $\tau = \tau^*$ is seen to satisfy the equation

$$\int_0^\tau \left[c_m \left(h_m(\tau) - h_m(t) \right) + c_e (h_e(\tau) - h_e(t)) \right] dt = c_p. \qquad (10.20)$$

Combining (10.19) and (10.20) yields the minimal maintenance cost rate

$$K(\tau^*) = c_m h_m(\tau^*) + c_e h_e(\tau^*). \qquad (10.21)$$

In particular, if $p(t) \equiv p$, $0 < p \leq 1$, then $H_m(t) = (1/p - 1)H_e(t)$, $t \leq \tau$, so that equations (10.19) to (10.21) become

$$K(\tau) = \frac{1}{\tau} \left\{ \left[c_m(1-p)/p + c_e \right] H_e(\tau) + c_p \right\},$$

$$\tau h_e(\tau) - H_e(\tau) = pc_p / \left[(1-p)c_m + pc_e \right],$$

$$K(\tau^*) = \left[c_m(1-p)/p + c_e \right] h_e(\tau^*).$$

Policy 11 contains as special cases Policy 4 ($p(t) \equiv 1$) and Policy 6 ($p(t) \equiv 0$), respectively. Under Policy 4, every failure is removed by a replacement, even if a minimal repair would be sufficient. In view of $c_m < c_e$, this cannot be economical. On the other hand, Policy 6 requires that principally every failure can be removed by a minimal repair. Policy 11 has eliminated these disadvantages. For generalizations see *Sheu* ([275], [276]).

Policy 12   The system is maintained according to the failure type (Policy 2). After $n-1$ successive minimal repairs a preventive replacement is done at the time point of the nth failure [34].

Let Y (Z) denote as before the time (number of minimal repairs) between two neighboring type 2-failures under Policy 2. Then the random cycle length under Policy 12 is $X(n) = \min\{Y, T_n\}$, and the maintenance cost rate has structure

$$K_{12}(n)$$
$$= \frac{\left[c_m \mathbb{E}(Z \,|\, Y < T_n) + c_e \right] P(Y < T_n) + \left[(n-1)c_m + c_p \right] P(Y \geq T_n)}{\mathbb{E}(X(n))}.$$

To determine the ingredients of this formula, note that

$$\mathbb{E}(X(n)) = \int_0^\infty P(X(n) > t)dt = \int_0^\infty P(Y > t, T_n > t)dt,$$

where

$$P(Y > t, T_n > t) = \sum_{k=1}^n P(Y > t, T_{k-1} \leq t < T_k)$$

and, by (9.57),

$$\Pr(Y > t, T_{k-1} \leq t < T_k) =$$

$$= \int_t^\infty \int_0^t \int_0^{x_{k-1}} \cdots \int_0^{x_3} \int_0^{x_2} \prod_{i=1}^{k-1} \overline{p}(x_i)\lambda(x_i)dx_i \ f(x_k)dx_k, \ k \geq 2.$$

Hence, for all $k = 1, 2, ...$, using (9.59),

$$\Pr(Y > t, T_{k-1} \leq t < T_k) = \frac{1}{(k-1)!} \left[\int_0^t \overline{p}(x)\lambda(x)dx \right]^{k-1} \overline{F}(t).$$

Analogously one obtains for $n = 2, 3, ...$ (obviously, $P(Y \geq T_1) = 1$),

$$\Pr(Y \geq T_n) = \frac{1}{(n-2)!} \int_0^\infty \left[\int_0^t \overline{p}(x)\lambda(x)dx \right]^{n-2} \overline{p}(t)dF(t).$$

Finally, for $n = 2, 3, ...$,

$$\mathbb{E}(Z \mid Y < T_n) = \sum_{k=1}^{n-2} k \frac{P(Z = k)}{P(Y < T_n)},$$

where the probability distribution of Z is given by (9.60). In particular, for $p(t) \equiv p$,

$$\Pr(Y \geq T_n) = \overline{p}^{n-1}, \ \Pr(Y < T_n) = 1 - \overline{p}^{n-1},$$

and, with $\mathbb{E}(L_k)$ given by (9.56),

$$\mathbb{E}(X(n)) = \sum_{k=0}^{n-1} \overline{p}^k \mathbb{E}(L_{k+1}).$$

10.4 Exercises

Exercise 10.1 *Solve Equation (10.2) for τ on condition that the system lifetime L is uniformly distributed over the interval $[0, T]$.*

Exercise 10.2 *Show that the left-hand side of Equation (10.2) is strictly increasing in τ if $\lambda(t)$ is strictly increasing in τ.*

Exercise 10.3 *When following an age replacement policy (Policy 1) with replacement interval τ, assume that the times for emergency and preventive replacements are not negligibly small, but require the constant times d_e and d_p, respectively, $0 < d_p < d_e$. Determine the long-run availability $A(\tau)$ under this policy and an equation for the corresponding optimal τ.*

Exercise 10.4 *Consider the age replacement policy for a two-unit system described in Section 10.1. Given*

$$F(t) = 1 - e^{-t^2}, \quad F_e = 1 - e^{-5t}, \quad F_p(t) = 1 - e^{-10t}, \quad t \geq 0,$$

determine approximately the optimal replacement interval $\tau = \tau^*$. (Make sure that the assumptions are fulfilled.)

Exercise 10.5 *Under otherwise the same assumptions as in Policy 4, a system is preventively replaced with a new one at time points τ, 2τ, ... But when a failure occurs in between, nothing is done. A financial loss $v(t)$ arises if t is the time between the failure and the next preventive replacement. Determine the corresponding cost rate and an equation for the optimal replacement interval $\tau = \tau^*$.*

Exercise 10.6 *Exercise 10.5 is modified as follows: If a failure occurs, it is removed by a minimal repair. If a second failure occurs, nothing is done and a loss of $v(t)$ arises if t is the time between the second failure and the next preventive replacement. Determine the corresponding cost rate and an equation for the optimal replacement interval $\tau = \tau^*$.*

Exercise 10.7 *The time L_n between system failure $(n-1)$ and n has the expected value $\mathbb{E}L_n = 0.9^n$, $n = 1, 2, ...$ With regard to the cost rate, determine the optimal number of minimal repairs between two replacements under Policy 9 if $c_e = 10$ and $c_m = 1$.*

Chapter 11

System Replacement Based on Cost Limits

11.1 Introduction

Replacement policies based on cost limits are widely acknowledged as particularly user friendly and efficient strategies for organizing the maintenance of technical systems. Different from the maintenance policies considered so far, repair cost limit replacement policies explicitly take into account that repair costs are random variables. Likely the first paper suggesting the use of repair cost limits and demonstrating their application was *Gardent and Nonant* [132], followed by *Drinkwater and Hastings* [111], *Hastings* [152].

Policy 13 (*repair cost limit replacement policy*) After a failure of the system at age t, the necessary repair cost is estimated. The system is replaced by an equivalent new one if the repair cost exceeds a given level $c(t)$. Otherwise a minimal repair is done.

In this section, the theoretical basis for the analysis of Policy 13 and its modifications is the 2-failure type-model developed in Section 9.3.2. Let C_t be the random repair cost of the system if it fails at time t. Then the 2-failure-type model applies to Policy 13 in the following way: A system failure is of type 1 or of type 2 if

$$C_t \leq c(t) \text{ or } C_t > c(t),$$

respectively. Thus, if $R_t(x) = \Pr(C_t \leq x)$ is the cdf of C_t and $\overline{R}_t(x) = 1 - R_t(x)$, then the respective probabilities of type 1 and type 2-failures are

$$1 - p(t) = R_t(c(t)) \text{ and } p(t) = \overline{R}_t(c(t)). \tag{11.1}$$

Formally, Policy 13 is identical to Policy 2 with $p(t)$ given by (11.1). But, different from Policies 10 to 12, the corresponding cost rate has now to be minimized with regard to $p(t)$ via $c(t)$. Note that it may be possible to equivalently write $\overline{R}_t(c(t))$ as $\overline{R}(\widetilde{c}(t))$ with a transformed "repair cost limit function" $\widetilde{c}(t)$ and a time-independent repair cost distribution function R.

As before, let c_e be the cost of a replacement after a type 2-failure. It makes sense to assume that

$$0 < c(t) < c_e, \quad R_t(x) = \begin{cases} 1 & \text{if } x \geq c_e \\ 0 & \text{if } x \leq 0 \end{cases}. \tag{11.2}$$

Three popular repair cost distributions satisfying this condition (but not depending on t) are:

a) *Power distribution:*

$$R(x) = 1 - \left(\frac{c_e - x}{c_e}\right)^{\alpha}; \quad 0 \leq x \leq c_e, \ \alpha > 1. \tag{11.3}$$

b) *Density $r(x)$ of the beta distribution over $[0, 1]$:*

$$r(x) = \begin{cases} \dfrac{1}{B(a,b)} x^{a-1}(1-x)^{b-1}; & 0 \leq x \leq 1; \ a, b > 0, \\ 0 & \text{otherwise,} \end{cases} \tag{11.4}$$

where $B(a,b) = \int_0^1 x^{a-1}(1-x)^{b-1}dx$ is the *beta-function* over $[0, 1]$ with parameters a and b.

c) *Truncated exponential distribution:*

$$R(x) = \frac{1 - e^{-x/\mu}}{1 - e^{-c_e/\mu}}, \quad 0 \leq x \leq c_e, \ R(x) = 1 \text{ for } x > c_e. \tag{11.5}$$

11.2 Constant Repair Cost Limit

For the sake of comparison, let us next consider the case that the repair cost limit and the random repair cost do not depend on t:

$$c(t) = c, \ C_t = C, \ R_t(x) = R(x) \text{ for all } t \geq 0. \tag{11.6}$$

1. An obvious assumption about the expected cost of a minimal repair is $c_m = E(C \mid C < c)$, since between replacements all repair costs are less than c:

$$c_m = \frac{1}{R(c)} \int_0^c x dR(x) = \frac{1}{R(c)} \left[\int_0^c \overline{R}(x)dx - c\overline{R}(c) \right]. \tag{11.7}$$

With this c_m, $p = \overline{R}(c)$ and $\overline{G}(t) = \left[\overline{F}(t)\right]^{\overline{R}(c)}$, the corresponding cost rate is now given by (10.18):

$$K_{13}(c) = \frac{\dfrac{1}{\overline{R}(c)} \displaystyle\int_0^c \overline{R}(x)dx + c_e - c}{\displaystyle\int_0^\infty \left[\overline{F}(t)\right]^{\overline{R}(c)} dt}. \tag{11.8}$$

Weibull distributed lifetime

Let $F(t)$ be given by (10.11) with $\beta > 1$. Then the expected time between two neighboring replacements is

$$\mathbb{E}(Y) = \int_0^\infty \overline{G}(t)dt = \theta\Gamma(1 + 1/\beta)\left[\overline{R}(c)\right]^{-1/\beta}.$$

Hence,

$$K_{13}(c) = \frac{\int_0^c \overline{R}(x)dx + (c_e - c)\overline{R}(c)}{\theta\Gamma(1 + 1/\beta)\left[\overline{R}(c)\right]^{(\beta-1)/\beta}}. \tag{11.9}$$

The optimal repair cost limit is solution of the equation $dK_{13}(c)/dc = 0$ or

$$\frac{1}{\overline{R}(c)}\int_0^c \overline{R}(x)dx = \frac{c_e - c}{\beta - 1}. \tag{11.10}$$

Since $\beta > 1$, there exists a unique solution $c = c^*$ with $0 < c^* < c_e$. In case of the power distribution (11.3),

$$c^* = \left[1 - \left(\frac{\beta - 1}{\beta + 1}\right)^{1/(\alpha+1)}\right].$$

For numerical results under the Weibull assumption, see Table 3.1.

	$\beta = 2$		$\beta = 3$		$\beta = 4$	
b	c^*	K_{13}^*	c^*	K_{13}^*	c^*	K_{13}^*
4	0.355	83.42	0.263	93.47	0.211	96.78
6	0.309	76.87	0.235	90.02	0.192	94.79
8	0.275	71.60	0.213	86.92	0.176	92.88

Table 11.1: Numerical results for Policy 13

If C has the truncated exponential distribution (11.5), then c^* is solution of

$$\frac{(1 - \mu c e^{-\mu c_e})e^{\mu c} - 1}{1 - e^{-\mu(c_e - c)}} = \frac{\mu}{\beta - 1}(c_e - c).$$

For μ being sufficiently large, i.e. the truncation has no significant effect, then this equation approximately becomes (*Park* [248])

$$e^{\mu c} - 1 = \frac{\mu}{\beta - 1}(c_e - c).$$

Remark 11.1 *If we simply assume that a minimal repair costs on average c_m units, not explicitly depending on the repair cost limit, then, by (10.18), the cost rate under Policy 13 is*

$$\widehat{K}_{13}(c) = \frac{c_e + \frac{1 - \overline{R}(c)}{\overline{R}(c)}c_m}{\theta\Gamma(1 + 1/\beta)\left[\overline{R}(c)\right]^{-1/\beta}}. \tag{11.11}$$

In this case, the cost rate depends on c only via $\overline{R}(c)$. The value of $y = \overline{R}(c)$ minimizing $\widehat{K}_{13}(c)$ is

$$y^* = \overline{R}(c^*) = \frac{\beta - 1}{k - 1} \quad \text{with } k = c_e/c_m.$$

Then, for any $\overline{R}$ with inverse function $\overline{R}^{-1}$,

$$c^* = \overline{R}^{-1}\left(\frac{\beta - 1}{k - 1}\right). \tag{11.12}$$

By assumption, $k > 1$ and $\beta > 1$. Moreover, β must satisfy the two conditions

$$1 < \beta < k \quad \text{and} \quad \beta < (k - 1)\overline{R}(c_m) + 1. \tag{11.13}$$

The first one results from $0 < y^* < 1$, and the second one makes sure that $c_m < c^*$. Since k is large, (11.13) should usually be fulfilled. Application of this c^* yields the cost rate

$$\widehat{K}_{13}(c^*) = \frac{\beta c_m}{\theta \Gamma(1 + 1/\beta)} \left(\frac{k - 1}{\beta - 1}\right)^{1 - 1/\beta}. \tag{11.14}$$

For $\beta = 2$ (Rayleigh distribution), $\Gamma(1.5) = \sqrt{\pi/4}$ so that

$$\widehat{K}_{13}(c^* \,|\, \beta = 2) = \frac{4 c_m}{\theta} \sqrt{\frac{k - 1}{\pi}}.$$

On condition $1 < \beta < k$, the cost rate $\widehat{K}_{13}(c^*)$ is smaller than the cost rate K belonging to the corresponding renewal process policy, i.e. after every failure the system is replaced by an equivalent new one (usually most uneconomically):

$$K = c_e / \int_0^\infty \overline{F}(t) dt. \tag{11.15}$$

In our case, $K = c_e/\theta \Gamma(1 + 1/\beta)$.

Policy 14 (combined repair cost limit-age replacement policy) The system is maintained according to Policy 13 with a constant repair cost limit c. But when there is no replacement within τ time units after the previous replacement, a preventive replacement is done at time τ.

Formally, this policy is a special case of Policy 10: In K_{10} given by (10.15), substitute p with $\overline{R}(c)$ and c_m with (11.7) to obtain

$$K_{14}(c, \tau) = \frac{\left\{\frac{1}{\overline{R}(c)} \int_0^c \overline{R}(x) dx + c_e - c\right\} G(\tau) + c_p \overline{G}(\tau)}{\int_0^\tau \overline{G}(t) dt}, \tag{11.16}$$

where $\overline{G}(t) = \left[\overline{F}(t)\right]^{\overline{R}(c)}$. Then, for fixed c, equation (10.16) becomes

$$\overline{R}(c)\lambda(\tau) \int_0^\tau \overline{G}(t)dt - G(\tau) = \frac{\overline{R}(c)c_p}{(c_e - c_p - c)\overline{R}(c) + \int_0^c \overline{R}(x)dx}. \qquad (11.17)$$

A solution τ^* of this equation is a function of c: $\tau^* = \tau^*(c)$. A sufficient condition for the existence of a finite $\tau^*(c)$ for all c with $0 < c < c_e - c_p$ is $\lim_{t\to\infty} \lambda(t) = \infty$. If an optimal vector (τ^*, c^*) with regard to $K_{14}(c, \tau)$ exists, it can be determined as follows:

1. For fixed c, determine $\tau^*(c)$ as the solution of Equation (11.17).
2. Determine the optimal vector (c^*, τ^*) by minimizing $K_{14}(\tau^*(c), c)$ with regard to c.

Example 11.2 *Let $F(t)$ be given by the Weibull-distribution (10.11) with scale parameter (time unit) $\theta = 1$ and $\beta = 2, 3,$ or 4, respectively. The cost unit is given by $c_e = 1$, $c_p = 0.75$, and the repair costs have the beta-distribution (11.4) with $a = 2$ and $b = 4, 6$ or 8, respectively. We will compare the cost rates of policies 13 and 14 under constant repair cost limits. Tables 1.1 and 1.2 show the corresponding c^* and τ^* and the procentual cost rates $K(c^*)$ and $K(c^*, \tau^*)$ with regard to the cost rate (11.15) of the renewal process policy. Table 11.1 indicates that with increasing variance of the repair costs, Policy 13 becomes more and more efficient ($K_{13}^* = K_{13}(c^*)$). On the other hand, its maintenance cost rate increases with increasing β, i.e. applying Policy 13 is more rewarding for slowly aging systems than for fast aging ones. This is plausible, since aging increases the intensity of minimal repairs (failures). Table 11.2 shows that enriching Policy 13 by a preventive maintenance interval τ leads to a further significant reduction of the cost rate ($K_{14}^* = K_{14}(c^*, \tau^*)$). But this reduction also depends heavily on the ratio c_p/c_e.*

	$\beta = 2$			$\beta = 3$			$\beta = 4$		
b	c^*	τ^*	K_{14}^*	c^*	τ^*	K_{14}^*	c^*	τ^*	K_{14}^*
4	0.414	1.92	81.85	0.359	1.22	89.85	0.332	1.04	91.39
6	0.397	1.98	73.96	0.354	1.25	84.68	0.331	1.07	87.72
8	0.389	2.08	67.48	0.349	1.30	80.04	0.329	1.10	84.28

Table 11.2: Numerical results for Example 11.2

11.3 Time-Dependent Repair Cost Limits

The useful life of a system with increasing failure rate $\lambda(t)$, $t \geq 0$, and integrated failure rate $\Lambda(t)$ is partitioned into $k = 1, 2, ..., n$ time intervals of length y (for convenience: *years*). During year k, the repair cost distribution function is $R_k(x) = \Pr(C_k \leq x)$, and the repair cost limit c_k is applied. If there is no (emergency) replacement in $[0, ny]$, a preventive replacement is done at time ny. Under these assumptions, in what follows, essentially Policy 14 is applied with a piece-wise constant repair cost limit function $c(t)$, emergency (preventive) replacement costs c_e (c_p), and $\tau = ny$. In terms of our previous notation,

$$\left.\begin{array}{l} C_t = C_k, \ c(t) = c_k \\ p(t) = p_k = \overline{R}_k(c_k) \end{array}\right\} \text{ for } (k-1)y \le t < ky, \quad k = 1, 2, ..., n. \qquad (11.18)$$

By (11.7), the expected cost of a minimal repair in year k is

$$c_{m,k} = \frac{1}{\overline{p}_k} \int_0^{c_k} x \, dR_k(x) = \frac{1}{\overline{p}_k} \left[\int_0^{c_k} \overline{R}_k(x) dx - c_k p_k \right]. \qquad (11.19)$$

Note that every replacement brings the failure rate back to its initial value $\lambda(+0)$. In addition, we will assume that the occurrence of system failures in a specific year is independent of the occurrence of system failures in all the other years. Let

$$\Lambda_k(t) = \Lambda(t) - \Lambda((k-1)y), \quad (k-1)y \le t \le ky, \quad k = 1, 2, ...$$

Then, according to (9.58), the distribution function of the length Y of a renewal cycle can be written as

$$\overline{G}(t) = e^{-p_1 \Lambda_1(t)}, \ 0 \le t < y, \quad \overline{G}(t) = 0, \ t > ny,$$

$$\overline{G}(t) = e^{-\sum_{i=1}^{k-1} p_i \ \Lambda_i(iy) - p_k \Lambda_k(t)} \text{ for } (k-1)y \le t < ky$$

and $k = 2, 3, ..., n$. Hence, the probability π_k that a replacement occurs in year k is given by

$$\pi_1 = 1 - e^{-p_1 \Lambda_1(y)},$$

$$\pi_k = \left(1 - e^{-p_k \Lambda_k(ky)}\right) e^{-\sum_{i=1}^{k-1} p_i \ \Lambda_i(iy)}, \quad k = 2, 3, ..., n,$$

and the expected cycle length $\mathbb{E}(Y)$ is

$$\mathbb{E}(Y) = \int_0^y \overline{G}(t) dt = \sum_{k=1}^n \frac{\pi_k}{1 - e^{-p_k \Lambda_k(ky)}} \int_0^y e^{-p_k \Lambda(t + (k-1)y)} dt.$$

By (9.66) and (9.67), depending on whether there is a replacement in year k or not, the expected number of minimal repairs in year k is

$$\mathbb{E}_k(Z \,|\, Y < y) = \overline{p}_k \left[\frac{1}{p_k} - \frac{\Lambda_k(ky)}{e^{p_k \Lambda_k(ky)} - 1} \right],$$

$$\mathbb{E}_k(Z \,|\, Y \ge y) = \overline{p}_k \Lambda_k(ky), \quad k = 1, 2, ..., n.$$

Let r_k be the expected total cost of minimal repairs in $[0, ky]$ on condition that an emergency replacement takes place in year k. Then,

$$r_1 = \bar{p}_1 \Lambda_1(y)c_{m,1},$$

$$r_k = \sum_{i=1}^{k-1} \bar{p}_i \Lambda_i(iy)c_{m,i} + \bar{p}_k \left[\frac{1}{p_k} - \frac{\Lambda_k(ky)}{e^{p_k \Lambda_k(ky)} - 1} \right] c_{m,k}, \quad k = 2, 3, ...$$

where $c_{m,i}$ is given by (11.19). Hence, the cost rate under the repair cost limit vector $\mathbf{c} = (c_1, c_2, ..., c_n)$ and a preventive replacement interval of length $\tau = ny$ is

$$K(\mathbf{c}, n) = \frac{\left(1 - \sum\limits_{k=1}^{n} \pi_k\right)\left(\sum\limits_{k=1}^{n} \bar{p}_k \Lambda_k(ky)c_{m,k} + c_p\right) + \sum\limits_{k=1}^{n} \pi_k(r_k + c_e)}{\mathbb{E}(Y)}. \quad (11.20)$$

Special Cases

1. If there is no preventive replacement scheduled (i.e. $n \to \infty$), the cost rate becomes

$$K(\mathbf{c}) = \frac{\sum_{k=1}^{\infty} \pi_k r_k + c_e}{\mathbb{E}(Y)}.$$

2. If the failure rate is piecewise constant,

$$\lambda(t) = \lambda_k, \quad (k-1)y \le t < ky, \quad k = 1, 2, ..., \quad (11.21)$$

then formula (11.20) becomes

$$K(\mathbf{c}, n) = \frac{\sum\limits_{k=1}^{n} \pi_k(r_k + c_e) + \left(1 - \sum\limits_{k=1}^{n} \pi_k\right)\left(\sum\limits_{k=1}^{n} \bar{p}_k \lambda_k y c_{m,k} + c_p\right)}{\sum\limits_{k=1}^{n} \pi_k / p_k \lambda_k}$$

$$(11.22)$$

with

$$\pi_1 = 1 - e^{-p_1 \lambda_1 y}, \quad \pi_k = \left(1 - e^{-p_k \lambda_k y}\right) e^{-\sum_{i=1}^{k-1} p_i \lambda_i y}, \quad k = 2, 3, ..., n.$$

$$r_1 = c_{m,1} \bar{p}_1 \lambda_1 y,$$

$$r_k = \sum_{i=1}^{k-1} c_{m,i} \bar{p}_i \lambda_i y + c_{m,k} \bar{p}_k \left(\frac{1}{p_k} - \frac{\lambda_k y}{e^{p_k \lambda_k y} - 1} \right), \quad k = 2, 3, ..., n.$$

This case has been considered in [41].

Minimizing $K(\mathbf{c}, n)$ with regard to $\mathbf{c}$ and n can be principally done by nonlinear programming. In the following example, a stochastic search procedure had been applied to determine the optimal vector $\mathbf{c}^* = (c_1^*, c_2^*, ..., c_n^*)$ with n given.

k	λ_k	μ_k	p_k	$c_{m,k}$	r_k	π_k	c_k^*
1	2.0	250	0.001	248	498	0.001	6120
2	3.0	300	0.001	298	1390	0.001	4750
3	3.5	320	0.001	318	2503	0.001	3720
4	4.0	340	0.001	338	3851	0.001	2930
5	4.3	350	0.001	348	5345	0.004	2420
6	4.5	370	0.005	360	6141	0.022	1950
7	4.7	380	0.015	355	7769	0.066	1590
8	5.0	390	0.030	348	9424	0.124	1370
9	5.0	400	0.051	336	11051	0.174	1190
10	5.0	440	0.110	320	12529	0.253	970
11	5.0	450	0.145	303	13875	0.177	870
12	5.0	480	0.269	248	14955	0.124	630
13	5.0	490	0.783	58	15521	0.043	120
14	5.0	520	0.841	44	15577	0.001	90
15	5.0	520	0.891	30	15607	0.001	60

Table 11.3: Numerical results for Example 11.3

Example 11.3 *We assume that the failure rate of the system in year k is λ_k, and that the time unit is "year" ($y = 1$). If there is no emergency replacement in $[0, n]$, after $n = 15$ years a preventive replacement is done. The replacement cost of a system is $c_p = c_e = \$6900$. In year k, the random cost C_k of a repair after a system failure has an exponential distribution with expected value μ_k, $k = 1, 2, ..., 15$. Table 11.3 shows the corresponding numerical parameters and the optimal vector $\mathbf{c}^*$. The minimal cost rate is*

$$K(\mathbf{c}^*) = 2050 \; [\$/year].$$

The mean time between replacements when applying vector $\mathbf{c}^*$ is $E(Y) = 6.5$ years. On the other hand, when the replacement decision is made on the basis of the average cost development, then we have to compare the cost rates

$$K(k) = \frac{6900 + \sum_{i=1}^{k} \lambda_i \mu_i}{k}, \quad k = 1, 2, ..., 15.$$

We find that the minimal value of $K(k)$ is assumed at $k = 9$:

$$K(9) = 2187 \; [\$/year].$$

The replacement interval $\tau^* = 9$ years is called the economic lifetime *of the system. Thus, applying optimal repair cost limits instead of the economic lifetime reduces the cost rate by*

$$\frac{2187 - 2050}{2187} [100\%] = 6.3\%.$$

The data set of this example is due to Drinkwater and Hastings *[111], who had to develop a cost-efficient maintenance plan for a fleet of military vehicles. Under otherwise the same assumptions, they determined the optimal repair cost limit vector* $\mathbf{c}^*$ *by a step-wise improvement procedure. The difference between our and their optimal vector* $\mathbf{c}^*$ *is negligibly small.*

A linear time-dependent repair cost limit was also considered by *Berg et al.* [47]. For any time-dependent repair cost limit, *Beichelt* [39] used formula (10.17) to derive an approximative cost rate by applying an appropriate average value for the repair cost. If $c(t)$ and $R_t(x)$ depend on a continuous time parameter $t \in [0, \infty)$ and with $c_m = c_m(t)$ given by (11.7), derivation of the minimal cost rate is principally possible by applying calculus of variation. *Jiang, Cheng, and Makis* [161] used optimal stopping theory to prove a key result, namely that a repair cost limit policy is optimal when the repair cost limit function is allowed to depend on time and the number n of the failure. If the repair cost do not depend on n, then there exists a single optimal repair-cost-limit function for all n. For discrete-time repair cost limit replacement policies with repairs not necessarily minimal and a fairly comprehensive literature survey, see [160].

11.4   Cumulative Repair Cost Limit Replacement Policies

When following a repair cost limit replacement policy, the decision to repair or to replace a failed system depends on the cost of a single repair. With increasing system age, repairs may occur more and more frequently, so that the repair cost rate, and, therefore, the total repair costs, will increase fairly fast, even if the costs of all repairs stay below the repair cost limit. Hence, it seems to be more cost-efficient to make the replacement decision on the basis of the *cumulative* or *total repair cost* $C(t)$ in $[0, t]$, where t refers to the time point of the last replacement.

Policy 15   The system is replaced with an equivalent new one as soon as the total repair cost $C(t)$ hits or exceeds a given level c.

Probably the first papers suggesting this approach to system replacement are [31], [32].

In comparison to the repair cost limit replacement policy (Policy 13), Policy 15 has four major advantages:

1. Applying policy 15 does not require information on the underlying lifetime distribution of the system. In particular, sophisticated classifications of failures become unnecessary. The assumption of "minimal repairs" can be dropped. Only the cost implications of failures matter. Note that, in case of complex systems like excavators, belt conveyors, trucks and the like, it does

not really make sense to schedule preventive maintenance based on a "lifetime distribution."

2. Apart from the actual repair costs, costs due to monitoring, servicing, stock keeping, downtime costs, personnel costs etc. can be taken into account.

3. In view of the simple structure of Policy 15 and the fact that the required input, namely the maintenance cost development in time (maybe including follow-up costs), is usually known, Policy 15 is principally easy to implement and can serve as a basic policy for scheduling the cost-optimal overhaul cycles of whole industrial plants.

4. In practical applications, decision-making via Policy 15 can be based on the running maintenance cost data.

Of course, instead of limiting the total repair cost $C(t)$, functionals of $C(t)$, for instance the repair cost rate $R(t) = C(t)/t$, $t > 0$, may be the basis for replacement decisions. This leads to

Policy 16   The system is replaced with an equivalent new one as soon as the repair cost rate $R(t) = C(t)/t$ hits or exceeds a given level r.

Let $L_C(x)$ and $L_R(y)$ be the first passage times of the processes $\{C(t), t \geq 0\}$ and $\{R(t), t > 0\}$ with regard to levels x and y, respectively:

$$L_C(x) = \inf_{t \in (0,\infty)} \{t, \ C(t) \geq x\}, \quad L_R(y) = \inf_{t \in (0,\infty)} \{t, \ R(t) \geq y\}.$$

Then, from the strong law of the great numbers, the cost rates arising from Policies 15 and 16 are

$$K_{15}(c) = \frac{c + c_e}{\mathbb{E}\left(L_C(c)\right)}, \tag{11.23}$$

$$K_{16}(c) = r + \frac{c_e}{\mathbb{E}\left(L_R(r)\right)}, \tag{11.24}$$

where c_e is the cost of a replacement. As usual, we assume that the lengths of the replacement cycles are independent random variables. If the processes $\{C(t), t \geq 0\}$ and $\{R(t), t > 0\}$ have nondecreasing sample paths (that is a matter of course for $\{C(t), t \geq 0\}$), then

$$C(t) \leq c \Leftrightarrow L_C(c) \geq t, \quad R(t) \leq r \Leftrightarrow L_R(r) \geq t.$$

Hence,

$$F_{C(t)}(x) = \Pr(C(t) \leq x) = \Pr(L_C(x) \geq t) = 1 - F_{L_C(x)}(t),$$

$$F_{R(t)}(y) = \Pr(R(t) \leq y) = \Pr(L_R(y) \geq t) = 1 - F_{L_R(y)}(t),$$

so that

$$\mathbb{E}\left(L_C(c)\right) = \int_0^\infty F_{C(t)}(c)dt, \quad \mathbb{E}\left(L_R(r)\right) = \int_0^\infty F_{R(t)}(r)dt. \tag{11.25}$$

In what follows, we will compare in a series of examples the cost rates arising from Policies 15 and/or 16 to the corresponding cost rates arising from the continuous-time economic lifetime approach (Policy 5), which requires the same data input. In makes no sense to compare Policies 15 and 16 to Policy 13 since the latter needs additional data input, namely information on the system lifetime.

Example 11.4 Let $C(t) = A_2 t^2 + A_3 t^3$, where A_2 and A_3 are independent, exponentially with respective parameters λ_2 and λ_3 distributed random variables. Then $F_{C(t)}(x)$, $x \geq 0$, is given by the convolution

$$F_{C(t)}(x) = \int_0^x \left(1 - e^{-\frac{\lambda_2(x-u)}{t^2}} \right) \frac{\lambda_3}{t^3} e^{-\frac{\lambda_3 u}{t^3}} \, du$$

$$= 1 - e^{-\frac{\lambda_3 x}{t^3}} - \frac{\lambda_3}{\lambda_3 - \lambda_2 t} \left(e^{-\frac{\lambda_2 x}{t^2}} - e^{-\frac{\lambda_3 x}{t^3}} \right).$$

Let $c_e = 20$, $\lambda_2 = 2$, and $\lambda_3 = 3$. Then the corresponding cost rate (11.23) is

$$K_{15}(c) = \frac{c + 20}{\int_0^\infty \left[1 - e^{-\frac{3c}{t^3}} - \frac{3}{3 - 2t} \left(e^{-\frac{2c}{t^2}} - e^{-\frac{3c}{t^3}} \right) \right] dt}.$$

The optimal values are (slightly rounded)

$$c^* = 12.6, \quad \mathbb{E}(L_C(c^*)) = 3.4750, \quad K_{15}(c^*) = 9.3814.$$

Comparison to Policy 5   The trend function of $\{C(t),\ t \geq 0\}$ is

$$\mathbb{E}(C(t)) = \mathbb{E}(A_2)t + \mathbb{E}(A_3)t^2 = \frac{1}{\lambda_2}t^2 + \frac{1}{\lambda_3}t^3$$

$$= \frac{1}{2}t^2 + \frac{1}{3}t^3.$$

Hence, the average cost rate when applying a replacement interval of length τ is

$$K_5(\tau) = \frac{20 + \frac{1}{2}\tau^2 + \frac{1}{3}\tau^3}{\tau}.$$

The optimal values are

$$\tau^* = 2.8764 \quad \text{and} \quad K_5(\tau^*) = 11.149.$$

Thus, applying Policy 15 instead of Policy 5 reduces the cost rate by

$$[100\%][K_5(\tau^*) - K_{15}(c^*)]/K_5(\tau^*) = 15.9\%.$$

Comparison to Policy 16 *It seems to be obvious that Policies 15 and 16 are equivalent from the cost point of view. But this is not the case as the following calculation result shows. The repair cost rate belonging to $C(t)$ is $R(t) = A_2t + A_3t^2$. Under the above assumptions on A_2 and A_3, its distribution function is*

$$F_{R(t)}(y) = 1 - e^{-\frac{3y}{t^2}} - \frac{3}{3-2t}\left(e^{-\frac{2y}{t}} - e^{-\frac{3y}{t^2}}\right).$$

Since the sample paths of $\{R(t),\ t \geq 0\}$ are nondecreasing, the corresponding cost rate (11.24) is

$$K_{16}(r) = r + \frac{20}{\displaystyle\int_0^\infty \left[1 - e^{-\frac{3r}{t^2}} - \frac{3}{3-2t}\left(e^{-\frac{2r}{t}} - e^{-\frac{3r}{t^2}}\right)\right] dt}.$$

The optimal values are

$$r^* = 3.49, \quad \mathbb{E}(L_R(r^*) = 3.65, \quad K_{16}(r^*) = 8.97.$$

Thus, applying Policy 16 decreases the cost rate by another 4.4% compared to Policy 15 [40]. The same observation can be made if $C(t) = A_1t + A_2t^2 + A_3t^3$ where the A_i are independent, normally distributed random variables, $\mathbb{E}(A_i) > 3\sqrt{Var(A_i)}$, $i = 1, 2, 3$. Hence, an interesting problem arises: For given $\{C(t),\ t \geq 0\}$, limiting what functional(s) of $C(t)$ reduces the cost rate most?

Example 11.5 *Let $C(t) = c_0\left[e^{D(t)} - 1\right]$, where $\{D(t),\ t \geq 0\}$ is the Brownian motion process with drift with positive drift parameter μ and volatility σ^2, i.e. $D(t) = \mu t + B(t)$, where $\{B(t),\ t \geq 0\}$ is the Brownian motion process with $Var(B(1)) = \sigma^2$, $0 < c_0 \leq c_e$, and $L_D(d)$ be the first passage time of $\{D(t),\ t \geq 0\}$ with regard to level d: $L_D(d) = \min\{t,\ D(t) = d\}$. Since $\mathbb{E}(L_D(d)) = d/\mu$ and*

$$c_0\left[e^{D(t)} - 1\right] = c \iff D(t) = \ln\left(\frac{c + c_0}{c_0}\right),$$

it follows $\mathbb{E}(L_C(c)) = \dfrac{1}{\mu}\ln\left(\dfrac{c + c_0}{c_0}\right)$. Hence,

$$K_{15}(c) = \frac{c_e + c}{\ln\left(\dfrac{c + c_0}{c_0}\right)}\mu.$$

The optimal level c^* satisfies the equation

$$\ln\left(\frac{c + c_0}{c_0}\right) = \frac{c_e + c}{c_0 + c}.$$

A unique solution $c = c^*$ exists and the corresponding minimal cost rate is $K_{15}(c^*) = (c^* + c_0)\mu$. Moreover,

$$K_5(\tau) = \frac{c_0 \left[e^{(\mu + \sigma^2/2)\tau} - 1\right] + c_e}{\tau}$$

so that the economic lifetime $\tau = \tau^*$ satisfies the equation

$$\frac{c_0}{c_e}\left[(\mu + \sigma^2/2)\tau - 1\right]e^{(\mu + \sigma^2/2)\tau} = \frac{c_e - c_0}{c_e}.$$

Let, in particular, $\mu = 0.1$ and $\sigma^2 = 0.02$. Then,

$$c^* = 25.911, \quad K_{15}(c^*) = 3.5911,$$
$$\tau^* = 11.622, \quad K_5(\tau^*) = 3.9502.$$

Thus, the savings when applying the optimal repair cost rate instead of the economic lifetime are $\Delta K = (3.9502 - 3.5911)/3.9502\ [100\%] = 9.1\%$. If under otherwise the same assumptions the volatility is $\sigma^2 = 0.08$, then the corresponding savings would be 28.6%. Generally, it can be observed that with increasing variability of the cumulative repair cost process $\{C(t),\ t \geq 0\}$, the efficiency of applying Policy 15 compared to applying Policy 5 is increasing as well. From the modeling point of view, it is a formal disadvantage of the process $\{C(t),\ t \geq 0\}$ that, although its trend function is strictly increasing, its sample paths do not have this property. However, the replacement decision as prescribed by Policy 15 and its comparison to the economic lifetime approach does not at all depend on properties of the sample paths, but only on the behavior of $\mathbb{E}(L_C(c))$ and on the trend function $m(t) = \mathbb{E}(C(t))$. Hence, the results obtained are relevant for all those stochastic cumulative repair cost developments, where the pair "expected first passage time" and '"trend function" exhibit at least approximately the same behavior as the corresponding pair resulting from the given or other functionals of the Brownian motion. In view of this, Brownian motion based cost models seem to be more realistic than Brownian motion based wear models.
This and other $C(t)$ derived from the Brownian motion were discussed in [36].

Example 11.6 *Let $C(t)$ have a Weibull distribution with distribution function $F_{C(t)}(x)$ and density $f_{C(t)}(x)$:*

$$F_{C(t)}(x) = 1 - e^{-\left(\frac{x}{\lambda t^\alpha}\right)^\beta}, \quad f_{C(t)}(x) = \frac{\beta}{\lambda^\beta t^{\alpha\beta}x^{\beta-1}}e^{-\left(\frac{x}{\lambda t^\alpha}\right)^\beta}, \quad x \geq 0,$$

where $\alpha > 1$, $\beta \geq 2$, $\lambda > 0$. The cumulative cost process $\{C(t), t \geq 0\}$ is assumed to have increasing sample paths. (Note that in what follows we do not need the assumption that $\{C(t), t \geq 0\}$ is a stochastic process in the strict *mathematical sense.) Then, for a given total repair cost limit, making use of (11.25),*

$$\mathbb{E}\left(L_C(c)\right) = \int_0^\infty F_{C(t)}(c)dt = \int_0^\infty \int_0^c f_{C(t)}(x)dxdt.$$

By changing the order of integration in the second integral,

$$\mathbb{E}\left(L_C(c)\right) = k_1 c^{1/\alpha} \quad \text{with} \quad k_1 = (1/\lambda)^{1/\alpha} \, \Gamma\left(1 - 1/\alpha\beta\right),$$

where $\Gamma(y)$ is the Gamma-function:

$$\Gamma(y) = \int_0^\infty x^{y-1} e^{-x} dx, \quad y > 0.$$

The corresponding cost rate is

$$K_{15}(c) = \frac{1}{k_1} \frac{c_e + c}{c^{1/\alpha}}.$$

The optimal limit c^* and the minimal cost rate are

$$c^* = \frac{c_e}{\alpha - 1}, \quad K_{15}(c^*) = \frac{\alpha}{k_1} \left(\frac{c_e}{\alpha - 1}\right)^{(\alpha - 1)/\alpha}.$$

Since $\mathbb{E}(C(t)) = k_2 t^\alpha$ with $k_2 = \lambda \Gamma\left(1 + 1/\beta\right)$, the cost rate under Policy 5 is

$$K_5\left(\tau\right) = \frac{c_e + k_2 \tau^\alpha}{\tau}.$$

The optimal values are

$$\tau^* = \left(\frac{c_e}{k_2(\alpha - 1)}\right)^{1/\alpha}, \quad K_5(\tau^*) = \alpha k_2^{1/\alpha} \left(\frac{c_e}{\alpha - 1}\right)^{(\alpha - 1)/\alpha}.$$

The inequality $K_{15}(c^*) < K_5(\tau^*)$ is equivalent to

$$1 < \Gamma(1 + 1/\beta) \, [\Gamma(1 - 1/\alpha\beta)]^\alpha. \tag{11.26}$$

$[\Gamma(1 - 1/\alpha\beta)]^\alpha$ is a decreasing function in α, $\alpha \geq 1, \beta \geq 2$. For $\alpha \downarrow 1$, the right-hand side of this inequality tends to

$$\Gamma(1 + 1/\beta) \, \Gamma(1 - 1/\beta) = \frac{\pi/\beta}{\sin(\pi/\beta)} > 1,$$

since β is finite and the function $f(x) = x/\sin x$ assumes its minimum in $[0, \pi/2]$ at $x = 0$. Hence, (11.26) is always true, so that Policy 15 is superior to Policy 5. In particular, if $1.1 \leq \alpha \leq 5$ and $\beta = 2$, then average cost savings between 31% and 4.2% are achieved by applying the optimum cumulative repair cost limit c^* instead of the economic lifetime τ^*.

Other models along this line can be found in [37]. Surely, from the mathematical point of view it also makes sense to consider processes $\{C(t), t \geq 0\}$ the sample paths of which have positive increments (jumps) at the time points of repairs. In this case explicit results can only be obtained when making use of asymptotic or approximative distributions of the first passage time of jump processes or of Monte Carlo simulation.

Combined Age-Total Repair Cost Limit Replacement Policy

Analogous to Policy 14, we now supplement the total repair cost limit replacement Policy 15 with preventive replacements based on system age to be able to meet given reliability criteria.

Policy 17   The system is replaced as soon as the total repair cost reaches or exceeds level c or after τ time units, whichever occurs first [37].

Following Policy 17, the probability distribution function of the cycle length $Y = \min\{L_C(c), \tau\}$ and the expected cycle length is

$$\Pr(Y \le t) = \begin{cases} 1 - F_{C(t)}(c) & \text{for } 0 \le t \le \tau, \\ 1 & \text{for } \tau < t \end{cases} \quad , \quad \mathbb{E}(Y) = \int_0^\tau F_{C(t)}(c)dt.$$

Hence, the cost rate has structure

$$K_{17}(c,\tau) = \frac{[c_p + E(C(\tau)\,|C(\tau) \le c)]F_{C(\tau)}(c) + (c_e + c)\overline{F}_{C(\tau)}(c)}{\displaystyle\int_0^\tau F_{C(t)}(c)dt}.$$

Since

$$\mathbb{E}(C(\tau)\,|C(\tau) < c) = \frac{1}{F_{C(\tau)}(c)}\left[\int_0^c \overline{F}_{C(\tau)}(x)dx - c\overline{F}_{C(\tau)}(c)\right],$$

we have

$$K_{17}(c,\tau) = \frac{\displaystyle\int_0^c \overline{F}_{C(\tau)}(x)dx - (c_e - c_p)F_{C(\tau)}(c) + c_e}{\displaystyle\int_0^\tau F_{C(t)}(c)dt}. \tag{11.27}$$

This formula becomes (11.23) if τ tends to infinity. If τ is fixed, then $K_{17}(c,\tau)$ has to be minimized with regard to c for given τ.

Example 11.7 *Let $c_e = 20$ and $C(t)$ have a Rayleigh distribution:*

$$F_{C(t)}(x) = 1 - e^{-x^2/t^4}, \ t > 0.$$

1. *If $\tau = \infty$, then*

$$K_{17}(c,\infty) = K_{15}(c) = \frac{c + 20}{\displaystyle\int_0^\infty (1 - e^{-\frac{c^2}{t^4}})dt}.$$

The optimal values are $c^* = 20$ and $K_{17}(c^*, \infty) = 7.2990$.
2. If $c_e = c_p = 20$, then (11.27) becomes

$$K_{17}(c, \tau) = \frac{\displaystyle\int_0^c e^{-\frac{x^2}{\tau^4}} \, dx + 20}{\displaystyle\int_0^\tau (1 - e^{-\frac{c^2}{t^4}}) \, dt}.$$

The optimal values are $c^* = 20$, $\tau^* = \infty$, and $K_{17}(c^*, \tau^*) = 7.2990$.
3. If $0 < c_p < c_e = 20$, then

$$K_{17}(c, \tau) = \frac{\displaystyle\int_0^c e^{-\frac{x^2}{\tau^4}} \, dx - (20 - c_p)(1 - e^{-\frac{c^2}{\tau^4}}) + 20}{\displaystyle\int_0^\tau (1 - e^{-\frac{c^2}{t^4}}) \, dt}.$$

For $c_p = 16$, the optimal values are $c^* = 20$, $\tau^* = \infty$, and $K_{17}(c^*, \tau^*) = 7.299$. Thus, age replacement is not cost-efficient. For $c_p = 14$, applying a finite repair cost limit is not cost-efficient: $c^* = \infty$, $\tau^* = 4.1$, and $K_{17}(c^*, \tau^*) = 7.048$.

Recent papers, which combine the total repair cost limit approach with established maintenance policies, are [90, 193, 236, 194, 85, 89, 277]. However, when including lifetime parameters in the model, the advantages of the cumulative repair cost limit approach listed above are lost.

11.5   Exercises

Exercise 11.1 *Derive equation (11.10) for the the optimal repair cost limit $c = c^*$ in case of the Weibull distribution with form parameter β and discuss the behavior of c^* for $\beta \to 1$ and $\beta \to \infty$, respectively.*

Exercise 11.2 *Let $c_e = 10$ and $c_m = 1$, the random repair cost C be uniformly distributed over the interval $[0, 8]$, and $F(t) = 1 - e^{-t^\beta}$, $t \geq 0$. Compare the differences between the minimal cost criteria (11.9) and (11.11) in dependence on β. Interpret the results.*

Exercise 11.3 *A system is replaced according to policy 17. The cumulative maintenance cost $C(t)$ in $[0, t]$ has distribution function (Frechèt distribution)*

$$F_t(x) = P(C(t) \leq x) = e^{-(t^\alpha/x)^\beta}, \quad \alpha > 1, \; \beta > 1, \; x > 0, \; t \geq 0.$$

(1) Determine the cost-optimal limit for $C(t)$.
(2) Compare the corresponding minimal cost rate with the one achieved by applying the economic lifetime.

Chapter 12

Maintenance Models with General Degree of Repair

12.1 Imperfect Repair

Apart from Policy 5 and the cumulative repair cost limit replacement policies, all maintenance policies considered so far were based on maintenance actions, which only comprise replacements ("good as new"-concept) and minimal repairs ("bad as old"-concept). In many cases, this is a satisfactory approximation to real-life maintenance management. While it may makes sense, doing replacements of aging systems both in case of emergency (corrective) and preventive maintenance, preventive minimal repairs of an operating system do not make any sense. Replacements of an aging system by an equivalent new one reduce the failure rate to its smallest value $\lambda_{\min}$. In this chapter, we will assume that $\lambda(t)$ is nondecreasing in $[0, \infty)$ so that $\lambda_{\min} = \lambda(0)$. After a minimal repair, $\lambda(t)$ is on the same level as immediately before the repair (failure). Therefore, replacements and minimal repairs can be considered extreme maintenance measures. This motivates the definition of the *degree of a repair* as the ratio

$$d_r(t) = \frac{\lambda(t) - \lambda_r(t)}{\lambda(t) - \lambda(0)}, \tag{12.1}$$

where $\lambda(t) = \lambda(t-0)$ is the value of the failure rate immediately before a failure at time t and $\lambda_r(t) = \lambda_r(t + 0)$ is the value of the failure rate immediately after a repair done at time t, $\lambda(0) \le \lambda_r(t) \le \lambda(t)$. In this case, $d_r = 0$ $(d_r = 1)$ refers to a minimal repair (replacement). λ_r defines the *virtual age* v of the system after a repair at time t via $\lambda_r(t) = \lambda(v)$, $0 \le v \le t$. Thus, from the reliability (survival) point of view, a repair at time t with degree $d_r(t)$ reduces the age of the system by $t - v$ time units, and, for any calender time point t, the value of the failure rate is actually determined by the virtual age of the system at time t. (Obviously, the virtual age of a technical system is the

273

analog to the *biological age* of a human being.) The case that a repair enables the system to continue its work, but actually increases its failure rate, may happen in practice, but we will exclude this atypical situation. Hence, when pursuing a maintenance policy with general repairs, every repair reduces the virtual age of the system, which it had immediately before the failure (repair) so that the virtual age of the system at calender time t depends on all repairs up to time t. Assigning to a minimal repair the degree $d_r = 0$ makes sense, since such a repair has zero influence on the (virtual) age and, therefore, on the survival probability of the system. However, in the literature $1 - d_r$ is frequently defined as degree of repair.

It is now widely used terminology to call a repair *imperfect* if $0 \le d_r(t) < 1$ and *perfect* if $d_r(t) = 1$ (see e.g. *Wang and Pham* [303]). Thus, the imperfect repair concept includes our two-failure-type model applied in Chapter 10 and the maintenance policies 10 to 12 derived from it. In the literature, the two failure type model is now called "$(p(t), q(t))$-modeling" and started with the papers [26], [54], and [55]. The special case $p(t) \equiv p$, $q(t) \equiv 1 - p = q$ is used in [29], [20], and [72]. For first summaries see the monographs [28], [42]. For more recent summaries see [148],[124] , [125], [303], [238], [268]. The (p, q)-model had been applied to preventive repairs as well: A preventive repair is minimal with probability p (as mentioned before, in this case the preventive repair is useless) or renews the system with probability $q = 1 - p$. Maybe the first papers along this line were [84], [235], [233].

The terms *imperfect repair* and *perfect repair* are likely to be a bit disturbing to maintenance engineers/mechanics, since, from their layman point of view, they may believe to have carried out a perfect *imperfect repair* and a *perfect repair* not that perfect. Hence, in what follows we prefer to use the term *general repair* if the degree of a repair can assume any value between 0 and 1. Two simple analytical models for quantifying the development in time of the system failure rate, when following a general repair policy, will be considered. In both models, let $\{t_1, t_2, ...\}$ be the sequence of repair (failure) time points, and $\{a_1, a_2, ...\}$ be a sequence of real numbers with $0 \le a_n \le 1$, $n = 1, 2, ...$ To avoid routine modifications to the models, all repairs are assumed to take negligibly small times.

Model 1

Firstly we assume that a repair only aims at removing or reducing the wear between the current and the previous failure. At the time point of the first failure, the system failure rate has value $\lambda(t_1)$. The value of the system failure rate after the first repair is $\lambda_1 = \lambda(0) + a_1[\lambda(t_1) - \lambda(0)]$. (Thus, a repair can not lead to a failure rate smaller than the initial value $\lambda(0)$ which belongs to a new system.) This decrease of the failure rate defines the virtual age v_1 of the system immediately after the first repair by $\lambda_1 = \lambda(v_1)$. At the time point of the second failure, the value of the failure rate is $\lambda(v_1 + t_2 - t_1)$. Since the second repair only affects the damage which occurred between the first and the second failure, the system failure rate immediately after the second repair is $\lambda_2 = \lambda(v_1) + a_2[\lambda(v_1 + t_2 - t_1) - \lambda(v_1)]$. Thus, the virtual age v_2 of the system after the second repair is given by $\lambda_2 = \lambda(v_2)$. Proceeding in this way yields the system failure rate $\lambda_{\mathbf{a}}(t)$ under the $\mathbf{a} = \{a_1, a_2, ...\}$-repair policy as

$$\lambda_{\mathbf{a}}(t) = \begin{cases} \lambda(t) & \text{for } 0 \le t < t_1, \\ \lambda(t - t_n + v_n) & \text{for } t_n \le t < t_{n+1}, \ n = 1, 2, ..., \end{cases} \quad (12.2)$$

where the virtual age v_n of the system after the nth repair is given by

$$\lambda_n = \lambda(v_n) = \lambda(v_{n-1}) + a_n[\lambda(t_n - t_{n-1} + v_{n-1}) - \lambda(v_{n-1})],$$
$$n = 1, 2, ..., \ t_0 = 0, \ v_0 = 0.$$

One can expect that the sequence $\{a_1, a_2, ...\}$ is decreasing with inreasing n, since with increasing system age the wear damage leading to failures will become more and more severe. Note that $a_n = 1$ implies that the nth repair is minimal, whereas $a_n = 0$ means that the nth repair restores the "reliability state" of the system to the one after the previous repair.

Model 2

Secondly, we assume that a repair at any time point not only influences the interval since the previous repair, but affects the whole preceding repair process in the following way: Next, let $\lambda(0) = 0$. Then the value of the system failure rate after the first repair is $\lambda_1 = a_1\lambda(t_1)$, and this decrease of the failure rate defines the virtual age v_1 of the system after the first repair by $\lambda_1 = \lambda(v_1)$. Immediately before the second repair, the system failure rate has value $\lambda(v_1 + t_2 - t_1)$, and after the second repair it has value $\lambda_2 = a_2\lambda(v_1 + t_2 - t_1)$. Thus, the virtual age v_2 of the system after the second repair by $\lambda_2 = \lambda(v_2)$. Proceeding in this way shows that the system failure rate $\lambda_{\mathbf{a}}(t)$ under the $\mathbf{a} = \{a_1, a_2, ...\}$-repair policy has again structure (12.2), but now the virtual age v_n of the system after the nth repair is given by

$$\lambda_n = \lambda(v_n) = a_n\lambda(t_n - t_{n-1} + v_{n-1}), \ n = 1, 2, ..., \ t_0 = 0, \ v_0 = 0. \quad (12.3)$$

If $\lambda(0) > 0$ and it is not possible that a repair can achieve a failure rate less than $\lambda(0)$, then Equation (12.3) has to be replaced with

$$\lambda_n = \lambda(v_n) = \lambda(0) + a_n[\lambda(t_n - t_{n-1} + v_{n-1}) - \lambda(0)],$$
$$n = 1, 2, ..., \ t_0 = 0, \ v_0 = 0.$$

Note that the function $\{\lambda_{\mathbf{a}}(t), \ t \ge 0\}$ defined by (12.2) is a sample path of a stochastic process since the failure times t_n of the system are random variables. *Chan and Shaw* [83] suggested a general repair model similar to (12.2) on condition $a_n = a$. *Guo et al.* [149] proposed a general repair model based on the expected number of repairs in $[0, t)$.

12.2   Virtual Age

In the previous section, we introduced the virtual age of a system via its failure rate. But the first ones who coined the term "virtual age" of a technical system and used it for maintenance optimization were *M. Kijima, H. Morimura*, and *Y. Suzuki* in their breakthrough paper [180]. To outline their approach, let L_1 be the lifetime of a system with the IFR-distribution function $F(t)$ and density $f(t)$, $t \geq 0$. All repairs are assumed to take negligibly small times. At the time L_1 of the first failure a general repair is done. It reduces the calender age L_1 of the system by $L_1 - A_1 L_1$ time units, i.e. the virtual age of the system after the repair is $V_1 = A_1 L_1$, where A_1 is a random variable in the range $0 \leq A_1 \leq 1$. Now, let the corresponding random variables after the nth failure (general repair) be L_{n+1}, A_n, and V_n, $n = 1, 2, ...$, where L_{n+1}, by formula (1.3), has the conditional distribution function

$$F_{L_{n+1}}(t \,|\, V_n = x) = \Pr(L_{n+1} \leq t \,|\, V_n = x) = \frac{F(t+x) - F(x)}{\overline{F}(x)}, \quad n = 1, 2, ...$$

$$(12.4)$$

and the A_n are random variables in the range $[0, 1]$. Thus, L_{n+1} on condition $V_n = x$ is the residual lifetime of a system which has survived the interval $[0, x]$. (V_n is the virtual age of the system immediately after the nth general repair.)

Kijima [179] suggests two models for the development of the virtual age in time:

Model 1

The nth repair affects only the wear or other damages to the system, which occurred after the $(n-1)^{th}$ failure. Thus, the nth repair reduces the calender lifetime L_n to the virtual lifetime $A_n L_n$, so that the virtual lifetime of the system after the nth repair is

$$V_n = V_{n-1} + A_n L_n, \quad n = 1, 2, ..., \quad V_0 = 0, \qquad (12.5)$$

or, equivalently,

$$V_n = A_1 L_1 + A_2 L_2 + \cdots + A_n L_n, \quad n = 1, 2, ...$$

Note that $D_r(n) = 1 - A_n$ is the random degree of the nth repair. Compared to (12.1), it does not explicitly depend on the failure time, but on the number of the repair. If the A_n are identically distributed as A with

$$A = \begin{cases} 0 \text{ with probability } p \\ 1 \text{ with probability } 1 - p \end{cases},$$

then formula (12.5) yields the virtual age of a system which is maintained according to Policy 2 with $p(t) \equiv p$.

Model 2

A repair affects all the wear or other damage to the system, which has occurred since the beginning of its operation. In this case, the virtual age of the system after the nth repair is

$$V_n = A_n(V_{n-1} + L_n), \ n = 1, 2, ..., \ V_0 = 0, \tag{12.6}$$

or, equivalently

$$V_n = A_1 A_2 \cdots A_n L_1 + A_2 A_3 \cdots A_n L_2 + \cdots + A_n L_n, \ n = 1, 2, ...$$

In some situations, other virtual age models may even be more adequate, for instance

Model 3

The nth repair affects the wear or other damage to the system as in Model 1. In addition, it affects the wear or other damage to the system, which occurred between the $(n-2)^{th}$ failure and the $(n-1)^{th}$ failure with degree of repair $D_r(n-1) = 1 - B_{n-1}$, where B_{n-1} is a random variable with range $[0, 1]$, $n = 2, 3, ...$ Hence, the virtual age V_n of the system after the nth repair is

$$V_0 = 0, \ V_1 = A_1 L_1,$$
$$V_n = A_1 B_1 L_1 + A_2 B_2 L_2 + \cdots + A_{n-1} B_{n-1} L_{n-1} + A_n L_n, \ n = 2, 3, ...$$

It can be expected that the nth repair reduces L_n on average more than L_{n-1}. This is e.g. the case if A_i is stochastically smaller than B_i, i.e. if

$$F_{A_i}(t) \ge F_{B_i}(t), \ i = 1, 2, ...$$

As much as Models 1 to 3 are motivated by real-life maintenance processes, as difficult is their use in optimizing even simple maintenance policies with general repairs with regard to the cost rate. Hence [180] assume

$$A_n = a = \text{const.}, \ 0 < a \le 1, \ n = 1, 2, ..., \tag{12.7}$$

and consider for Model 1 a block replacement policy: preventive renewal after τ time units. Nevertheless, the derivation of the corresponding cost rate is a nontrivial, tedious exercise. For random A_i, tractable explicit formulas for the cost rates do not exist. At least under assumption (12.7), recursive formulas for the distribution function of the virtual lifetime $F_{V_n}(t) = \Pr(V_n > t)$ and its density $f_{V_n}(t) = dF_{V_n}(t)/dt$ can easily be derived:

Model 1 From (12.4) and (12.5), letting $\overline{F}_{V_n}(t) = 1 - F_{V_n}(t)$,

$$\overline{F}_{V_1}(t) = \overline{F}(t/a),$$

$$\overline{F}_{V_n}(t) = \overline{F}_{V_{n-1}}(t) + \int_0^t \frac{\overline{F}\left(x + \dfrac{t-x}{a}\right)}{\overline{F}(x)} f_{V_{n-1}}(x)\,dx, \ n = 2, 3, ...$$

By differentiation, for $n = 2, 3, ...,$

$$f_{V_1}(t) = \frac{1}{a} f\left(\frac{t}{a}\right), \quad f_{V_n}(t) = \frac{1}{a} \int_0^t \frac{f\left(x + \dfrac{t-x}{a}\right)}{\overline{F}(x)} f_{V_{n-1}}(x) dx.$$

Model 2   From (12.4) and (12.6), under assumption (12.7),

$$\overline{F}_{V_1}(t) = \overline{F}(t/a),$$

$$\overline{F}_{V_n}(t) = \overline{F}_{V_{n-1}}(t/a) + \int_0^{t/a} \frac{\overline{F}(t/a)}{\overline{F}(x)} f_{V_{n-1}}(x) dx, \quad n = 2, 3, ...$$

By differentiation,

$$f_{V_1}(t) = \frac{1}{a} f(\frac{t}{a}), \quad f_{V_n}(t) = \frac{1}{a} f\left(\frac{t}{a}\right) \int_0^{t/a} \frac{1}{\overline{F}(x)} f_{V_{n-1}}(x) dx, \quad n = 2, 3, ...$$

For the distribution function and density of V_n under Model 3 on condition $A_n = a$, $0 < a \leq 1$, $B_n = b$, $0 < b \leq 1$, $n = 1, 2, ...,$ see Exercise 12.2.

A generalization of Model 2 under assumption (12.7) has been proposed by *Dagpunar* [102]: Given a "repair functional" $g(\cdot)$ and negligibly small repair times, he assumes that the virtual age of the system after the nth repair is

$$V_n = g(V_{n-1} + L_n), \quad n = 1, 2, ... \tag{12.8}$$

If $g(v) = av$, then (12.8) becomes (12.6) with $A_n = a$. Given $V_0 = r$, $r \geq 0$, let $N_r(t)$ be the random number of failures in $(0, t]$ and $H_r(t) = \mathbb{E}N_r(t)$. Analogously to (9.14), conditioning on the first failure at time x, $0 \leq x \leq t$, yields the integral equation

$$H_0(t) = \int_0^t \left[1 + H_{g(x)}(t - x)\right] dF(x). \tag{12.9}$$

Dagpunar also obtained asymptotic results for the virtual age of the system.
Since the publication of the paper *[180]* quite a few papers have made use of the virtual age concept to determine and minimize the maintenance cost under various general repair policies. A particularly good paper from the theoretical and application point of view is [204].

The Liu, Makiš, Jardine Policy

A system starts operating at time $t = 0$. Overhauls are scheduled at time points $\tau, 2\tau, ... N\tau$. At time point $(N + 1)\tau$, a replacement of the system is carried out. Between overhauls and in the intervals $(0, \tau)$ and $(N\tau, (N+1)\tau)$, the system is minimally repaired. Repairs, overhauls and replacements are assumed to take only negligibly small times. Repairs have no influence on the virtual age of the system. Replacements turn the virtual age back to 0. Hence, in this policy only overhauls are genuine general repairs. These assumptions considerably simplify the theoretical treatment of the policy and facilitate its practical application.

Let $V_i(\tau)$ be the virtual age of the system at the end of the overhaul period $((i-1)\tau, i\tau]$, $i = 1, 2, ..., N$, i.e. immediately after the overhaul at time $i\tau$, $i = 1, 2, ..., N$, and $V_0(\tau) = 0$. Then, under assumption (12.7), for model 1,

$$V_n(\tau) = V_{n-1}(\tau) + a\tau, \; n = 1, 2, ..., N, \; \text{or}$$
$$V_n(\tau) = an\tau, \; n = 1, 2, ..., N,$$

and for model 2,

$$V_n(\tau) = a[V_{n-1}(\tau) + \tau], \; n = 1, 2, ..., N, \; \text{or}$$
$$V_n(\tau) = a\frac{1 - a^n}{1 - a}\tau, \; n = 1, 2, ..., N.$$

For both models 1 and 2, with increasing τ, the difference between the calendar age and the virtual age of a system between two neighbouring overhauls $(1 - a)\tau$ is increasing. This may not be realistic in view of technical or economical restrictions to the overhaul capacity. Hence, Liu et al. propose the following virtual age model for $0 < a < 1$ and $n = 0, 1, ..., N$:

$$V_n(\tau) = \begin{cases} an\tau & \text{if } \tau \leq \tau_m/(1-a), \\ n(\tau - \tau_m) & \text{if } \tau > \tau_m/(1-a). \end{cases}$$

Thus, τ_m is the maximum amount of time by which an overhaul can reduce the calendar age of the system.

To derive the corresponding maintenance cost rate, let $c_o(V)$, and $c_m(V)$ be the cost of an overhaul and a minimal repair depending on the virtual age V at the time of the maintenance action, respectively, and c_p be the cost of a (preventive) replacement. Then the total overhaul cost per replacement cycle is

$$C_o(\tau, N) = \sum_{n=1}^{N} c_o[V_{n-1}(\tau) + \tau].$$

(Immediately before the nth overhaul the virtual age of the system is given by $V_{n-1}(\tau) + \tau$.) The total (minimal) repair cost per replacement cycle is

$$C_m(\tau, N) = \sum_{n=0}^{N} \int_{V_n(\tau)}^{V_n(\tau) + \tau} c_m(t)\lambda(t)dt.$$

Hence, the (long-run) maintenance cost rate is

$$K(\tau, N) = \frac{c_p + C_o(\tau, N) + C_m(\tau, N)}{(N+1)\tau}.$$

Under some additional assumptions, Liu et al. derive the vector (τ^*, N^*), which minimizes $K(\tau, N)$. Of course, the optimal vector (τ^*, N^*) heavily depends on the structure of the time-dependency of $C_o(V)$ and $C_m(V)$.

A combination of the virtual age approach with the repair-cost-limit replacement method is presented in [162]. Its authors *Jiang, Makiš and Jardine* assume that the degree of a repair depends on the virtual age and on the repair cost. An attempt to combine the decrease of the failure rate after a repair with the corresponding repair cost had been made in [21]. Since an exact cost analysis for maintenance policies with general repairs is not possible, or is at least complicated, *Scarsini and Shaked* [273] constructed bounds on the profit generated by a system subjected to general repairs. For a statistical analysis and an industrial application of the Kijima et al. virtual age approach see [134]. A preventive maintenance model based on the virtual age of a system, which is subjected to shocks, has been analyzed by *Sheu et al.* [274].

Malik [215] was probably the first one who modeled the aging behavior of a technical system via its virtual age (without using this term). His model is essentially given by formula (12.5) with constant $A_n = a$. Another and even more general approach to virtual age than the one of Kijima et al. was proposed and analyzed by *Uematsu and Nishida* [299].

12.3   Geometric Time Scaling

Lam Yeh [195] introduced a powerful generalization of the well-known alternating renewal process for application in maintenance:

Let $\{(L_n, R_n), \; n = 1, 2, ...\}$ be an alternating renewal process (Definition 9.7), where the L_n and R_n are identically distributed as L and R, respectively, and a and b are constants with $0 < a < 1$ and $b > 1$. To take into account that the operating periods of a system tend to get shorter and shorter, and the repair periods longer and longer with increasing system age, let $X_n = a^{n-1} L_n$ and $Y_n = b^{n-1} R_n$ be the respective new lengths of the operating and repair periods. If $F(t) = P(L \leq t)$, $f(t) = dF(t)/dt$, and $\lambda(t)$ are distribution function, density, and failure rate of L, then the corresponding functions of X_n are

$$F_n(t) = F(a^{-(n-1)}t),$$
$$f_n(t) = a^{-(n-1)} f(a^{-(n-1)}t), \qquad (12.10)$$
$$\lambda_n(t) = a^{-(n-1)} \lambda(a^{-(n-1)}t), \quad t \geq 0, \; n = 1, 2, ...$$

In addition, expected value and variance of X_n are

$$\mathbb{E}X_n = a^{n-1} \mathbb{E}L, \quad Var(X_n) = a^{2n-2} Var(L). \qquad (12.11)$$

Hence, for any n, if $F(t)$ is *IFR*, then $F_n(t)$ is *IFR* as well. (If $G(t) = \Pr(R \leq t)$, then analogous formulas hold for Y_n.) Lam Yeh used his process

$\{(X_n, Y_n), \; n = 1, 2, ...\}$ to model the sequence of the operating-repair intervals of a system, and analyzed two replacement policies:

1. The system is replaced as soon as its total operating time reaches a given level (total operating time limit repair policy).
2. For given n, the system is replaced after its nth failure.

Stadje and Zuckerman [284] generalized the Lam Yeh process by assuming that the operating times are *IFR and* get stochastically shorter and shorter, whereas the repair times are *NBUE* (new better than used in expectation) and get stochastically longer and longer. They discuss the cost efficiency of two more maintenance policies, dealing extensively with that operating time process $\{X_n\}$, which satisfies (12.10). The optimization approach used in [284] has in turn been made use of in [196]. In a subsequent paper, Lam Yeh (see [197]) applied his process to the reliability analysis of a two-unit series system. Obviously, the Lam Yeh process $\{(X_n, Y_n), \; n = 1, 2, ...\}$ is also a suitable model for health insurance if X_n (Y_n) is interpreted as the length of the nth period, in which the insured is healthy (sick). Surely, with increasing n, these periods will stochastically tend to become shorter and shorter (longer and longer).

Given an ordinary renewal process $\{L_n, \; n = 1, 2, ...\}$ with the L_n distributed as L, *Wang and Pham* [302] defined a quasi-renewal process in the following way:

Definition 12.1 *The random counting process* $\{N_a(t), \; t \geq 0\}$ *is said to be a quasi-renewal process with a positive parameter a if its interarrival times between the $(n-1)^{th}$ event and the nth event have structure* $X_n = a^{n-1} L_n$, $n = 1, 2, ..., \; X_0 = 0$.

Thus, characteristics of a quasi-renewal process are given by (12.10) and (12.11). Let

$$T_n = \sum_{i=1}^{n} X_i, \quad n = 1, 2, ..., \; T_0 = 0.$$

The maintenance actions at times T_n, $n \geq 1$, have the character of general repairs if $a < 1$. Analogously to (9.13), we get from (9.2)

$$H_a(t) = \mathbb{E}N_a(t) = \sum_{n=1}^{\infty} F^{*(n)}(t)$$

with $F^{*(n)}(t) = F_1 * F_2 * \cdots * F_n(t)$, $n = 1, 2, ...$ Since the value of the failure rate after every repair is $\lambda(0) = \lambda_{min}$, the degree of all repairs as given by (12.1) is always 1, although the repairs do not have the character of renewals if $a < 1$. In this case, a more suitable degree of repair might be $d_n = \mathbb{E}X_{n+1}/\mathbb{E}X_n$, $n = 1, 2, ...$Then, in view of (12.11), $d_n = a$ for all $n = 1, 2, ...$

Wang and Pham [302], [303] have discussed a large number of general repair models based on a quasi-renewal process. Here we consider modifications of our Policies 9 and 13. Note that under Policy 9* (just as in case of Policy 9), the emergency replacement is partially a preventive replacement.

Policy 9* The first $n-1$ failures are removed by repairs. At the time point of the nth failure, an emergency replacement is carried out.

If c_r is the cost of a repair (no longer a minimal one), then the corresponding cost rate is

$$K_{9*}(n) = \frac{c_e + (n-1)c_r}{\sum_{i=1}^{n} \mathbb{E}X_i},$$

or, by (12.11),

$$K_{9*}(n) = \frac{c_e + (n-1)c_r}{\dfrac{1-a^n}{1-a}\mathbb{E}L}.$$

For given numerical model parameters, minimizing $K_{9*}(n)$ with regard to n is a simple exercise, since $\mathbb{E}L$ does not depend on n. For instance, if $c_e = 10$, $c_r = 2$, and $a = 0.9$, the optimal n is $n^* = 8$, and the minimal cost rate is $K_{9*}(8) = 4.214/\mathbb{E}L$.

Policy 13* After a system failure, the necessary repair cost is estimated. The system is replaced by an equivalent new one if the repair cost exceeds a given level c. Otherwise a repair is done.

With the assumptions/notations (11.6) and (11.7) we have the cost rate

$$K_{13*}(c) = \frac{c_e + \left[\dfrac{1}{R(c)}\left(\int_0^c \overline{R}(x)dx - c\overline{R}(c)\right)\right]\mathbb{E}(N-1)}{\mathbb{E}T},$$

where N is the number of that repair at which the repair cost exceeds for the first time level c, and T is the random time between neighboring replacements. N has a geometric distribution with parameter $p = \overline{R}(c)$:

$$\Pr(N = n) = R(c)^{n-1}\overline{R}(c), \quad n = 1, 2, ...; \quad \mathbb{E}N = 1/\overline{R}(c).$$

$\mathbb{E}T$ is given by

$$\mathbb{E}T = \sum_{n=1}^{\infty} [\mathbb{E}T_n] \Pr(N = n)$$

$$= \sum_{n=1}^{\infty} [\mathbb{E}X_1 + \mathbb{E}X_2 + \cdots + \mathbb{E}X_n] \Pr(N = n)$$

so that, by (12.11),

$$\mathbb{E}T = \frac{\mathbb{E}L}{1 - aR(c)}.$$

Hence, the cost rate has the simple structure

$$K_{13*}(c) = \left(\frac{1 - aR(c)}{R(c)}\int_0^c \overline{R}(x)dx + (c_e - c)\overline{R}(c)\right)\frac{1}{\mathbb{E}L}.$$

For instance, if $c_e = 10$, $a = 0.9$, and the random repair cost C is uniformly distributed over the interval $[0, 10]$, then the optimal repair cost limit is $c^* = 6.40$ and the minimum cost rate is $K_{13^*}(c) = 6.652/\mathbb{E}L$.

The following general repair process may sometimes be more appropriate for modeling the maintenance of technical systems than the quasi-renewal process: Let $\{a_0, a_1, a_2, ...\}$ and $\{b_0, b_1, b_2, ...\}$ be two sequences of real numbers with properties

$$0 = b_0 \leq b_1 \leq b_2 \leq \cdots \quad \text{and} \quad 1 = a_0 \leq a_1 \leq a_2 \leq \cdots$$

The failure rate of the system after the nth repair is assumed to be given by

$$\lambda_n(t) = b_n + \lambda(a_n t); \quad n = 0, 1, 2, ..., \tag{12.12}$$

In (12.12), t refers to the time of the previous repair and $\lambda(t)$ is as usual the failure rate of a new system $(n = 0)$. Then the time X_n between the $(n-1)^{th}$ and the nth repair has survival probability

$$\overline{F}_{n-1}(t) = \exp\left(-b_{n-1}t - \frac{1}{a_{n-1}}\int_0^{a_{n-1}t}\lambda(x)dx\right), \quad n = 1, 2, ...$$

In particular, if $b_n = 0$ for all n,

$$\overline{F}_{n-1}(t) = \left[\overline{F}(a_{n-1}t)\right]^{1/a_{n-1}}, \quad n = 1, 2, ... \tag{12.13}$$

Example 12.2 *A system is maintained according to Policy 9* with $c_e = 10$, $c_r = 2$, and the probability distribution of the X_n is given by*

$$\lambda(t) = 2t, \ t \geq 0, \ a_n = (1.2)^{n-1}, \ and \ b_n = 0.1(n-1), \quad n = 1, 2, ...$$

Therefore,

$$F_{n-1}(t) = \Pr(X_n \leq t) = e^{-0.1(n-1)t-(1.2)^{n-1}t^2}, \ t \geq 0, \ n = 1, 2, ...,$$

and

$$K_{9^*}(n) = \frac{c_e + c_r(n-1)}{\sum_{i=1}^n \mathbb{E}X_i}$$

$$= \frac{10 + 2(n-1)}{\sum_{i=1}^n \left(\int_0^\infty e^{-0.1(i-1)t-(1.2)^{i-1}t^2}dt\right)}, \quad n = 1, 2, ...$$

The optimal values are $n^* = 7$ and $K_{9^*}(7) = 5.052$. Thus, the optimum number of repairs between replacements is $n^* - 1 = 7$. For the sake of comparison: $K_{9^*}(6) = 5.091$ and $K_{9^*}(8) = 5.064$. Hence, the cost rate is fairly insensitive with regard to small deviations from the optimal n.

12.4   Exercises

Exercise 12.1 *Under Model 1 (formula (12.2)) consider the $\mathbf{a} = (a_1, a_2, ...)$-repair policy with $a_n = (0.9)^n$, $n = 1, 2, ...$, for a system with failure rate $\lambda(t) = 0.4t^3$, $t \geq 0$. Determine the corresponding sequences of the degrees of repairs given by formula (12.1) and the virtual ages for $n = 1, 2, 3$.*

Exercise 12.2 *Determine distribution function and density of V_n under Model 3 (Section 12.2) on conditions*

$$A_n = a, \; 0 < a \leq 1, \; B_n = b, \; 0 < b \leq 1, \; n = 1, 2, ...$$

Exercise 12.3 *A system is maintained according to Policy 9* based on a quasi-renewal process with $a = 0.95$ and survival function $F(t) = e^{-0.1t^3}$, $t \geq 0$, of L. Given $c_e = 20$ and $c_r = 5$, determine the optimal number of repairs n^* between two neighboring replacements and the corresponding minimal cost rate. How sensitive is the cost rate $K_{9*}(n)$ with regard to deviations from n^*?*

Exercise 12.4 *With otherwise the same input as in Exercise 12.3 assume that the repair costs have a uniform distribution over the interval $[0, 20]$.*
(1) Apply Policy 13* and determine the optimal repair cost limit c^* and the corresponding minimal cost rate $K_{13*}(c^*)$.
(2) Apply Policy 9* with the same expected repair cost as under (1) and compare $K_{9*}(n^*)$ to $K_{13*}(c^*)$.

Exercise 12.5 *A system with failure rate $\lambda(t) = 0.2t^2$ is maintained according to policy 9* based on the repair Policy (12.12). Let $c_e = 10$ and $c_r = 2$, $b_n = 0.1n$, and $a_n = (1.1)^n$, $n = 1, 2, ...$ Determine the optimum number of repairs between two neighboring replacements.*

Chapter 13

Inspection of Systems

13.1   Basic Problem and Notation

Application of the maintenance policies considered so far requires that the state of the system, operating or failed, is known at any time point. However, this may not be true when operating for instance fully automated systems, sensors or more complex monitoring systems. In such cases, inspections are indispensable maintenance measures. A standard situation, first formulated by *R.E. Barlow and L.C. Hunter* [15], is the following one: An undetected failure causes an economic loss which increases in time, whereas inspections are costly, too. What is the most cost-efficient way to schedule inspections in time? This chapter deals with this situation and some of its recent ramifications.

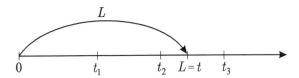

Figure 13.1: Illustration of the inspection policy

More exactly, a system starts operating at time $t = 0$. Distribution function, density, and expected value of the system lifetime L be $F(t)$, $f(t)$, and μ with μ finite. An inspection costs c units. If between the occurrence of a failure till its detection by an inspection a time t elapses, then a loss cost $v(t)$, $v(+0) = 0$, arises. Function $v(t)$ is assumed to be differentiable and strictly increasing in $[0, \infty)$. Inspections are scheduled at time points $t_1, t_2, ...$ with $0 < t_1 < t_2 < \cdots$. The sequence $P = \{t_1, t_2, ...\}$, shortly: $P = \{t_k\}$, is called an *inspection policy* or simply *policy*. If $B = \sup\{t, F(t) < 1\} = \infty$, then the sequence $\{t_1, t_2, ...\}$ is supposed to have property $\lim_{k \to \infty} t_k = \infty$, i.e. P is a point process. But if $B < \infty$, or if we anyway monitor the system

285

only over a finite time period $[0, T]$, $T \leq B$, then, for a finite integer n, the t_k are such that $0 < t_1 < t_2 < \cdots < t_n = T$. The differences $t_{k+1} - t_k = \tau_k$, $k = 0, 1, ..., t_0 = 0$, are the *(lengths of the) inspection intervals*. Thus, we can equivalently characterize an inspection policy by its inspection intervals: $P = \{\tau_0, \tau_1, ...\} = \{\tau_k\}$. P is called *strictly periodic* if $\tau_k = \tau$ for all $k = 0, 1, ...$ Notation: $P = P^{(\tau)}$. If, in case $T < \infty$, P prescribes exactly n inspections (the nth one by agreement at time T), we will write $P_n = \{t_1, t_2, ..., t_n\}$ or $P_n = \{\tau_0, \tau_1, ..., \tau_{n-1}\}$. In this chapter, inspections are assumed to only take negligibly small times.

13.2   Inspection without Replacement

13.2.1   Known Lifetime Distribution

For given $P = \{t_k\}$, let the system failure occur at a time t with $t_k < t \leq t_{k+1}$. Then the corresponding conditional total cost caused by downtime and inspections is $(k+1)c + v(t_{k+1} - t)$. For instance, the situation depicted in Figure 13.1 leads to the loss $3c + v(t_3 - t)$. Hence, if $T = \infty$, the expected total loss cost till detecting the failure is

$$K(P, F) = \sum_{k=0}^{\infty} \int_{t_k}^{t_{k+1}} [(k+1)c + v(t_{k+1} - t)] dF(t). \qquad (13.1)$$

If $T < \infty$, then

$$K(P_n, F) = \sum_{k=0}^{n-1} \int_{t_k}^{t_{k+1}} [(k+1)c + v(t_{k+1} - t)] dF(t) + nc\overline{F}(T). \qquad (13.2)$$

Next, for given F, the aim is to determine a cost-optimal policy $P^* = \{t_k^*\}$ with regard to $K(P, F)$:

$$K^*(F) = K(P^*, F) = \min_P K(P, F).$$

$K^*(F)$ is finite as the following elementary evaluation shows: For any strictly periodic inspection policy $P^{(\tau)} = \{k\tau\}$,

$$K(P^{(\tau)}, F) = \sum_{k=0}^{\infty} \int_{k\tau}^{(k+1)\tau} [(k+1)c + v((k+1)\tau - t)] dF(t)$$

$$\leq \sum_{k=0}^{\infty} \int_{k\tau}^{(k+1)\tau} [(k+1)c + v(\tau)] dF(t)$$

$$\leq \frac{c}{\tau} \sum_{k=0}^{\infty} (k\tau)[F((k+1)\tau) - F(k\tau)] + c + v(\tau),$$

so that

$$K^*(F) \leq K(P^{(\tau)}, F) \leq \left(\frac{\mu}{\tau} + 1\right) c + v(\tau) < \infty. \qquad (13.3)$$

An optimum inspection policy $P = \{t_k\}$ satisfies the necessary conditions $\partial K(\{t_k\}, F)/\partial t_k = 0$, $k = 1, 2, ...$, or

$$v(t_{k+1} - t_k) = \int_0^{t_k - t_{k-1}} \frac{f(t_k - t)}{f(t_k)} dv(t) - c, \quad k = 1, 2, ... \quad (13.4)$$

In particular, for a linear loss function $v(t) = at$, $a > 0$, this system of equations becomes

$$t_{k+1} - t_k = \frac{F(t_k) - F(t_{k-1})}{f(t_k)} - \frac{c}{a}, \quad k = 1, 2, ... \quad (13.5)$$

Hence, if the first optimal inspection time point t_1^* is known, the other ones can be recursively calculated by (13.4). An additional effort has to be made to determine $t_1 = t_1^*$. This task is facilitated by assuming that $f(t)$ is a PD_2-density.

Definition 13.1 *A probability density $f(t)$ is said to be a* Pólya-density *of order 2 (PD_2) if for all $h \geq 0$ the ratio $f(t - h)/f(t)$ is increasing in t, or, equivalently, the ratio $f(t + h)/f(t)$ is decreasing in t.*

It is easy to show that $f(t)$ is PD_2 iff for all $h \geq 0$

$$\frac{f(t)}{F(t + a) - F(t)} \quad (13.6)$$

is increasing in t for all real $a \neq 0$. The ratio (13.6) shows that the PD_2- property is closely related to the *IFR*-concept. Indeed, if $f(t)$ is PD_2, then $F(t)$ is *IFR*, i.e. the failure rate of L, namely $\dfrac{f(t)}{F(\infty) - F(t)} = \dfrac{f(t)}{1 - F(t)}$, is increasing in t. But the opposite need not be true. *Barlow et. al.* [16] were the first to exploit the properties of PD_2-densities for characterizing the optimum inspection policy in case of a linear loss function $v(t) = at$, $t \geq 0$. *Beichelt* [22], [24] did the same for any $v(t)$ with properties specified above.

To formulate the results, let $P(t_1)$ be the sequence $\{t_i\}$ generated by (13.4) starting with t_1. By agreement, this sequence breaks off at t_m if the right-hand side of (13.4) is 0 or negative for $k = m$.

Theorem 13.2 *Let $f(t)$ be PD_2 and $\mu < \infty$.*
1) If $T < \infty$, then the first optimal inspection time point t_1^* is the smallest t_1 with property that $P(t_1)$ breaks off at t_{n^*} with $t_{n^*} = T$, i.e. n^* is the optimal number of inspections.
2) If $T = \infty$, then t_1^* is the smallest time point with property that $\lim\limits_{k \to \infty} t_k = \infty$ [24].
3) If $t_1 > t_1^*$, there exists a positive integer m such that the inspection intervals τ_k in $P(t_1)$ start increasing at $k = m$.
4) In both cases, $T < \infty$ and $T = \infty$, the optimum inspection intervals are decreasing.

Already the first two properties of the optimal strategy $P^* = \{t_k^*\}$ listed in this theorem would be sufficient to determine the optimal inspection policy with regard to any desired degree of accuracy. It seems to be a good idea to start with a t_1 satisfying

$$c = \int_0^{t_1} v(t_1 - t)dF(t), \qquad (13.7)$$

since in this case the cost of an inspection is equal to the expected loss cost caused by a possible failure in $[0, t_1]$. This t_1 is increased/decreased till the generated sequence $P(t_1)$ has the characteristic properties of the optimal policy listed in the theorem. This is no longer a computational problem, so that the construction of "nearly optimal" inspection policies has lost some of its importance, see, e.g. *Munford and Shahani* [231], [232] and *Tadikamalla* [290]. In case of frequent inspections, *Keller* [177] proposed an inspection model based on the *inspection density* $d(t)$, which is the number of inspections per unit time at time t, and constructed an approximative policy. His idea was used by *Kaio and Osaki* [165] for constructing a nearly optimal policy. In two subsequent papers, *Kaio and Osaki* [166], [167] compared "exact" and approximative optimal inspection policies of the Barlow-Hunter type.

In case $B = \infty$, another way to determine an optimal policy with a desired degree of accuracy is to truncate $F(t)$ with regard to a sufficiently large finite inspection period $[0, T]$.

Example 13.3 *Let L be uniformly distributed over the interval $[0, T]$. Then, for an integer n with $n \leq n^*$, the system of equations (13.4) becomes*

$$v(\tau_{k+1}) = v(\tau_k) - c, \quad k = 0, 1, 2, ..., n - 2, \qquad (13.8)$$

$$\sum_{k=0}^{n-1} \tau_k = T.$$

Proposition: *The optimal number of inspections n^* is the largest integer n satisfying*

$$\sum_{k=0}^{n-1} v^{-1}(kc) < T, \qquad (13.9)$$

where $v^{-1}(\cdot)$ is the inverse function of $v(t)$. To prove this proposition, we show that condition (13.9) on n is necessary and sufficient for the existence of a solution of (13.8). By adding up all $n - 1$ equations of (13.8) we get $v(\tau_{n-1}) = v(\tau_0) - (n-1)c$. Hence, $v(\tau_0) > (n-1)c$. By adding up the last $n - 2$ equations we get analogously $v(\tau_1) > (n-2)c$. Generally, $v(\tau_k) > (n-k-1)c$ or, equivalently, $\tau_k > v^{-1}((n-k-1)c)$ is true for $k = 0, 1, ..., n-1$. Since $\sum_{k=0}^{n-1} \tau_k = T$, condition (13.9) is necessary. Now, let (13.9) be true. Starting with a $t_1 = \tau_0$ satisfying $v(\tau_0) > nc$ we generate the sequence $P(\tau_0) = \{\tau_0, \tau_1, ..., \tau_{n-1}\}$ from (13.8). It fulfills the condition $\sum_{k=0}^{n-1} \tau_k > \sum_{k=0}^{n-1} v^{-1}(kc)$. Moreover, in view of the continuity of $v(t)$, τ_0 can be chosen in such a way that $\sum_{k=0}^{n-1} \tau_k = T$. This proves the proposition.

In case of the linear loss function $v(t) = at$, the optimal number of inspections is the largest integer n satisfying

$$(n-1)n < 2cT/a,$$

and the optimal inspection time points are

$$t_k = k\left[\frac{T}{n} + \frac{c}{2a}(n-k)\right], \quad k = 1, 2, ..., n. \tag{13.10}$$

Example 13.4 *Let the system lifetime L have an exponential distribution with parameter λ: $F(t) = 1 - e^{-\lambda t}$, $t \geq 0$. In this case, the optimal inspection policy is strictly periodic [24]. From (13.1),*

$$K(P^{(\tau)}, F) = \frac{1}{1 - e^{-\lambda \tau}}\left[c + e^{-\lambda \tau}\int_0^\tau v(t)e^{+\lambda t}dt\right]. \tag{13.11}$$

The optimal inspection interval $\tau = \tau^*$ satisfies the equation

$$v(\tau) = \int_0^\tau e^{+\lambda t}dv(t) - c. \tag{13.12}$$

Under our assumptions, (13.12) always has a positive solution. For a linear loss function $v(t) = at$, (13.11) and (13.12) become

$$K(P^{(\tau)}, F) = \frac{c + a\tau}{1 - e^{-\tau/\mu}} - a\mu, \tag{13.13}$$

$$e^{\tau/\mu} - \tau/\mu = 1 + c/a\mu. \tag{13.14}$$

Example 13.5 *Let $F(t) = 1 - e^{-(t/10)^2}$, $t \geq 0$, $v(t) = t^2$, and $c = 1$. Then (13.4) becomes*

$$\tau_k^2 = 2\int_0^{\tau_{k-1}}\left(1 - \frac{t}{t_k}\right)e^{(2\,t\,t_k - t^2)/100}tdt - 1, \quad k = 1, 2, ..., \tau_0 = t_1.$$

Inspecting will be maintained over a time interval of length $T = 10$. We start calculating $P(t_1)$ with $t_1 = 5$, because this t_1 is close to the solution of (13.7). However, by theorem 13.2, the optimum first inspection time point t_1^* must be smaller, since $P(5) = \{5, 7.99, 10.77, ...\}$. More accurately, we need to determine the smallest t_1 with the property that there is an integer n so that $P(t_1) = \{t_1, t_2, ..., t_n = 10\}$ and $\tau_n = 0$. This property has the time point $t_1^* = 3.8246$ (slightly rounded), so that the optimal inspection policy is

$$P^* = P(t_1^*) = \{3.8246, \; 5.9233, \; 7.6171, \; 8.9794, \; 9.9511, 10\}.$$

The corresponding cost is $K(P^*, F) = 2.5823$. Note that the system has the expected lifetime $\mathbb{E}(L) = \sqrt{\pi/4} \approx 8.8623$.

For the sake of comparison, the optimal strictly periodic inspection strategy has, under the same assumptions, the inspection interval $\tau^* = 2.5$, and the corresponding loss cost is $K(P^{(\tau^*)}) = 4.9514$.

13.2.2   Unknown Lifetime Distribution

In this section, the lifetime distribution of the system is assumed to be unknown. Hence, instead of $K(P, F)$ we optimize timing of inspections with regard to the criterion

$$K(P) = \max_{F \in \mathbf{F}} K(P, F),$$

where $\mathbf{F}$ is the set of all lifetime distribution functions, i.e. $F(+0) = 0$ for all $F \in \mathbf{F}$. Hence, our aim is to determine $P = P^*$ such that

$$K(P^*) = \min_{P} \max_{F \in \mathbf{F}} K(P, F).$$

We call P^* *minimax (inspection) policy.* Derman [103] has found P^* assuming a linear loss function $v(t) = at$. (But, in addition, he assumed that an inspection does not detect a failure with probability p.) Minimizing $K(P)$ only makes sense if $K(P)$ is finite. Hence, we assume that the inspection process is restricted to the finite interval $[0, T]$ with the last inspection occurring at T. Let

$$g_k(t, t_{k+1}) = (k+1)c + v(t_{k+1} - t), \quad t_k < t \le t_{k+1}, \quad k = 0, 1, \ldots$$

The following elementary evaluation shows that there is indeed a distribution in $\mathbf{F}$ which, for any given P, maximizes $K(P, F)$:

$$K(P, F) = \sum_{k=0}^{n-1} \int_{t_k}^{t_{k+1}} g_k(t, t_{k+1}) dF(t) + nc\overline{F}(T)$$

$$\le \sum_{k=0}^{n-1} g_k(t_k, t_{k+1}) \left[F(t_{k+1}) - F(t_k) \right] + nc\overline{F}(T)$$

$$\le \max_{k=0,1,\ldots,n-1} g_k(t_k, t_{k+1}).$$

This estimation is sharp, so that

$$K(P) = \max_{k=0,1,\ldots,n-1} g_k(t_k, t_{k+1}).$$

Theorem 13.6 *There exists a unique minimax inspection policy. It coincides with the optimal policy $P_{n^*}^* = \{t_1^*, t_2^*, \ldots, t_{n^*}^* = T\}$ with regard to $K(P, F)$ if F is the distribution function of an in $[0, T]$ uniformly distributed lifetime L.*

    Proof The system of equations (13.8) is obviously equivalent to the following one:

$$g_0(0, t_1) = g_1(t_1, t_2) = \cdots = g_{n-1}(t_{n-1}, T). \tag{13.15}$$

Hence, as we have seen in example 13.3, $v(t_1^*) < n^*c$, where n^* is the largest integer n satisfying (13.9). Therefore, a policy P_m with $m > n^*$ can never be the minimax policy, since

$$K(P_{n^*}^*) - K(P_m) \le v(t_1^*) + c - mc < n^*c - (m-1)c \le 0.$$

Now we show that if $P'_n = \{t'_k\}$, $0 < n \leq n^*$ are solutions of (13.15) and P_n is any policy with n inspections, then there is $K(P'_n) \leq K(P_n)$. Let us assume that there is a policy $\overline{P}_n = \{\overline{t}_k\}$ different from P'_n with property

$$K(\overline{P}_n) < K(P'_n). \tag{13.16}$$

Then let $\overline{t}_j$ be the smallest of the $\overline{t}_k$ with $\overline{t}_j \neq t'_j$. But it can only be $\overline{t}_j < t'_j$, since $\overline{t}_j > t'_j$ would imply $K(\overline{P}_n) \geq g_{j-1}(\overline{t}_{j-1}, \overline{t}_j) > g_{j-1}(t'_{j-1}, t'_j) = K(P'_n)$, contradictory to the assumption (13.16). Similarly, we conclude that $\overline{t}_{j+1} < t'_{j+1}$, since $\overline{t}_{j+1} > t'_{j+1}$ would again lead to a contradiction to (13.16). Proceeding in this way, we finally obtain $\overline{t}_{n-1} < t'_{n-1}$ so that $K(\overline{P}_n) \geq g_{n-1}(\overline{t}_{n-1}, T) > g_{n-1}(t'_{n-1}, T) = K(P'_n)$, contradictory to (13.16). Hence there is no policy $\overline{P}_n$ which yields smaller "minimax costs" than P'_n. To complete the proof, the intuitively obvious fact that $K(P'_n) \leq K(P'_{n-1})$, $n = 1, 2, ..., n^*$, can be shown analogously.

In particular, for $v(t) = at$ the minimax policy is given by (13.10), and the corresponding minimax costs are

$$K(P^*_{n^*}) = \frac{aT}{n^*} + \frac{n^* + 1}{2} c,$$

where n^* is the largest integer satisfying (13.9). This is the result of *Derman* [103] given that an inspection detects a failure with probability $p = 1$.

Example 13.7 *Let $v(t) = \sqrt{t}$, $t \geq 0$, $c = 1$ and $T = 100$. By (13.9), n^* is the largest integer satisfying*

$$\sum_{k=0}^{n-1} k^2 < 100.$$

Hence, $n^* = 7$. The corresponding system of equations for the minimax policy (13.15) can be written as

$$\sqrt{t^*_{k+1} - t^*_k} = \sqrt{t^*_1} - 1, \quad k = 1, 2, ..., 7 \quad \text{with } t^*_7 = 100.$$

The solution is

$$t^*_1 = 38.5, \ t^*_2 = 65.5, \ t^*_3 = 83.1, \ t^*_4 = 93.3, \ t^*_5 = 98.2, \ t^*_6 = 99.6.$$

The minimax cost is $K(P^*_7) = 7.25$.

Leung [200] constructs a minimax inspection policy based on the model of *Keller* [176]. In addition, he provides a fairly comprehensive survey of papers on cost-optimal inspection (many of them are not cited here). Another, more recent, survey is given by *Osaki* [245].

13.2.3 Partially Known Lifetime Distribution

Let us now assume that the only piece of information we have about the probability distribution of the system lifetime L is its expected value $\mu = \mathbb{E}(L) < \infty$. In addition, we will assume that $\sup\{t, \ F(t) < 1\} = \infty$. Thus, we know that the distribution function F of L is a member of the family

$$\mathbf{F}_\mu = \left\{ F \in \mathbf{F}, \ \sup\{t, \ F(t) < 1\} = \infty, \ \mu = \int_0^\infty (1 - F(t))\, dt \right\}.$$

In this case its makes sense to consider the cost criterion

$$K_\mu(P) = \max_{F \in \mathbf{F}_\mu} K(P, F).$$

An inspection policy P_μ^* minimizing $K_\mu(P)$ is called a *partial minimax (inspection) policy:*

$$K_\mu(P_\mu^*) = \min_P \max_{F \in \mathbf{F}_\mu} K(P, F).$$

Let $\mathbf{F}_\mu^2$ be that subset of $\mathbf{F}_\mu$, the elements of which have only two points of increase. Then, as a corollary of a more general result of *Hoeffding* [156],

$$\max_{F \in \mathbf{F}_\mu} K(P, F) = \max_{F \in \mathbf{F}_\mu^2} K(P, F).$$

After some elementary calculations, this relationship leads to the following representation of $K_\mu(P)$:

$$K_\mu(P) = \max_{\{0 \le r \le m < \ s\}} G_{rs}(P)$$

with $P = \{t_k\}$, $t_m \le \mu < t_{m+1}$, and

$$G_{rs}(P) = \frac{t_s - \mu}{t_s - t_r} g_r(t_r, t_{r+1}) + \frac{\mu - t_r}{t_s - t_r} g_s(t_s, t_{s+1}).$$

For a strictly periodic inspection policy $P = P^{(\tau)}$ we have for all pairs (r, s) with $0 \le r \le m < s$ always $G_{rs}(P^{(\tau)}) = c + v(\tau) + c\mu/\tau$ so that

$$K_\mu(P^{(\tau)}) = c + v(\tau) + c\mu/\tau. \tag{13.17}$$

The inspection interval τ_μ^* of the strictly periodic policy $P^{(\tau)}$, which minimizes $K_\mu(P^{(\tau)})$, is therefore a solution of the equation

$$\tau^2 v'(\tau) = \mu c. \tag{13.18}$$

By assumption, $v(t)$ is strictly increasing. Hence, there exists a positive solution of (13.18). For $v(t) = at$, the solution is $\tau_\mu^* = \sqrt{\mu c/a}$ so that

$$K_\mu(P^{(\tau_\mu^*)}) = 2\sqrt{\mu a c} + c. \tag{13.19}$$

Theorem 13.8 *(Beichelt [23]) The partial minimax inspection policy P_μ^* is strictly periodic. Its inspection interval τ_μ^* is a solution of (13.18).*

Proof. Let us assume there exists a policy $P = \{t_k\}$ satisfying

$$K_\mu(P) < K_\mu(P^{(\tau_\mu^*)}). \tag{13.20}$$

With this P and the corresponding inspection intervals $\tau_k = t_{k+1} - t_k$ we write $G_{rs}(P)$ in the form

$$G_{rs}(P) = (r+1)c + v(\tau_r) + (\mu - t_r)cw_{rs}, \quad 0 \le r \le m < s,$$

where

$$w_{rs} = \frac{(s-r)c + v(\tau_s) - v(\tau_r)}{(t_s - t_r)c}.$$

Let

$$w_r = \sup_{\{s,\ s>m\}} w_{rs}, \text{ and } w = \min_{\{r,\ 0 \le r \le m\}} w_r.$$

Without loss of generality, P can be assumed to be such that $0 < w < \infty$. (Otherwise, $G_{rs}(P)$ would be infinite, whereas a strictly periodic policy always yields a finite G_{rs} for all pairs (r,s), $0 \le r \le m < s$. The case $w > 0$ is evident anyway.) With $\tau = 1/w$, formulas (13.17) and (13.20) yield for all $0 \le r \le m$

$$(r+1)c + v(\tau_r) + (\mu - t_r)cw < c + v(1/w + \mu cw.$$

Starting with $r = 0$, we successively conclude that

$$\tau_r < 1/w, \quad 0 \le r \le m. \tag{13.21}$$

Let r_0 be defined by $w_{r_0} = w$. Then we have by definition of w_r,

$$\frac{(s-r_0)c + v(\tau_s) - v(\tau_{r_0})}{(t_s - t_r)c} \le w, \quad s > m.$$

Using (13.21), it can be seen by induction that $\tau_s < \tau_{r_0}$, $s > m$. But this implies

$$w = \sup_{\{s,\ s>m\}} w_{r_0 s} \ge 1/\tau_{r_0},$$

contradictory to (13.21). Hence, a policy P satisfying (13.20) cannot exist. This proves the theorem. Moreover, the proof shows that any partial minimax policy must be strictly periodic. Its inspection interval is a solution of Equation (13.18). ∎

Partial minimax inspection policies have been obtained on condition that one or two percentiles of the lifetime distribution are known, see *Roeloffs* [257] and *Kander and Raviv* [168] for linear loss function, and *Beichelt* [24] for arbitrary loss function. *Koh* [184] determined the partial minimax policy with the expected lifetime known on condition $T < \infty$.

13.3 Inspection with Replacement

13.3.1 Basic Model

An obvious extention of the model considered in the previous section is to replace the system with a new one as soon as a failure is detected [69]. In what follows, we will consider this situation under the following assumptions:
1) The system is replaced by an equivalent new one in a constant time d.
2) The replacement cost is c_e.
3) The inspection-replacement process is maintained unlimitedly with the same inspection policy $P = \{t_k\}$ after every replacement. (The time point, at which the replacement of a failed system is accomplished, becomes the new origin $t = 0$.)
4) The loss function is linear: $v(t) = at$. This assumption allows us to make use of results of the previous section.
5) If μ is the expected system lifetime, then

$$(c + c_e)/\mu < a. \tag{13.22}$$

Otherwise, the loss cost per unit downtime of the system a would be smaller or equal than the expected loss cost per unit time which arises by "ideal inspection and replacement", i.e. a failure of the system is immediately detected and replacement takes only a negligibly small time. But then inspection and replacement would be uneconomical anyway.

Under these assumptions, the expected (long-run) loss cost per unit time has structure

$$Q(P, F) = \frac{U(P, F)}{V(P, F)} \tag{13.23}$$

where $U(P, F)$ are the expected loss cost per replacement cycle, and $V(P, F)$ is the expected length Y of a replacement cycle (time span between the begin of the operation of new systems). Thus, $U(P, F) = K(P, F) + ad + c_e$, where $K(P, F)$ is given by (13.1), and $V(P, F) = Z(P, F) + d$ with $Z(P, F)$ being the expected time till the detection of a failure in a cycle:

$$Z(P, F) = \sum_{k=0}^{\infty} t_{k+1} \left[F(t_{k+1}) - F(t_k) \right]. \tag{13.24}$$

To be able to characterize the optimal inspection policy with regard to $Q(P, F)$, let for a constant λ

$$D(P, F, \lambda) = U(P, F) - \lambda V(P, F).$$

Theorem 13.9 For all λ, $0 \leq \lambda \leq a$, there exist a unique inspection policy P_λ with property $D(P_\lambda, F, \lambda) = \min_P D(P, F, \lambda)$. If, in addition, for a $\lambda = \lambda^*$, $D(P_{\lambda^*}, F, \lambda^*) = 0$, then P_{λ^*} is the only optimal inspection policy:

$$Q(P_{\lambda^*}, F) = \min_P Q(P, F).$$

Proof. The equation $D(P_{\lambda^*}, F, \lambda^*) = 0$ yields

$$\lambda^* = \frac{U(P_{\lambda^*}, F)}{V(P_{\lambda^*}, F)} = Q(P_{\lambda^*}, F). \qquad (13.25)$$

According to the definition of P_{λ^*}, we have for all P the relationship $0 = D(P_{\lambda^*}, F, \lambda^*) \leq D(P, F, \lambda^*)$. Hence,

$$\lambda^* \leq \frac{U(P, F)}{V(P, F)} = Q(P, F).$$

From this and (13.25) we get $\lambda^* = Q(P_{\lambda^*}, F) \leq Q(P, F)$ for all P. To finish the proof, let us assume that there is a policy P_0, $P_0 \neq P_{\lambda^*}$, so that

$$Q(P_0, F) = \min_P Q(P, F). \qquad (13.26)$$

Then define λ_0 by $D(P_0, F, \lambda_0) = 0$. By definition of P_λ, $D(P_{\lambda_0}, F, \lambda_0) < D(P_0, F, \lambda_0) = 0$. This implies $Q(P_{\lambda_0}, F) < \lambda_0 = Q(P_0, F)$, contradictory to (13.26). $D(P, F, \lambda)$ is easily verified to be

$$D(P, F, \lambda) = \sum_{k=0}^{\infty} \int_{t_k}^{t_{k+1}} [(k+1)c + (a - \lambda)(t_{k+1} - t)]dF(t)$$
$$+ (a - \lambda)d + c_e - \mu\lambda.$$

With regard to its dependence on P, $D(P, F, \lambda)$ has for $0 \leq \lambda < a$ the same structure as $K(P, F)$. Hence, by theorem 13.2, there exists P_λ, and the same theorem (see also comments made after it) can be used to determine P_λ if the role of (13.5) is taken over by the system of equations

$$t_{k+1} - t_k = \frac{F(t_k) - F(t_{k-1})}{f(t_k)} - \frac{c}{a - \lambda}, \quad 0 \leq \lambda < a, \ k = 1, 2, \ldots$$

Obviously, $D(P_0, F, 0) > 0$, and, by assumption (13.22), $D(P_a, F, a) = c + c_e - a\mu < 0$. Since $D(P_\lambda, F, \lambda)$ is a continuous, strictly decreasing function in λ, $0 \leq \lambda \leq a$, there exists exactly one $\lambda = \lambda^*$ with property $D(P_{\lambda^*}, F, \lambda^*) = 0$. ∎

Theorem 13.9 implies the following algorithm for determining the optimal inspection policy P_{λ^*} with regard to $Q(P, F)$:

Algorithm
1. For a λ with $0 < \lambda < a$ determine P_λ.
2.1 If $D(P_\lambda, F, \lambda) > 0$, increase λ and go to 1.
2.2 If $D(P_\lambda, F, \lambda) < 0$, decrease λ and go to 1.
2.3 If $D(P_\lambda, F, \lambda) = 0$, then P_λ is optimal.

Partial Minimax Inspection Policy

The partial minimax policy P_μ^* for known expected value μ is defined as

$$Q(P_\mu^*) = \min_P \max_{F \in \mathbf{F}_\mu} Q(P, F).$$

It can be determined by using $D(P, \lambda) = \max_{F \in \mathbf{F}_\mu} D(P, F, \lambda)$ and results obtained in Section 13.2.3 [26], [30]: P_μ^* is strictly periodic with inspection interval $\tau_{\lambda^*} = \sqrt{c\mu/(a - \lambda^*)}$, $0 \le \lambda^* < a$, where

$$\lambda^* = Q(P_\mu^*) = \frac{1}{\mu + d} \left\{ c + ad + c_e - \frac{2\mu c}{\mu + d} \right.$$

$$\left. +2 \sqrt{\frac{\mu c}{\mu + d} \left(a\mu - c - c_e + \frac{\mu c}{\mu + d} \right)} \right\}.$$

Partial minimax inspection policies for systems with replacement if one or two percentiles of $F(t)$ are known, have been obtained by *Roeloffs* [258] for linear and by *Beichelt* [25] for general loss functions. *Wang* [304] analyzed an inspection-replacement model with two types of inspections and repairs. Similarly, *Taghipour and Banjevic* [291] optimized the inspection and renewal of a system in case of minimal repairs or renewals after the detection of failures.

13.3.2   Maximum Availability under Cost Restrictions

Apart from minimizing the long-run cost per unit time $Q(P, F)$, the long-run availability of the system is of interest. Under the assumptions and with the notation of the previous section, it is given by

$$A(P, F) = \frac{\mu}{Z(P, F) + d}.$$

Surely, $A(P, F)$ tends to its maximum possible value $\mu/(\mu+d)$ if the lengths of all inspection intervals tend to 0. But in this case the loss cost per replacement cycle (not taking into account c_e) $K(P, F)$ would tend to infinity. Hence, to compromise, the solution of the following problem is important: An inspection policy P^* has to be found so that

$$A(P^*, F) = \max_{P \in \mathbf{P}_0} A(P, F),$$

where, for a given positive cost limit K_0 with $\min_P K(P, F) \le K_0$,

$$\mathbf{P}_0 = \{P, \ K(P, F) = K_0\}.$$

This problem can be written in the equivalent form

$$\begin{aligned} Z(P, F) &\to \min \\ K(P, F) - K_0 &= 0 \end{aligned} \tag{13.27}$$

It is obvious to solve this problem by the Lagrange multiplier method. Subjected to our problem, we formulate its principle in the following lemma.

Lemma 13.10 *Let* $M(P, F, \lambda) = Z(P, F) + \lambda[K(P, F) - K_0]$. *For any non-negative constant* λ *there exist an inspection policy* $P(\lambda)$ *with property*

$$M(P(\lambda), F, \lambda) = \min_{P} M(P, F, \lambda).$$

If there is a $\lambda^* \in \Lambda = \{\lambda, \ P(\lambda) \in \mathbf{P}_0\}$ *so that* $M(P(\lambda^*), F, \lambda^*) = \min_{\lambda \in \Lambda} M(P(\lambda), F, \lambda)$, *then* $P^* = P(\lambda^*)$ *is solution of (13.27)*.

From (13.1) and (13.24) we get

$$M(P, F, \lambda) = \sum_{k=0}^{\infty} \int_{t_k}^{t_{k+1}} [(k+1)\lambda c + (1 + \lambda a)(t_{k+1} - t)]dF(t) + \mu - K_0.$$

Since $M(P, F, \lambda)$ and $K(P, F)$ have the same functional structures, for any $\lambda \geq 0$ a unique $P(\lambda)$ exists and can be determined by theorem 13.2 if the role of (13.5) is taken over by

$$t_{k+1} - t_k = \frac{F(t_k) - F(t_{k-1})}{f(t_k)} - \frac{\lambda c}{1 + \lambda a}, \quad k = 1, 2, \dots$$

As proved in [27], $K(S(\lambda), F)$ is a continuous, strictly decreasing function in λ with $\lim_{\lambda \to 0} K(P(\lambda)) = \infty$ and $\lim_{\lambda \to \infty} K(P(\lambda), F) = \min_{P} K(P, F)$. Hence, there exists exactly one $\lambda = \lambda^*$ with property $K(P(\lambda^*), F) = K_0$. Thus, we have the following algorithm for determining the solution $P(\lambda^*)$ of (13.27):

Algorithm
1. Select $\lambda > 0$ and calculate $P(\lambda)$.
2. If $K(P(\lambda), F) < K_0$, decrease λ and go to 1.
3. If $K(P(\lambda), F) > K_0$, increase λ and go to 1.
4. If $K(P(\lambda), F) = K_0$, $P(\lambda)$ is the solution of (13.27).

Example 13.11 *Let* $F(t) = 1 - e^{-t/\mu}$, $t \geq 0$. *Then, as pointed out in example 13.4, the optimal policy with regard to* $K(P, F)$, *and, hence, with regard to* $M(P, F, \lambda)$ *as well, is strictly periodic. From (13.13), the inspection interval* τ^* *of the solution* $P(\lambda^*) = P^{(\tau^*)}$ *of (13.27) is, therefore, determined by the equation*

$$\frac{c + a\tau}{1 - e^{-\tau/\mu}} - a\mu = K_0. \tag{13.28}$$

There exist two solutions of (13.28). (Mind that $\min_{\tau} K(P^{(\tau)}, F) < K_0$.) *Since* $Z(P, F) = \tau/(1 - e^{-\tau/\mu})$ *is increasing in* τ, τ^* *is the smallest one.*

Remark 13.12 *Different to the previous sections, problem (13.27) also makes sense if $a = 0$.*

Remark 13.13 *The condition $K(P, F) - K_0 = 0$ can be replaced with $Q(P, F) = Q_0$, $Q_0 < a$, since the latter is equivalent to $U(P, F) - Q_0 V(P, F) = 0$.*

Partial minimax solution If the expected value μ is the only information about the lifetime distribution of the system, then we consider instead of (13.27) the following problem [30]:

$$Z(P) = \max_{F \in \mathbf{F}_\mu} Z(P, F) \to \min$$
$$K(P) - K_0 = \max_{F \in \mathbf{F}_\mu} K(P, F) - K_0 = 0, \tag{13.29}$$

where now K_0 is the minimax cost limit, which, by (13.19), must satisfy

$$2\sqrt{\mu a c} + c < K_0. \tag{13.30}$$

Based on
$$M(P, \lambda) = \max_{F \in \mathbf{F}_\mu} M(P, F, \lambda)$$

one can show that the policy P_{μ, K_0}, which is solution of (13.29), is strictly periodic with inspection interval

$$\tau^*_{\mu, K_0} = \frac{w}{2} - \sqrt{\frac{w^2}{2} - \frac{\mu c}{a}} \quad \text{with} \quad w = (K_0 - c)/a. \tag{13.31}$$

In view of (13.30), the radicand in (13.31) is positive. In case of $a = 0$, the inspection interval (13.31) is not applicable. It has to be replaced with

$$\tau^*_{\mu, K_0} = \sqrt{\mu c/(K_0 - c)}.$$

Biswas et al. [53] determined the availability of a periodically inspected system with replacement subjected to an imperfect repair policy.

13.4   Exercises

Exercise 13.1 *Let L be uniformly distributed over the interval $[0, 100]$, $c = 5$, and $v(t) = \sqrt{t}$, $t \geq 0$. Determine the cost-optimal inspection policy P^* and compare $K(P^*, F)$ to $K(P^{\tau^*}, F)$, where P^{τ^*} is the optimal strictly periodic inspection policy.*

Exercise 13.2 *Let L have an exponential distribution with distribution function $F(t) = 1 - e^{t/100}$, $c = 5$, and $v(t) = \sqrt{t}$, $t \geq 0$. Determine the optimal inspection policy.*

Exercise 13.3 *Verify formula (13.17).*

Exercise 13.4 *Given $c = 1$ and $v(t) = t$, what is the partial minimax inspection policy with regard to an unknown expected lifetime $\mu = E(L)$ if L has distribution function $F(t) = 1 - e^{t/\mu}$, $t \geq 0$?*

Bibliography

[1] J. A. Abraham. An improved algorithm for network reliability. *IEEE Transactions on Reliability*, R-28:58–61, April 1979.

[2] Martin Aigner. *Combinatorial Theory*. Springer Verlag, 1979.

[3] Saieed Akbari, Saeid Alikhani, and Yee-Hock Peng. Characterization of graphs using domination polynomials. *European Journal of Combinatorics*, 31:1714–1724, 2010.

[4] Ian Anderson. *Combinatorics of Finite Sets*. Dover Publications, Mineola, 1987.

[5] George E. Andrews. *The Theory of Partitions*. Cambridge University Press, 1984.

[6] Bernhard Anrig and Frank Beichelt. Disjoint sum forms in reliability theory. *South African Journ. of Operations Research*, 16:75–86, 2000.

[7] Dan Archdeacon, Charles J. Colbourn, Isidoro Gitler, and J. Scott Provan. Four-terminal reducibility and projective-planar wye-delta-wye-reducible graphs. *Journal of Graph Theory*, 33:83–93, 2000.

[8] Jorge Luis Arocha and Bernardo Llano. Mean value for the matching and dominating polynomials. *Discussiones Mathematicae Graph Theory*, 20:57–69, 2000.

[9] Alexandru O. Balan and Lorenzo Traldi. Preprocessing minpaths for sum of disjoint products. *IEEE Transactions on Reliability*, 52(3):289–295, September 2003.

[10] Michael O. Ball. Complexity of network reliability computation. *Networks*, 10:153–165, 1980.

[11] Michael O. Ball, Charles J. Colbourn, and J. Scott Provan. Network reliability. Technical Report TR 92-74, University of Maryland, Systems Research Center, June 1992.

[12] Michael O. Ball and J. Scott Provan. Calculating bounds on reachability and connectedness in stochastic networks. *Networks*, 13(2):253–278, 1983.

[13] Michael O. Ball and J. Scott Provan. Disjoint products and efficient computation of reliability. *Operations Research*, 36(5):703–715, Sept/Oct 1988.

[14] Jørgen Bang-Jensen and Gregory Z. Gutin. *Digraphs: Theory, Algorithms and Applications*. Springer, 2 edition, 2009.

[15] R.E. Barlow and L.C. Hunter. Mathematical models for system reliability. Engineering Report EDL-E35, Sylvania Electric Products Inc., Mountain View, California, August 1959.

[16] R.E. Barlow, L.C. Hunter, and F. Proschan. Optimum checking procedures. *Journ. Soc. Industr. Appl. Mathem.*, 4:1078–1095, 1963.

[17] Richard Barlow and Larry Hunter. Optimum preventive maintenance policies. *Operations Research*, 8(1):90–96, 1960.

[18] Richard E. Barlow. *Engineering Reliability*. ASA-SIAM Series on Statistics and Applied Probability 2. SIAM, 1998.

[19] Richard E. Barlow and Frank Proschan. *Statistical Theory of Reliability and Life Testing: Probability Models*. To Begin With, 1981.

[20] F. Beichelt and K. Fischer. On a basic equation of reliability theory. *Microelectronics and Reliability*, 19:367–369, 1979.

[21] F. Beichelt and A. Kröber. Maintenance policies with time-dependent repair cost limits. *Economic Quality Control-Journal and Newsletter for Quality and Reliability*, 7:79–84, 1992.

[22] Frank Beichelt. Optimale Inspektionsstrategien bei beliebiger Verlustfunktion. *Biometrische Zeitschrift*, 13:384–395, 1971.

[23] Frank Beichelt. Optimale Inspektionsstrategien bei bekanntem Erwartungswert der Lebenszeit. *Zeitschrift für Angew. Mathem. und Mechanik (ZAMM)*, 52:431–433, 1972.

[24] Frank Beichelt. Über eine Klasse von Inspektionsmodellen der Zuverlässigkeitstheorie. *Mathematische Operationsforschung und Statistik*, 5:281–298, 1974.

[25] Frank Beichelt. Minimax inspection strategies for replaceable systems with partial information on the lifetime distribution. *Mathematische Operationsforschung und Statistik*, 6:479–492, 1975.

[26] Frank Beichelt. A general preventive maintenance policy. *Mathematische Operationsforschung und Statistik*, 7:927–932, 1976.

[27] Frank Beichelt. Maximum availability under cost restrictions. *Zeitschrift für Angewandte Mathematik und Mechanik (ZAMM)*, 57:77–78, 1977.

[28] Frank Beichelt. *Effektive Planung prophylaktischer Maßnahmen in der Instandhaltung*. Verlag Technik, Berlin, 1978.

[29] Frank Beichelt. A new approach to repair cost limit replacement policies. In *Transactions of the 8th Prague Conference on Information Theory, Statistical Decision Functions, and Random Processes*, volume C, pages 31–37, 1978.

[30] Frank Beichelt. The availability of replaceable single unit systems in case of cost restrictions and partial information on the lifetime distribution. *Mathematische Operationsforschung und Statistik, Ser. Optimization*, 12:453–461, 1981.

[31] Frank Beichelt. Minimax inspection strategies for single unit systems. *Naval Research Logistics Quarterly*, 28:375–381, 1981.

[32] Frank Beichelt. A replacement policy based on limits for the repair cost rate. *IEEE Transactions on Reliability*, R-31:401–403, 1982.

[33] Frank Beichelt. Zuverlässigkeitstheoretische Importanz von Elementen stochastischer Netzstrukturen. *Nachrichtentechnik/Elektronik*, 37:342–343, 1987.

[34] Frank Beichelt. A unifying treatment of replacement policies with minimal repair. *Naval Research Logistics*, 40:51–67, 1993.

[35] Frank Beichelt. *Zuverlässigkeits- und Instandhaltungstheorie*. B.G. Teubner Stuttgart, 1993.

[36] Frank Beichelt. Total repair cost limit replacement policies. *South African Journal of Operations Research (ORION)*, 13:37–44, 1997.

[37] Frank Beichelt. A replacement policy based on limiting the cumulative maintenance cost. *Internat. Journ. on Reliability and Quality Management*, 18:76–83, 2001.

[38] Frank Beichelt. *Stochastic Processes in Science, Engineering and Finance*. CRC Press, 2006.

[39] Frank Beichelt. Advances in repair cost limit replacement policies. *Communications in Dependability and Quality Management*, 13:5–13, 2010.

[40] Frank Beichelt. Repair cost limit maintenance policies and random marked point processes. In M. Kempf, S. Rommel, J.E. Fernandez, and Subramanian A., editors, *Proc. of the 16th Annual Int. Conf. on Industrial Engineering-Theory, Applications and Practice*, pages 57–62, 2011.

[41] Frank Beichelt and Klaus Fischer. General failure model applied to preventive maintenance policies. *IEEE Transactions on Reliability*, R-29:39–41, 1980.

[42] Frank Beichelt and Peter Franken. *Zuverlässigkeit und Instandhaltung (Reliability and Maintenance)*. Verlag Technik, Berlin, 1983.

[43] Frank Beichelt and Lutz Sproß. An effective method for reliability analysis of complex systems. *J. Inf. Process. Cybern.*, 23:227–235, 1987.

[44] Frank Beichelt and Lutz Sproß. Bounds on the reliability of binary coherent systems. *IEEE Transactions on Reliability*, R-38:425–427, 1989.

[45] Frank Beichelt and Lutz Sproß. Comment on "An improved Abraham-method for generating disjoint sums'.'. *IEEE Trans. Reliab.*, 38(4):422–424, 1989.

[46] Frank Beichelt and Peter Tittmann. A generalized reduction method for the connectedness probability of stochastic networks. *IEEE Transactions on Reliability*, 40(2):198–204, June 1991.

[47] M. Bienvenu Berg, M. and R. Cleroux. Age-replacement policy with age-dependent minimal repair. *INFOR*, 24:26–32, 1986.

[48] Bo Bergman. On some new reliability importance measures. In J. Quirk, editor, *Proceedings of the IFAC SAFECOMP'85*, pages 61–64, 1985.

[49] Daniel Bienstock. Some lattice-theoretic tools for network reliability analysis. *Mathematics of Operations Research*, 13(3):467–478, 1988.

[50] Z.W. Birnbaum and S.C. Saunders. A new family of life distributions. *Journal of Appied Probability*, 6:319–327, 1969.

[51] Zygmund William Birnbaum. On the importance of different components in a multicomponent system. In P. R. Krishnaiah, editor, *Multivariate Analysis II*, pages 581–592. Academic Press, New York, 1969.

[52] Zygmund William Birnbaum, James D. Esary, and S.C. Saunders. Multicomponent systems and structures and their reliability. *Technometrics*, 3:55–77, 1961.

[53] J. Biswas, A., Sarkar, and S. Sarkar. Availability of a periodically inspected system, maintained under an imperfect repair policy. *IEEE Transactions on Reliability*, 52:311–318, 2003.

[54] H.W. Block, W.S. Borges, and T.H. Savits. Age-dependent minimal repair. *Journal of Applied Probability*, 22:370–385, 1985.

[55] H.W. Block, W.S. Borges, and T.H. Savits. A general age replacement model with minimal repair. *Naval Research Logistics*, 35:365–372, 1988.

[56] Hans L. Bodlaender. A partial k-arboretum of graphs with bounded treewidth. *Theoretical Computer Science*, 209:1–45, 1998.

[57] Hans L. Bodlaender and Arie M. C. A. Koster. On the Maximum Cardinality Search lower bound for treewidth. In J. Hromkovič, M. Nagl, and B. Westfechtel, editors, *Proc. 30th International Workshop on Graph-Theoretic Concepts in Computer Science WG 2004*, pages 81–92. Springer-Verlag, Lecture Notes in Computer Science 3353, 2004.

[58] Hans L. Bodlaender, Arie M. C. A. Koster, and Thomas Wolle. Contraction and treewidth lower bounds. Technical Report UU-CS-2004-34, Dept. of Computer Science, Utrecht University, Utrecht, The Netherlands, 2004.

[59] Hans L. Bodlaender and Thomas Wolle. A note on the complexity of network reliability problems. Technical Report UU-CS-2004-001, Utrecht University, Department of Information and Computing Sciences, 2004.

[60] Frank T. Boesch, Appajosyula Satyanarayana, and Charles L. Suffel. On residual connectedness network reliability. Reliability of computer and communication networks, Proc. Workshop, New Brunswick, NJ/USA 1989, DIMACS Ser. Discret. Math. Theor. Comput. Sci. 5, 51-59 (1991)., 1991.

[61] F.T. Boesch, A. Satyanarayana, and C.L. Suffel. A survey of some network reliability analysis and synthesis results. *Networks*, 54:99–107, 2009.

[62] P.J. Boland and Samaniego. The signature of a coherent system and its applications in reliability. In *Mathematical Reliability: An Expository Perspective*, International Series in Operations Research & Management Science, chapter 1, pages 1–29. Kluwer Academic Publishers, Boston, 2004.

[63] Béla Bollobás. *Extremal Graph Theory*. Academic Press, London, 1978.

[64] Adrian Bondy and U.S.R. Murty. *Graph Theory*. Springer, 2008.

[65] Ulrik Brandes. A faster algorithm for betweenness centrality. *The Journal of Mathematical Sociology*, 25(2):163–177, 2001.

[66] Ulrik Brandes and Thomas Erlebach. *Network Analysis*. Number 3418 in LNCS. Springer, 2005.

[67] Andreas Brandstädt, Van Bang Le, and Jeremy P. Spinrad. *Graph Classes: A Survey*. SIAM Monographs on Discrete Mathematics and Applications. Society for Industrial and Applied Mathematics, 1999.

[68] Timothy B. Brecht and Charles J. Colbourn. Multiplicative improvements in network reliability bounds. *Networks*, 19:521–529, 1989.

[69] D.N. Brender. A surveillance model for recurrent events. Technical report, IBM Watson Research Center, 1963.

[70] Gunnar Brinkmann and Bradan McKay. Posets on up to 16 points. *Order*, 19:147–179, 2002.

[71] J. I. Brown and Xiaohu Li. The strongly connected reliability of complete digraphs. *Networks*, 45(3):165–168, May 2005.

[72] M. Brown and F. Proschan. Imperfect repair. *Journ. Appl. Probability*, 20:851–859, 1983.

[73] Adam L. Buchsbaum and Milena Mihail. Monte Carlo and Markov chain techniques for network reliability and sampling. *Networks*, 25(3):117–130, 1995.

[74] Fred Buckley and Frank Harary. *Distances in Graphs*. Addison-Wesley, 1990.

[75] Stéphane Bulteau and Gerardo Rubino. Evaluating network vulnerability with mincuts frequency vector. Technical Report 1089, Institut de Recherche en Informatique et Systèmes Aléatoires, Campus de Beaulieu, France, 1997.

[76] D.A. Butler. A complete ranking for components of binary coherent systems with extensions to multistate systems. *Naval Research Logistics Quarterly*, 26:565–578, 1979.

[77] John A. Buzacott and J.S.K. Chang. Cut set intersections and node partitions. *IEEE Transactions on Reliability*, R-33:385–389, 1984.

[78] H. Cancela and M. E. Khadiri. A recursive variance reduction algorithm for estimating communication network reliability. *IEEE Transactions on Reliability*, 44(4):595–602, 1995.

[79] H. Cancela and M. E. Khadiri. A series-parallel reductions in Monte Carlo network-reliability. *IEEE Transactions on Reliability*, 47(2):159–164, 1998.

[80] Héctor Cancela and Louis Petingi. On the characterization of the domination of a diameter-constrained network reliability model. *Discrete Appl. Math.*, 154(13):1885–1896, 2006.

[81] Bernard Carré. *Graphs and Networks.* Oxford Applied Mathematics & Computing Science Series. Clarendon Press, Oxford, 1980.

[82] Arthur Cayley. A theorem on trees. *Quart. J. Math.*, 23:376–378, 1889.

[83] J.-K. Chan and L. Shaw. Modeling repairable systems with failure rates that depend on age and maintenance. *IEEE Trans. on Reliability*, 42:566–571, 1993.

[84] P.K.W. Chan and T. Downs. Two criteria for preventive maintenance. *IEEE Trans. on Reliability*, 27:272–273, 1978.

[85] C.C. Chang, S.H. Sheu, and Y.L. Chen. Optimal number of minimal repairs before replacement on a cumulative repair-cost limit policy. *Computers & Industrial Engineering*, 59:603–610, 2010.

[86] Manoj K. Chari, Thomas A. Feo, and J. Scott Provan. The delta-wye approximation procedure for two-terminal reliability. *Operations Research*, 44:745–757, 1996.

[87] A. A. Chernyak. Residual reliability of P-threshold graphs. *Discrete Applied Mathematics*, 135(1-3):83–95, January 2004.

[88] D.T. Chiang and S. Niu. Reliability of a consecutive k-out-of-n : F system. *IEEE Transactions on Reliability*, R-35:65–67, 1981.

[89] Y.H. Chien, C.C. Chang, and S.H. Sheu. Optimal age replacement model with age-dependnet type of failureand random lead time based on a cumulative repair-cost limit policy. *Annals Oper. Research*, 181:723–744, 2010.

[90] Y.H. Chien and J.A. Chen. Optimal age-replacement model with minimal repair based on a cumulative repair cost limit and random lead time. In *Proceedings of the 2007 IEEE IEEM*, pages 636–639, 2007.

[91] J.C.R. Clapham. Economic life of equipment. *Oper. Res. Quart.*, 8:181–190, 1957.

[92] Charles J. Colbourn. *The Combinatorics of Network Reliability.* Oxford University Press, 1987.

[93] Charles J. Colbourn. Edge-packing of graphs and network reliability. *Discrete Appl. Math.*, 72:49–61, 1988.

[94] Charles J. Colbourn. Analysis and synthesis problems for network resilience. *Math. Comput. Modelling*, 17(11):43–48, 1993.

[95] Charles J. Colbourn, Daryl D. Harms, and Wendy J. Myrvold. Reliability polynomials can cross twice. *Journal of the Franklin Institute*, 330(3):629–633, May 1993.

[96] Charles J. Colbourn, A. Satyanarayana, C. Suffel, and K. Sutner. Computing residual connectedness reliability for restricted networks. *Discrete Appl. Math.*, 44(1-3):221–232, 1993.

[97] Derek G. Corneil and Udi Rotics. On the relationship between clique-width and treewidth. *SIAM Journal on Comput.ing*, 34:825–847, 2005.

[98] B. Courcelle, J. Engelfriet, and G. Rozenberg. Handle-rewriting hyper-graph grammers. *Journal of Computer and System Sciences*, 46:218–270, 1993.

[99] Bruno Courcelle and Stephan Olariu. Upper bounds to the clique width of graphs. *Discrete Applied Mathematics*, 101:77–114, 2000.

[100] Richard Cox. *Renewal Theory*. London: Methuen, 1962.

[101] Margaret B. Cozzens and Shu-Shih Y. Wu. Bounds of edge-neighbor-integrity of graphs. *Australasian Journal of Combinatorics*, 15:71–80, 1997.

[102] J. S. Dagpunar. Renewal-type equations for a general repair process. *Quality and Reliability Engineering International*, 13:235–245, 1997.

[103] C. Derman. On minimax surveillance schedules. *Naval Research Logistics Quarterly*, 8:415–419, 1961.

[104] Yefim Dinitz, Naveen Garg, and Michel X. Goemans. On the single-source unsplittable flow problem. *Combinatorica*, 19(1):17–41, 1999.

[105] Klaus Dohmen. Inclusion-exclusion and network reliability. *Electronic Journal of Combinatorics*, 5(R36):1–8, June 1998.

[106] Klaus Dohmen. *Improved Bonferroni Inequalities via Abstract Tubes*. Springer-Verlag, 2004.

[107] Klaus Dohmen and Peter Tittmann. Bonferroni-Galambos inequalities for partition lattices. *Electronic Journal of Combinatorics*, 11(1):9, 2004. Research paper R85.

[108] Klaus Dohmen and Peter Tittmann. Domination reliability. *The Electronic Journal of Combinatorics*, 19:1–15, 2012.

[109] Peter Doubilet. On the foundations of combinatorial theory. vii: Symmetric functions through the theory of distribution and occupancy. *Studies in Applied Mathematics*, 51:377–396, 1972.

[110] Rod G. Downey and Michael R. Fellows. *Parameterized Complexity*. Springer, 1999.

[111] R.W. Drinkwater and N.A.J. Hastings. An economic replacement model. *OR*, 18(2):121–138, 1967.

[112] M. Easton and C. Wong. Sequential destruction method for Monte Carlo evaluation of system reliability. *IEEE Transactions on Reliability*, 29(1):27–32, 1980.

[113] Tov Elperin, Ilya B. Gertsbakh, and Michael Lomonosov. Estimation of network reliability using graph evolution models. *IEEE Tranastions on Reliability*, 40:572–591, 1991.

[114] G.V. Epifanov. Reduction of a plane graph to an edge by star-triangle transformations. *Soviet Math. Doklady*, 166:13–17, 1966.

[115] S. Eryilmaz. Review of recent advances in reliability of consecutive k-out-of-n and related systems. *Proceedings of the Institution of Mechanical Engineers, Part O: Journal of Risk and Reliability*, 224(3):225–237, 2010.

[116] J.D. Esary and F. Proschan. Coherent structures of non-identical components. *Technometrics*, 5:191–209, 1963.

[117] J.D. Esary and F. Proschan. Relationships among some concepts of bivariate dependence. *The Annals of Mathematical Statistics*, 43(2):651–655, 1972.

[118] J.D. Esary, F. Proschan, and D.W. Walkup. Association of random variables with applications. *Ann. Math. Statist.*, 38:1466–1474, 1967.

[119] D.J.G. Farlie. The performance of some correlation coefficients for a general bivariate distribution. *Biometrika*, 47:307–323, 1960.

[120] William Feller. *An Introduction to Probability Theory and Its Applications*. John Wiley & Sons, third edition, 1968.

[121] Michael R. Fellows, Frances A. Rosamond, Udi Rotics, and Stefan Szeider. Clique-width is NP-complete. *SIAM Journal on Discrete Mathematics*, 23(2):909–939, 2009.

[122] Thomas A. Feo and J. Scott Provan. Delta-wye transformations and the efficient reduction of two-terminal planar graphs. *Operations Research*, 41(3):572–582, May-June 1993.

[123] J.K. Filus and L.Z. Filus. Some alternative approaches to system reliability modeling. In H. Pham, editor, *Recent Advances in Reliability and Quality Design*, pages 101–136. Springer, 2008.

[124] Maxim Finkelstein. *Failure Rate Modelling for Reliability and Risk*. Springer, 2008.

[125] M.S. Finkelstein. Modeling a process of non-ideal repair. In *Recent Advances in Reliability Theory*. Birkhäuser, 2000.

[126] G. Fishman. A comparison of four Monte Carlo methods for estimating the probability of $s-t$ connectedness. *IEEE Transactions on Reliability*, 35:145–155, 1986.

[127] George S. Fishman. *Monte Carlo: Concepts, Algorithms, and Applications*. Springer, 1996.

[128] Luigi Fratta and Ugo G. Montanari. Boolean algebra method for computing the terminal reliability in a communication network. *IEEE Transactions on Circuit Theory*, 20(3):203–211, 1973.

[129] Linton C. Freeman. A set of measures of centrality based on betweenness. *Sociometry*, 40(1):35–41, 1977.

[130] J.C. Fu and M.V. Koutras. Reliability bounds for coherent sructures with independent components. *Statistics and Probability Letters*, 22:137–148, 1995.

[131] J.P. Gadani. System effectiveness evaluation using star and delta transformations. *IEEE Tranastions on Reliability*, R-30:43–47, 1981.

[132] P. Gardent and L. Nonant. Entretien et renouvellement d'un parc de machines. *Revue Fr. Rech. Operat.*, 7:5–19, 1963.

[133] Michael R. Garey and David S. Johnson. *Computers and Intractability: A Guide to the Theory of NP-Completeness*. W.H. Freeman & Company, San Francisco, 1979.

[134] S. Gasmi, C.E. Love, and W. Kahle. A general repair, proportional-hazards framework to model complex repairable systems. *IEEE Trans. on Reliability*, 52:26–31, 2003.

[135] I. Gertsbakh and Y. Shpungin. *Models of Network Reliability: Analysis, Combinatorics, and Monte Carlo*. CRC Press, 2009.

[136] Ilya Gertsbakh. *Reliability Theory with Applications to Preventive Maintenance*. Springer Verlag, 2000.

[137] Ilya Gertsbakh and Yoseph Shpungin. *Network Reliability and Resilience*. SpringerBriefs in Electrical and Computer Engineering. Springer, Berlin Heidelberg, 2011.

[138] E.N. Gilbert. Random graphs. *The Annals of Mathematical Statistics*, 30:1141–1144, 1959.

[139] Omer Giménez, Petr Hlinený, and Marc Noy. Computing the Tutte polynomial on graphs of bounded clique-width. *SIAM J. Discrete Math.*, 20(4):932–946, 2006.

[140] W. Goddard and H.C. Swart. The integrity of a graph: Bounds and basics. *J. Combin. Math. Combin. Comput.*, 7:139–151, 1990.

[141] E.E. Gomory and T.C. Hu. Multi-terminal network flows. *Journal of the Society for Industrial and Applied Mathematics*, 9:551–570, 1961.

[142] Michel Gondran and Michel Minoux. *Graphs, Dioids and Semirings.* Springer, 2008.

[143] Valeri Gorlov and Peter Tittmann. A unified appoach to the reliability of recurrent structures. In Waltraud Kahle, Elart von Collani, Jürgen Franz, and Uwe Jensen, editors, *Advances in Stochastic Models for Reliability, Quality and Safety*, chapter 18, pages 261–273. Birkhäuser, 1998.

[144] Jonathan Gross and Jay Yellen. *Graph Theory and Its Applications.* CRC Press, Boca Raton, 1998.

[145] Jonathan L. Gross and Thomas W. Tucker. *Topological Graph Theory.* Dover Publications, 1987.

[146] Martin Grötschel, Clyde L. Monma, and Mechthild Stoer. Design of survivable networks. In Michael O. Ball, Thomas L. Magnanti, Clyde L. Monma, and George L. Nemhauser, editors, *Network Models*, volume 7 of *Handbooks in Operations Research and Management Science*. North-Holland, 1995.

[147] E.J. Gumbel. Bivariate exponential distributions. *Journal of the Americ. Statist. Assoc.*, 55:698–707, 1960.

[148] R. Guo, H. Ascher, and C.E. Love. Generalized models of repairable systems: A survey via stochastic process formalism. *South African Journ. of Operations Research (ORION)*, 16:87–128, 2000.

[149] R. Guo, H. Liao, W. Zhao, and A. Mettas. A new stochastic model for systems under general repairs. *IEEE Trans. on Reliability*, 56:40–49, 2007.

[150] Allan Gut. Convergence rates for record times and the associated counting process. *Stochastic Processes and their Applications*, 36:135–151, 1990.

[151] Mohammad Taghi Hajiaghayi and Mahdi Hajiaghayi. A note on the bounded fragmentation property and its applications in network reliability. *European Journal of Combinatorics*, 24:891–896, 2003.

[152] N.A.J. Hastings. The repair limit replacement method. *Operations Res. Quart.*, 20:337–349, 1969.

[153] Klaus Heidtmann. Inverting paths and cuts of two-state systems. *IEEE Transactions on Reliability*, 32(5):469–471, 1983.

[154] Klaus Heidtmann. Smaller sums of disjoint products by subproduct inversion. *IEEE Transactions on Reliability*, 38(3):305–311, August 1989.

[155] Klaus D. Heidtmann. Statistical comparison of two sum-of-disjoint-product algorithms for reliability and safety evaluation. In Stuart Anderson, Sandro Bologna, and Massimo Felici, editors, *SAFECOMP*, volume 2434 of *Lecture Notes in Computer Science*, pages 70–81. Springer, 2002.

[156] W. Hoeffding. The extrema of the expected value of a function of independent random variables. *Annals of Mathematical Statistics*, 26:268–275, 1955.

[157] Petter Holme, Beom Jun Kim, Chang No Yoon, and Seung Kee Han. Attack vulnerability of complex networks. *Physical Review E*, 65(5):056109, 2002.

[158] Arnljot Høyland and Marvin Rausand. *System Reliability Theory: Models and Statistical Methods*. Wiley Series in Probability and Mathematical Statistics. Wiley-Interscience, 1994.

[159] Toshiyuki Inagaki and Ernest J. Henley. Probabilistic evaluation of prime implicants and top-events for non-coherent systems. *IEEE Transactions on Reliability*, R-29(5):361–367, 1980.

[160] T. Dohi Iwamoto, K. and N. Kaio. Discrete repair cost limit replacement policies with/without imperfect repair. *Asia-Pacific Journ. of Operational Res.*, 6:735–751, 2008.

[161] X. Jiang, K. Cheng, and V. Makiš. On the optimality of repair-cost limit policies. *Journ. Appl. Prob.*, 35:936–949, 1998.

[162] X. Jiang, V. Makiš, and A.K.S. Jardine. Optimal repair/replacement policy for a general repair model. *Adv. Appl. Prob.*, 33:206–222, 2001.

[163] Harry Joe. *Multivariate Models and Dependence Concepts*. London: Chapman & Hall, 1997.

[164] K. Jogdeo. Concepts of dependence. In S. Kotz and N.L. Johnson, editors, *Encyclopedia of Statistical Sciences, Vol. 1*. New York: Wiley, 1982.

[165] N. Kaio and S. Osaki. Some remarks on optimal inspection policies. *IEEE Transactions on Reliability*, 33:277–279, 1984.

[166] N. Kaio and S. Osaki. Inspection policies: Comparisons and modifications. *Recherche Operationelle*, 22:387–400, 1988.

[167] N. Kaio and S. Osaki. Comparison of inspection policies. *Journ. of the Operational Res. Society*, 40:499–503, 1989.

[168] Z. Kander and A. Raviv. Maintenance policies when failure distribution of equipment is only partially known. *Naval Research Logistics Quarterly*, 21:419–429, 1974.

[169] Parmod K. Kapur, R.B. Garg, and S. Kumar. *Contributions to Hardware and Software Reliability*. Singapore: World Scientific, 1999.

[170] David Karger. A randomized fully polynomial time approximation scheme for the all terminal network reliability problem. In *Proceedings of the twenty-seventh annual ACM symposium on Theory of computing*, pages 11–17, Las Vegas, Nevada, United States, 1995. ACM.

[171] David Karger. Random sampling in cut, flow, and network design problems. *Mathematics of Operations Research*, 24(2):283–413, May 1999.

[172] David R. Karger. Minimum cuts in near-linear time. *Journal of the ACM*, 47(1):46–76, January 2000.

[173] Samuel Karlin. *Total Positivity, Vol. 1*. Stanford: Stanford University Press, 1968.

[174] Richard Karp and Michael Luby. Monte-Carlo algorithms for the planar multiterminal network reliability problem. *Journal of Complexity*, 1(1):45–64, October 1985.

[175] Richard Karp, MIchael Luby, and Neal Madras. Monte-Carlo approximation algorithms for enumeration problems. *Journal of Algorithms*, 10(3):429–448, 1989.

[176] J.B. Keller. Optimum checking schedules for systems subject to random failure. *Management Science*, 21:256–260, 1974.

[177] J.B. Keller. Optimum inspection policies. *Management Science*, 28:447–450, 1984.

[178] Alexander K. Kelmans. Connectivity of probabilistic networks. *Automation Remote Control*, 29:444–460, 1967.

[179] M. Kijima. Some results for repairable systems with general repair. *Journ. Applied Probability*, 26:89–102, 1989.

[180] M. Kijima, H. Morimura, and Y. Suzuki. Periodical replacement problem without assuming minimal repair. *Europ. Journ. of Operat. Res.*, 37:194–203, 1988.

[181] Gustav Robert Kirchhoff. Über die Auflösung der Gleichungen, auf welche man bei der Untersuchung der linearen Verteilung galvanischer Ströme geführt wird. *Ann. Phys. Chem.*, 72:497–508, 1847.

[182] Donald E. Knuth. *The Art of Computer Programming. Volume 2: Seminumerical Algorithms*, volume 2. Addison-Wesley, Reading, 2 edition, 1981.

[183] Subhash C. Kochar, Hari Mukerjee, and F. Samaniego. The 'signature' of a coherent system and its application to comparisons among systems. *Naval Research Logistic*, 46:507–523, 1999.

[184] S.P. Koh. *On minimax inspection schedules*. PhD thesis, Columbia University. School of Engineering and Applied Science, 1982.

[185] Dieter König and Volker Schmidt. *Zufällige Punktprozesse*. Stuttgart: Teubner, 1991.

[186] M.V. Koutras, S.G. Papastavridis, and K.I. Petakos. Bounds for coherent reliability structures. *Statistics and Probability Letters*, 26:285–292, 1996.

[187] Igor N. Kovalenko, Nickolaj Kuznetsov, and Philip A. Pegg. *Mathematical Theory of Reliability of Time Dependent Systems with Practical Applicatios*. Chichester: J. Wiley & Sons, 1997.

[188] Joseph B. Kruskal. The number of simplices in a complex. In Richard Bellmann, editor, *Mathematical Optimization Techiques*, pages 251–278. Cambridge University Press, 1963.

[189] K. Kumamoto, H., K. Inoue Tanaka, and E. Henley. Dagger-sampling Monte Carlo for system unavailability evaluation. *IEEE Transactions on Reliability*, 29(2):122–125, 1980.

[190] Way Kuo and Ming J. Zuo. *Optimal Reliability Modelling: Principles and Applications*. Wiley, New York, 2003.

[191] Chin-Diew Lai and Min Xie. A new family of positive dependence bivariate disriburions. *Statistics and Probability Letters*, 46:359–364, 2000.

[192] Chin-Diew Lai and Min Xie. *Stochastic ageing and dependence for reliability*. New York, Springer , 2006.

[193] M.T. Lai. A periodical replacement model based on cumulative repair cost limit. *Applied Stochastic Models in Business and Industry*, 26:455–464, 2007.

[194] M.T. Lai. Replacement model with cumulative repair cost limit for a two-unit system under failure rate interaction between units. In *Proceedings of Business and Information*, volume 8, Bangkok, 2011.

[195] Yeh Lam. A note on the optimal replacement problem. *Advances Appl. Probability*, 20:479–482, 1988.

[196] Yeh Lam. A repair replacement model. *Advances Appl. Probability*, 22:494–497, 1990.

[197] Yeh Lam and Yuan Lin Zhang. Analysis of a two-component series system with a geometric process model. *Naval Research Logistics*, 43:491–502, 1996.

[198] A. B. Lehman. Wye-delta transformations in probabilistic networks. *Journal of SIAM*, 11:773–805, 1962.

[199] E.L. Lehmann. Some concepts of dependence. *Annals of Mathematical Statistics*, 37:1137–1157, 1966.

[200] F.K. Leung. Inspection schedules when the lifetime distribution of a single-unit system is completely unknown. *European J. of Oper. Research*, 132:106–115, 2001.

[201] Yi-Kuei Lin and John Yuan. Flow reliability of a probabilistic capacitated-flow network in multiple node pairs case. *Computers & Industrial Engineering*, 45:417–428, 2003.

[202] Bernt Lindström. Determinants on semilattices. *Proc. Amer. Math. Soc.*, 20:207–208, 1969.

[203] Richard J. Lipton and Robert Endre Tarjan. A separator theorem for planar graphs. *SIAM J. Appl. Math.*, 36(2):177–189, April 1979.

[204] Xiao-Gao Liu, Viliam Makis, and Andrew K. S. Jardine. A replacement model with overhauls and repairs. *Naval Research Logistics*, 42:1063–1079, 1995.

[205] M. O. Locks. Recursive disjoint products, inclusion-exclusion, and min-cut approximations. *IEEE Transactions on Reliability*, R-29:368–371, 1980.

[206] M. O. Locks and J. M. Wilson. Nearly minimal disjoint forms of the Abraham reliability problem. *Reliability Engineering & System Safety*, 46:283–286, 1994.

[207] M.O. Locks. Inverting and minimalizing path sets and cut sets. *IEEE Transactions on Reliability*, R-27(2):107–109, 1978.

[208] M.O. Locks. A minimizing algorithm for sum of disjoint products. *IEEE Transactions on Reliability*, R-36(4):445–453, October 1987.

[209] M.V. Lomonosov and V. P. Polesskii. An upper bound for the reliability of information networks. *Problems of Information Transmission*, 7(4):337–339, 1971.

[210] Alfred J. Lotka. A contribution to the theory of selfrenewing aggregates with special reference to industrial replacement. *Annals of Mathematical Statistics*, 10:1–25, 1939.

[211] Tong Luo and K.S. Trivedi. An improved algorithm for coherent-system reliability. *IEEE Transactions on Reliability*, R-47:73–78, 1998.

[212] H. Makabe and H. Morimura. A new policy for preventive maintenance. *J. Oper. Res. Soc. Japan*, 5:110–124, 1963.

[213] H. Makabe and H. Morimura. On some preventive maintenance policies. *J. Oper. Res. Soc. Japan*, 5:17–47, 1963.

[214] H. Makabe and H. Morimura. Some considerations on preventive maintenance policies with numerical analysis. *J. Oper. Res. Soc. Japan*, 7:154–171, 1965.

[215] M. Malik. Reliable preventive maintenance scheduling. *AIIE Trans.*, 11:221–228, 1979.

[216] A.W. Marshall and I. Olkin. A multivariate exponential distribution. *Journal of the American Statistical Association*, 62:291–302, 1967.

[217] Klaus Matthes, Johannes Kerstan, and Joseph Mecke. *Unbegrenzt teilbare Punktprozesse (English edition 1978: Infinitely Divisible Point Processes. New York: Wiley)*. Berlin: Akademie-Verlag, 1974.

[218] F.C. Meng. Some further results on ranking the importance of system components. *Reliability Engineering and System Safety*, 47:97–101, 1995.

[219] F.C. Meng. Comparing the importance of system components by some structural characteristics. *IEEE Transactions on Reliability*, 44:59–64, 1996.

[220] F.C. Meng. Relationships of fussel-vesely and birnbaum-importance to structural importance in coherent systems. *Reliability Engineering and System Safety*, 67:55–60, 2000.

[221] F.C. Meng. Comparing birnbaum importance measure of system components. *Probab. in the Engin. and Informat. Sciences*, 18:247–255, 2004.

[222] F.C. Meng. On some structural importance of system components. *Journal of Data Science*, 7:277–283, 2009.

[223] Karl Menger. Zur allgemeinen Kurventheorie. *Fundamenta Mathematicae*, 10:96–115, 1927.

[224] Michael Mitzenmacher and Eli Upfal. *Probability and Computing*. Cambridge University Press, 2005.

[225] Bojan Mohar and Carsten Thomassen. *Graphs on Surfaces*. Johns Hopkins University Press, 2001.

[226] N.A. Mokhlis. Consecutive k-out-of-n systems. In N. Balakrishnan and C.R. Rao, editors, *Handbook of Statistics 20*, pages 237–280. Amsterdam: Elsevier, 2001.

[227] E.F. Moore and C.E. Shannon. Reliable circuits using less reliable relays. *J. Franklin Inst.*, 262:191–208, 1956.

[228] D. Morgenstern. Einfache Beispiele zweidimensionaler Verteilungen. *Mitteilungsblatt für Mathematische Statistik*, 8:234–235, 1956.

[229] H. Morimura. On some preventive maintenance policies for IFR. *Journal of the Oper. Res. Soc. Japan*, 12:94–124, 1970.

[230] Rajeev Motwani and Prabhakar Raghavan. *Randomized Algorithms*. Cambridge University Press, 1995.

[231] A.G. Munford and A.K. Shahani. A nearly optimal inspection policy. *Oper. Res. Quarterly*, 23:373–379, 1972.

[232] A.G. Munford and A.K. Shahani. An inspection policy for the Weibull case. *Oper. Res. Quarterly*, 24:453–458, 1973.

[233] D.N.P. Murthy and D.G. Nguyen. Optimal age-policy with imperfect preventive maintenance. *IEEE Trans. on Reliability*, 30:80–81, 1981.

[234] E.J. Muth. An optimal decision rule for repair vs replacement. *IEEE Transactions on Reliability*, R-26:179–181, 1977.

[235] T. Nakagawa. Optimum policies when preventive maintenance is imperfect. *IEEE Trans. on Reliability*, 28:331–332, 1979.

[236] T. Nakagawa and S. Mizutani. Periodic and sequential imperfect preventive maintenance policies for cumulative damage models. In H. Pham, editor, *Recent Advances in Reliability and Quality Design*, pages 85–98. Springer, 2008.

[237] Toshio Nakagawa. *Maintenance Theory of Reliability*. London: Springer-Verlag, 2005.

[238] Toshio Nakagawa. *Advanced Reliability Models and Maintenance Policies*. Springer Verlag, 2008.

[239] Hayao Nakazawa. Equivalence of a nonoriented line and a pair of oriented lines in a network. *IEEE Transactions on Reliability*, R-28:364 – 367, 1979.

[240] B. Natvig and J. Gåsemyr. New results on the Barlow-Proschan and Natvig measures of component importance in nonrepairable and repairable systems. *Methodology and Computing in Applied Probability*, 11(4):603–620, 2009.

[241] Bent Natvig. A suggestion of a new measure of importance of system components. *Stochastic Processes and their Applications*, 9:319–330, 1979.

[242] Bent Natvig. New light on measures of system components. *Scandinavian Journal of Statistics*, 12:43–54, 1985.

[243] Roger B. Nelson. *An Introduction to Copulas*. Springer, New York, second edition, 2006.

[244] M.E.J. Newman. *Networks: An Introduction*. Oxford University Press, 2010.

[245] S. Osaki. Optimal maintenance policies for stochastic deteriorating systems. In *Quality, Reliability, and Information Technology*, pages 1–9. New Delhi: Narosa, 2005.

[246] Christos H. Papadimitriou. *Computational Complexity*. Addison-Wesley, 1994.

[247] Christos H. Papadimitriou and Kenneth Steiglitz. *Combinatorial Optimization: Algorithms and Complexity*. Dover Publications, Mineola, N.Y., 1998.

[248] K.S. Park. Cost limit replacement policy under minimal repair. *Microelectronics and Reliability*, 23:347–349, 1983.

[249] Hoang Pham, editor. *Handbook of Reliability Engineering*. Springer, 2003.

[250] V.P. Polesskii. A lower boundary for the reliability of information networks. *Problems of Information Transmission*, 7:165–171, 1971.

[251] André Pönitz. *Über eine Methode zur Konstruktion von Algorithmen für die Berechnung von Invarianten in endlichen ungerichteten Hypergraphen*. PhD thesis, Technische Universität Bergakademie Freiberg, 2004.

[252] S. Rai and K. K. Agarwal. An efficient method for reliability evaluation of a general network. *IEEE Transactions on Reliability*, R-27:206–211, 1978.

[253] Suresh Rai, Malathi Veeraraghavan, and Kishor S. Trivedi. A survey of efficient reliability computation using disjoint products approach. *Networks*, 25:147–163, 1995.

[254] Lucia I.P. Resende. New expressions for the extended $\Delta - Y$ reductions. *Operations Research Letters*, 8:323–328, 1989.

[255] Neil Robertson and P.D. Seymour. Graph minors. II. Algorithmic aspects of tree-width. *Journal of Algorithms*, 7(3):309–322, September 1986.

[256] Jose Rodriguez and Lorenzo Traldi. (K, j)-domination and (K, j)-reliability. *Networks*, 30(4):293–306, 1997.

[257] R. Roeloffs. Minimax surveillance schedules with partial information. *Naval Research Logistics Quarterly*, 10:307–322, 1963.

[258] R. Roeloffs. Minimax surveillance schedules for replaceable units. *Naval Research Logistics Quarterly*, 12:461–471, 1965.

[259] Tomasz Rolzki, Hanspeter Schmidli, Volker Schmidt, and Jozef Teugels. *Stochastic Processes for Insurance and Finance*. John Wiley & Sons, 1999.

[260] A. Rosenthal and D. Frisque. Transformations for simplifying network reliability calculations. *Networks*, 7:97–111, 1977.

[261] Arnie Rosenthal. A computer scientist looks at reliability computations. In R.E. Barlow, J.B. Fussel, and D.D. Singpurwalla, editors, *Reliability and Fault Tree Ananlysis*, pages 133–152. SIAM, Philadelphia, 1975.

[262] Arnie Rosenthal. Computing the reliability of complex networks. *SIAM Journal on Applied Mathematics*, 32:384–393, 1977.

[263] Gian Carlo Rota. On the foundations of combinatorial theory I. Theory of Möbius functions. *Probability Theory and Related Fields*, 2(4):340–368, January 1964.

[264] A.M.A. Rushdi. Partially-redundant systems: Examples, reliability, and life expectancy. *International Magazine on Advances in Computer Science and Telecommunications*, 1:1–13, 2010.

[265] F. J. Samaniego. On closure of the IFR class under formation of coherent systems. *IEEE Transactions on Reliability*, R-34:60–72, 1985.

[266] F.J. Samaniego and M. Shaked. Systems with weighted components. *Statistics and Probability Letters*, 78:815–823, 2008.

[267] Francisco J. Samaniego. *System signatures and their applications in engineering reliability*, volume 110 of *International Series in Operations Research & Management Science*. Springer, New York, 2007.

[268] A. Sarkar, S.C. Panja, and B. Sarkar. Survey of maintenance policies for the last 50 years. *Int. Journ. of Software Engineering & Applications*, 2:1–33, 2011.

[269] Appajosyula Satyanarayana and Mark K. Chang. Network reliability and the factoring theorem. *Networks*, 13:107–120, 1983.

[270] Appajosyula Satyanarayana and A. Prabhakar. A new topological formula and rapid algorithm for reliability ananlysis of complex networks. *IEEE Transactions on Reliability*, R-27:82–100, 1978.

[271] Appajosyula Satyanarayana and Ralph Tindell. Chromatic polynomials and network reliability. *Discrete Mathematics*, 67(1):57–79, October 1987.

[272] Appajosyula Satyanarayana and R. Kevin Wood. A linear-time algorithm for computing K-terminal reliability in series-parallel networks. *SIAM J. Comput.*, 14:818–832, 1985.

[273] M. Scarsini and M. Shaked. On the value of an item subject to general repair or maintenance. *European Journ. Oper. Research*, 122:625–637, 2000.

[274] S.-H. Sheu, C.-C. Chang, and Y.-L. Chen. A generalized periodic preventive maintenance model with virtual age for a system subjected to shocks. *Communications in Statistics-Theory and Methods*, 39:2379–2393, 2010.

[275] S.H. Sheu. A generalized block replacement policy with minimal repair and general random repair cost for a multiunit system. *Journ. of the Operational Res. Soc.*, 42:331–341, 1991.

[276] S.H. Sheu. Optimal block replacement policies with multiple choice at failure. *J. Applied Probability*, 29:129–141, 1992.

[277] S.H. Sheu, C.C. Chang, Y.L. Chen, and Z.G. Zhang. A periodic replacement model based on cumulative repair-cost limit for a system subjected to shocks. *IEEE Transactions on Reliability*, 59:374–382, 2010.

[278] Douglas R. Shier. *Network Reliability and Algebraic Structures*. Clarendon Press, 1991.

[279] Akiyoshi Shioura, Akihisa Tamura, and Takeaki Uno. An optimal algorithm for scanning all spanning trees of undirected graphs. *SIAM Journal on Computing*, 26:678–692, 1994.

[280] Senju Shizuo. A probabilistic approach to preventive maintenance. *Journ. of the Operations Res. Society of Japan*, 1:49–55, 1957.

[281] F. Simon. Splitting the K-Terminal Reliability. *arXiv:1104.3444v1 [math.CO]*, pages 1–17, April 2011.

[282] Emanuel Sperner. Ein Satz über Untermengen einer endlichen Menge. *Mathematische Zeitschrift*, 27(1):544–548, 1928.

[283] Lutz Sproß. *Effektive Methoden der Zuverlässigkeitsanalyse stochastischer Netzstrukturen*. Dr. sc. techn. thesis, Hochschule für Verkehrswesen "Friedrich List, " Dresden, 1988.

[284] W. Stadje and D. Zuckerman. Optimal maintenance strategies for repairable systems with general degree of repair. *Journ. Applied Probability*, 28:384–396, 1991.

[285] Richard P. Stanley. *Enumerative Combinatorics, Volume I*, volume I. Cambridge University Press, 1997.

[286] Karl Stigman. *Stationary Marked Point Processes*. New York: Chapman & Hall, 1995.

[287] C. Stivaros and K. Sutner. Computing optimal assignments for residual network reliability. *Discrete Appl. Math.*, 75(3):285–295, 1997.

[288] Constantine Stivaros and Charles Suffel. Uniformly optimal networks in the residual node connectedness reliability model. Combinatorics, graph theory, and computing, Proc. 22nd Southeast Conf., Baton Rouge/LA (USA) 1991, Congr. Numerantium 81, 51-63 (1991)., 1991.

[289] K. Sutner, A. Satyanarayana, and C. Suffel. The complexity of the residual node connectedness reliability problem. *SIAM J. Comput.*, 20(1):149–155, 1991.

[290] P.R. Tadikamalla. An inspection policy for the gamma failure distribution. *Journ. Oper. Res. Society*, 30:77–80, 1979.

[291] S. Taghipour and D. Banjevic. Periodic inspection optimization models for a repairable system subject to hidden failures. *IEEE Transactions on Reliability*, 60:275–285, 2011.

[292] A. Tahara and T. Nishida. Optimal replacement policy for a minimal repair model. *J. Oper. Res. Soc. Japan*, 18:113–124, 1975.

[293] Peter Tittmann. Partitions and network reliability. *Discrete Applied Mathematics*, 95(1-3):445–453, July 1999.

[294] Peter Tittmann, Ilya Averbouch, and Johann A. Makowsky. The enumeration of vertex induced subgraphs with respect to the number of components. *European Journal of Combinatorics*, 32(7):954–974, October 2011.

[295] Lorenzo Traldi. On the star-delta transformation in network reliability. *Networks*, 23:151–157, 1993.

[296] Lorenzo Traldi. Non-minimal sums of disjoint products. *Reliability Engineering & System Safety*, 91(5):533–538, May 2006.

[297] William Tutte. A contribution to theory of chromatic polynomials. *Canadian Journal of Mathematics*, 6:26–40, 1954.

[298] William Tutte. *Graph Theory*. Addison-Wesley, 1984.

[299] K. Uematsu and T. Nishida. One-unit system with failure rate depending on repair. *Math. Japonica*, 32:139–147, 1987.

[300] Leslie G. Valiant. The complexity of enumeration and reliability problems. *SIAM J. Comput.*, 8:410–421, 1979.

[301] Richard M. Van Slyke and H. Frank. Network reliability analysis. I. *Networks*, 1:279–290, 1972.

[302] Hongzhou Wang and Hoang Pham. A quasi renewal process and its application in the imperfect maintenance. *Journ. of Systems Science*, 27:1055–1062, 1996.

[303] Hongzhou Wang and Hoang Pham. *Reliability and Optimal Maintenance*. London: Springer-Verlag, 2006.

[304] W. Wang. An inspection model for a process with two types of inspections and repairs. *Reliability Engineering and System Safety*, 94:526–533, 2009.

[305] Dominic J.A. Welsh. *Complexity: Knots, Colourings and Counting*. Cambridge University Press, 1993.

[306] Douglas B. West. *Introduction to Graph Theory*. Prentice Hall, 2 edition, 1999.

[307] G.E. Willmot and X.S. Lin. *Lundberg Approximations for Compound Distributions with Insurance Applications*. Springer, New York, 2001.

[308] Thomas Wolle. A framework for network reliability problems on graphs of bounded treewidth. Technical Report UU-CS-2003-026, Institute of Information and Computing Sciences, Utrecht University, 2003.

[309] R. Kevin Wood. Polygon-to-chain reductions and extensions for computing K-terminal reliability in an undirected network. Technical Report ORC 82-11, Operations Research Center, University of California, Berkeley, 1982.

[310] R. Kevin Wood. A factoring algorithm using polygon-to-chain reductions for computing K-terminal reliability. *Networks*, 15:173–190, 1985.

[311] R. Kevin Wood. Triconnected decomposition for computing K-terminal network reliability. *Networks*, 19:203–220, 1989.

[312] M. Xie. On some importance measures for system components. *Stochastic Processes and their Applications*, 25:273–208, 1987.

[313] Q. Zhang and Q. Mei. A sequence of diagnosis and repair for a 2-state repairable system. *IEEE Transactions on Reliability*, R-36(1):32–33, 1987.

[314] Lian-Chang Zhao and Jun-Chen Xu. An efficient minimizing algorithm for sum of disjoint products. *Microelectronics and Reliability*, 35(8):1157–1162, August 1995.

List of Symbols

$B(n)$	bell number
C_n	cycle of length n
$d(u,v)$	distance between u and v
$\deg v$	degree of vertex v
E	edge set
$\mathbb{E}$	expectation
F	distribution function, failure probability
$\overline{F}$	survival function
$F^{*(n)}$	n-fold convolution
$\hat{F}$	Laplace transform of F
G	graph
$G - e$	graph obtained from G by deletion of e
G/e	graph obtained from G by contraction of e
K_n	complete graph with n vertices
L	lifetime
lcm	least common multiple
$N(v)$	neighborhood of vertex v
$\mathbb{P}(X)$	partition lattice of X
p_e	reliability of an edge
P_n	path with n vertices
$\Pr(A)$	probability of the random event A
q_e	failure probability $q_e = 1 - p_e$
$R(G)$	all-terminal reliability of G
$R(G,K)$	K-terminal reliability of G
$R(G,p)$	reliability polynomial of G
$R_{st}(G)$	two-terminal reliability
$t(G)$	number of spanning trees of G
V	vertex set
$v(G)$	order (number of vertics) of G
V_n	state space of a binary system of order n
Var	variance
$\lfloor x \rfloor$	largest integer not greater than x
$\coprod_{i=1}^{n} z_i$	maximum (join) of binary variables
$\vee$	join (supremum)
$\wedge$	meet (infimum)
$*$	convolution
$\delta(x,y)$	Kronnecker function
$\Gamma(x)$	Gamma function
λ	failure rate
$\phi(\mathbf{z})$	structure function of a binary system

Index

Printed and bound by CPI Group (UK) Ltd, Croydon, CR0 4YY

21/10/2024

01777085-0013